THE SENSORY PHYSIOLOGY OF
AQUATIC MAMMALS

THE SENSORY PHYSIOLOGY OF AQUATIC MAMMALS

by

Alexander Ya. Supin
Vladimir V. Popov
Alla M. Mass

*Severtsov Institute of Ecology and Evolution
of the Russian Academy of Sciences*

KLUWER ACADEMIC PUBLISHERS
Boston / Dordrecht / London

Distributors for North, Central and South America:
Kluwer Academic Publishers
101 Philip Drive
Assinippi Park
Norwell, Massachusetts 02061 USA
Telephone (781) 871-6600
Fax (781) 681-9045
E-Mail <kluwer@wkap.com>

Distributors for all other countries:
Kluwer Academic Publishers Group
Distribution Centre
Post Office Box 322
3300 AH Dordrecht, THE NETHERLANDS
Telephone 31 78 6392 392
Fax 31 78 6546 474
E-Mail <services@wkap.nl>

 Electronic Services <http://www.wkap.nl>

Library of Congress Cataloging-in-Publication Data

Supin, Alexander Ya.
  The sensory physiology of aquatic mammals/ by Alexander Ya, Supin, Vladimir V.
Popov, Alla M. Mass
    p. cm.
  Includes bibliographical references (p.).
  ISBN 0-7923-7357-X (alk. Paper)
    1. Aquatic mammals—Physiology. 2. Senses and sensation. I. Popov, Vladimir V. II.
    Mass, A.M. (Alla Mikhailovna) III. Title.

QL713 .S86 2001
573.8'7195—dc21

                                                                2001022995

Printed on acid-free paper.   Printed in the United States of America

CONTENTS

PREFACE

This book is actually a product of efforts of many people, not only of the authors.

Wide investigations of marine mammals began in Russia (that time, in the former Soviet Union) in the 1960s when a few teams of enthusiasts founded experimental stations intended for keeping dolphins and seals in captivity and for performing experimental studies of these fascinating animals. It was a time when attention of many people throughout the world was attracted to dolphins and other marine mammals due to appearance of oceanariums and dolphinariums, which demonstrated unique capabilities of these animals. So scientists in many countries concentrated on studies of them. There was much to learn about the morphology, physiology, and psychology of marine mammals, and investigators spending their time and efforts on studies in this field were rewarded by a number of surprising findings.

The authors of this book represent one of such research teams focused on the neuro- and sensory physiology of marine mammals. A few decades of studies naturally resulted in the idea to summarize in a book both the results of these studies and a large body of data in adjacent fields. Our goal was to synthesize the many research findings and the present knowledge on sensory capabilities and mechanisms of sensory systems of aquatic mammals. We realize, however, that the appearance of this book was made possible due to the help and assistance of many colleagues.

First of all, it should be stressed that experimentation on marine mammals would be impossible without the collective participation of the entire research team. Therefore, it is our pleasant duty to note the participation of Vladimir Klishin, Tamara Ladygina, Olga Milekhina, Mikhail Tarakanov, and Mikhail Pletenko in various parts of the studies described herein.

Keeping large animals in captivity is expensive and effort-consuming. It would not have been possible for us if it were not for Administration of the

Institute of Ecology and Evolution of the Russian Academy of Science, headed initially by Professor Vladimir Sokolov and later by Professor Dmitriy Pavlov. They always supported our projects and ideas. In recent years, the Russian Foundation for Basic Research (RFBR) provided valuable support for our studies. Most of our studies were carried out in the Utrish Marine Station (the Black Sea coast) built purposefully for this type of investigation; the extraordinary efforts of the scientific supervisor of the Station, our friend Lev Mukhametov, made possible the successful functioning of this institution. We also received a lot of help from the Utrish Dophinarium.

Our understanding of the subject has been markedly enhanced due to discussions with many colleagues in national and international scientific meetings. For these opportunities, we are very thankful to organizers of these meetings, Whitlow Au, Nicolay Dubrovskiy, Ronald Kastelein, Patrick Moore, Paul Nachtigall, Eugeniy Romanenko, Jeanette Thomas, and many others.

We have to note that the idea to write this book was put forward by Michael Myslobodsky, who encouraged us to submit the book project to Kluwer Academic Publishers. The attitude of the Kluwer's staff, particularly Molly Taylor and Melinda Paul, who did not hesitate to deal with authors from a country so far from their home, and Rosemary Winfield, who carefully edited the manuscript, is greatly appreciated.

So there are many people who we are indebted to. We hope they are not disappointed by the final product.

Alexander Supin
Vladimir Popov
Alla Mass

ABBREVIATIONS

ABR Auditory brainstem response

ACF Autocorrelation function

ACR Auditory cortical response

AM Amplitude modulation

ANR Auditory nerve response

AP [Cochlear] Action potential

CCF Cross-correlation function

EFR Envelope-following response

EP Evoked potential

ERB Equivalent rectangular bandwidth

ERD Equivalent rectangular duration

ERG Eelectroretinogram

FM Frequency modulation

FFR Frequency following response

FRP Frequency resolving power

IID Interaural intensity difference

ILD Iinteraural latency difference

ISI Interstimulus interval

ITD Interaural time delay

MAA Minimal audible angle

MTF Modulation transfer function

PND Pposteronodal distance

RFR Rate-following response

RMS Root-mean-square

ROC Receiver operation characteristic

Roex Rounded exponential function

SAM Sinusoidal amplitude modulation

TSP Time separation pitch

Chapter 1
INTRODUCTION

1.1. GENERAL

Comparative studies of sensory systems in animals adapted to various living conditions reveal unknown abilities of sensory systems, aid our understanding of the mechanisms of perception and information processing, and extend our knowledge of animal behavior and orientation. It is no surprise that a number of researchers who worked for many years on traditional laboratory animals and made major contribution to basic neuro- and sensory physiology turn their attention to studies of exotic species that were not usual subject of study for physiologists. As well as yielding factual data on the functional basis of sensory systems, this approach provides a better understanding of several aspects of evolution. In particular, there is a great interest in the sensory systems of such a specific group of animals as aquatic mammals.

Aquatic mammals differ from most terrestrial ones not only in their life style in the aquatic environment. From the point of view of phylogeny, they are specific groups (orders) of mammals that evolved independently of other orders during tens millions of years. Some tasks that sensory systems encounter were evolutionarily solved in these animals in other ways than in terrestrial mammals. Therefore, they provide opportunities for fuller knowledge of mammalian sensory mechanisms.

There are several excellent books addressing the sensory systems of aquatic mammals (Busnel and Fish, 1980; Nachtigall and Moore, 1988; Thomas and Kastelein, 1990; Thomas et al., 1992; Kastelein et al., 1995). Unfortunately, all of them are contributed books composed of separated articles, so they do not present systematically the situation in the sensory physiology of aquatic mammals. The only authored book on a related topic was an outstanding work of Au (1993) summarizing data on the cetacean echolocation and hearing. However, because of the focus of that book, many other aspects of sensory systems of aquatic mammals remained unconsidered. Thus, a need remained to synthesize into a unified treatise much of the data on sensory physiology of aquatic mammals.

The present book summarizes some recent data on the physiology of the auditory, visual, and somatosensory systems of aquatic mammals. Apart from presenting the results of a quarter century of experience of the author team in the field of sensory physiology of aquatic mammals, it summarizes a considerable body of data existing in the literature on related topics.

It should be noted that the amount of available data markedly differs among different sensory systems and different groups of aquatic mammals. The hearing of cetaceans is investigated in the most detail since it has attracted the attention of many investigators due to the cetacean capability of echolocation. The vision of cetaceans was investigated to a lesser though significant extent, and their somatic sense was the least investigated. In pinnipeds, all these senses were studied significantly, though hearing was studied to a lesser extent than in cetaceans. Only a few studies are available on the hearing and vision of sirenians. Therefore, the length of the chapters and subchapters in this book is unequal. A separate chapter is devoted to the hearing of cetaceans, and it is the largest one. The chapters devoted to the hearing and vision of other aquatic mammals are smaller, and the chapter devoted to the somatic sense is the smallest one. This inequality reflects the real situation in studies of the different sensory systems of aquatic mammals.

1.2. AQUATIC MAMMALS AS SUBJECTS OF EXPERIMENTAL STUDIES

Before presenting the data on the aquatic mammal sensory systems, the term aquatic mammals needs to be defined. The lifestyle of many mammals is to some extent connected to their aquatic environment. However, usually only three mammalian orders are called the true aquatic mammals since all representatives of these orders are deeply adapted to the aquatic environment. These orders are cetaceans, pinnipeds, and sirenians (*Cetacea, Pinnipedia, Sirenia*). Two of these groups, cetaceans (whales and dolphins) and sirenians (sea cows) are completely aquatic mammals. This means that they spend all their life in water and never inhabit land. The third group, the pinnipeds (seals, sea lions, and walruses) is amphibious; all pinnipeds spend a part of their life in water and another part on land. Nevertheless, they are regarded as true aquatic mammals since inhabiting the aquatic environment is obligatory for all of them, and all their systems are deeply adapted to aquatic conditions.

Among completely aquatic mammals, cetaceans are studied to the greatest extent. They are deeply specialized animals evolved from primitive ancestral mammals. Apparently more than 50 million years ago their evolution paralleled that of terrestrial mammals, and their sensory systems (as well as all other ones) underwent deep modifications, which makes them extremely attractive subjects for comparative physiology.

The cetacean order is a prosperous group that includes two recent suborders, *Odontoceti* (toothed whales) and *Mysticeti* (baleen whales), 14 families, and a few dozens of species. Odontocetes (10 families, 65 species) inhabit aquatic niches throughout almost all the world, both fresh and marine water. All of them are efficient predators. Their body length is diverse, from approximately 1 m (small dolphins and porpoises) to 40 m (large sperm whales). Small odontocetes ("small" cetaceans may be of a size of a few meters) usually are called *dolphins* and *porpoises*, contrary to large *whales*. The term *dolphin*, although widely used, is not strictly defined. In the strict sense, only one of odontocete families is named dolphin, *Delphinidae*, but really this term is applied to many odontocete families, such as *Phocoenidae* (these small cetaceans are usually called *porpoises*), a few families of fresh-water dolphins (*Platanistidae, Pontoporidae, Iniidae, Lipotidae*), which previously were considered as a single family *Platanistidae*, and sometime to other odontocete families. The separation to whales and dolphins is arbitrary and based only on the body size. Middle-sized cetaceans, such as the beluga *Delphinapterus leucas* (*Monodontidae*) and the killer whale *Orcinus orca*, (*Delphinidae*), may be called both whales and dolphins.

The rather small size of many odontocetes (dolphins) has made it possible to keep them in captivity, so representatives of this suborder of cetaceans have been the subjects of many physiological experimental investigations.

All odontocetes are capable of echolocation. They get environmental information by analyzing echoes from self-generated sound signals. This imposes specific demands on their auditory system, which has been an object of active investigation over the last few decades.

All mysticetes (four families, 11 species) are large animals (from a few meters to tens meters in size) called *whales*. Keeping these animals in captivity is extremely rare, so they have not been subjects of extensive experimental studies. Mysticetes are not know to echolocate.

Contrary to cetaceans, another group of completely aquatic mammals, sirenians, is not numerous. These animals are threatened to extinction throughout the world, so only four species of two families, each for one genus, have survived today: the West Indian manatee *Trichechus manatus*, Amazonian manatee *T. inunguis*, West African manatee *T. Senegalensis*

(*Trichechidae*), and the dugong *Dugong dugong* (*Dugonidae*). All surviving sirenians are slowly moving herbivores inhabiting tropical marine waters along the coast and coastal rivers. Keeping manatees in a few marine parks provided opportunities for their experimental studies, but these studies are not numerous.

Contrary to cetaceans and sirenians, pinnipeds (*Pinnipedia*) are certain to spend a part of their life on land. Pinnipeds are a thriving group occurring along almost all coasts of the world, in some rivers, and in some inland lakes. This order includes three families: true seals (*Phocidae*), eared seals (*Otariidae*), and walruses (*Odobenidae*). True (earless) seals include 13 genera and 18 species. Eared seals (sea lions, fur seal) include six genera and 12 species. All of them are active fish-eating predators. Although spending part of their life on land, all true and eared seals are excellent swimmers and divers. The third family, *Odobenidae*, includes only one recent genus and species, the walrus *Odobenus rosmarus*. For foraging, walruses use their tusks to obtain bentic mollusks and other marine life.

Many representatives of all three pinniped families were kept in captivity and were used in experimental studies, particularly in the field of sensory physiology.

It should be noted that the lifestyles of many representatives of other mammalian orders (insectivores, rodents, carnivores, ungulates, and so on) are also more or less "aquatic"– for example, beavers (*Rodenita*: *Castoridae*), otters (*Carnivora*: *Mustelidae*), and hippopotamuses (*Artiodactila*: *Hippopotamidae*). However, the orders that they belong to contain mostly terrestrial species, and their anatomy and physiology are not deeply modified for the aquatic mode of life. All these animals are left out of scope of this book, which summarizes data on only cetaceans, pinnipeds, and sirenians.

1.3. THE PHYSICAL PROPERTIES OF WATER AS A SENSORY MEDIUM

Some information presented in this section is elementary. Nevertheless, it is presented because some physical properties of water might attract little attention of a physiologist who was used to study aerial hearing and vision in terrestrial animals. Therefore, a very short overview of differences between physical properties of water and air seems reasonable.

1.3.1. Acoustics

The different physical characteristics of air and water result in different relationship between sound intensity to other sound measures – first of all,
sound pressure. Sound intensity is the acoustic power flux density – that is,
the power spreading through a unit surface perpendicular to the direction of
sound propagation. It is

$$I = pv, \tag{1.1}$$

where I (W/m^2) is the power flux density, p (Pa) is the sound pressure, and v
(m/s) is the speed of moving particles. Since

$$p/v = \rho c, \tag{1.2}$$

where ρ (kg/m^3) is the medium density and c (m/s) is the sound speed, it follows

$$I = p^2/\rho c = v^2 \rho c. \tag{1.3}$$

The product ρc is the acoustic impedance for the medium.

This relation shows how differences in the physical properties of air and
water influence their acoustic properties. Acoustic impedances of air and
water are markedly different. In air, $c = 340$ m/s and $\rho = 1.3$ kg/m^3, so $\rho c =
442$ kg/s·m^2. In sea water, $c = 1530$ m/s and $\rho = 1030$ kg/m^3, so $\rho c =
1576 \times 10^6$ kg/s·m^2. Thus, the ratio of the water and air acoustic impedance is
3565. It means that one and the same sound pressure in water corresponds to
3565 times lower intensity than in air; one and the same intensity in water
corresponds to $3565^{1/2} = 59.7$ times higher sound pressure than in air.

A commonly adopted measure of sound intensity is its decibel (dB) level
relative to a certain reference intensity:

$$I_{dB} = 10 \log_{10}(I/I_{ref}) = 20 \log_{10}(p/p_{ref}), \tag{1.4}$$

where I and p are intensity and sound pressure of the measured sound and I_{ref}
and p_{ref} are those of a reference level. In aerial acoustics, the commonly
adopted reference level is 20 µPa (which corresponds to $9 \times 10^{-13} \approx 10^{-12}$
W/m^2), which is the standard hearing threshold for humans. In hydroacoustics, the use of this reference sound pressure would make no sense since in
water it corresponds to a much lower intensity, around 2.5×10^{-16} W/m^2,
which is well below any real hearing threshold. A commonly adopted refer-

ence level in hydroacoustics is, however, even lower sound pressure, 1 μPa (which corresponds to 6.5×10^{-19} W/m^2). So if a hearing threshold of an aquatic animal is, for example, 50 dB (*re* 1 μPa!), it does not mean that the animal is deaf: it is a very good threshold, less than 6.5×10^{-14} W/m^2 – much better than in humans.

Another significant sequence of the high acoustic impedance of water is that it is rather close to impedance of many body tissues. In air, the boundary between the medium (air) and tissues is not very transparent for sounds because of large difference of impedances; such a boundary reflects sounds. Therefore, a sound-conducting air-filled canal (the external auditory canal) is necessary to deliver sounds to the middle ear. On the other hand, the ability of the body surface to reflect sounds makes it possible to create a sound-focusing device, the pinna. In water, the boundary between the medium and many parts of the body is sound-transparent. This provides various ways to deliver sounds to the middle ear without the external auditory canal. However, a sound-reflective device like the pinna is not effective.

To be effective, any sound receiver must be impedance-matched to the medium. The impedance difference between air and water makes a sound-receiving device adapted to air not very effective in water, and vice versa, because of the impedance mismatch between the medium and receiver. In submerged humans, about a 30-dB hearing loss occurs (Wainwright, 1958; Hollien and Brandt, 1969). Similar hearing loss should occur for the water-adapted ear in air.

Therefore, properties of water as a sound-conducting medium dictate adaptation of the ear of aquatic mammals, mostly of completely aquatic mammals. It must be adapted to sound perception in conditions of much higher sound pressure but much less displacement of medium particles than the ear of terrestrial animals. This adaptation manifests itself in the anatomy and biomechanics of the ear.

1.3.2. Optics

From the point of view of visual physiology, the most important difference between air and water is the much higher refractive index of the latter. While the refractive index of air is close to that of a vacuum, very close to 1, the refractive index of water is 1.33 to 1.34. This difference is significant taking into account the range of refractive indices of biological tissues forming refractive structures of the eye.

The Newton's equation relates the refractive power of a spherical surface with its radius and refractive indices of the separated media:

$$f' = \frac{n'}{n'-n} r \,,$$ (1.5)

where f is the focal distance, n and n' are refractive indices of the media in front and in the rear of the surface, and r is the surface radius. Supposing that the refractive index of the media in front of the surface is $n = 1$ (air), that of some eye tissue $n' = 1.5$, and the radius $r = 10$ mm, this results in a focal distance $f = 30$ mm. Being comparable with the eye dimensions of large animals, this focal distance may provide focusing of an image at the eye retina. If the refractive index in front of the surface $n = 1.33$ (water), the same surface radius results in a focal distance of around 88 mm; to provide the 30-mm focal distance, the surface radius should be as short as 3.3 mm. Thus, to function in water, underwater eye optics require much higher convexity of refractive surfaces than those of aerial-adapted eyes. In particular, the eye lens of aquatic animals is always of much greater curvature (almost spherical) than rather thin (lenticular) lenses of terrestrial animals. As a rule, the spherical outer surface of the cornea serves as a main refractive device in terrestrial animals, and the lens serves only as a rather weak additional refractive device. This is not the case in aquatic animals: when the media in front of a surface (water) and in the rear of the surface (anterior chamber liquid behind the thin spherical cornea) have similar refractive indices, this surface does not function as a refractive device at all ($f = \infty$ if $n = n'$).

Another important optic feature of water is much higher light absorption than that of air. This is true for the absorption by water itself; however, in natural conditions, the main light absorption appears due to microparticles of both biogenic and nonbiogenic origin suspended in water. Due to intense light absorption, illumination conditions as a rule are at least mesopic at a depth of a few meters and scotopic at greater depths, even if bright daylight is above the water surface. If an animal inhabits a near-surface water layer, like aquatic mammals do, its visual system has to function in conditions of rapidly changing illumination within a wide range, from photopic conditions at the surface to mesopic or scotopic conditions at a depth.

Furthermore, absorption and scattering of light in any medium is wavelength-dependent. Since both absorption and scattering are much stronger in water than in air, this effect manifests itself in narrowing the light spectrum with increasing depth. In clear water, shorter (blue) wavelengths penetrate deeper; in turbid water, longer wavelengths may take advantage. In any case, the available spectrum becomes limited. This feature of the aquatic media may influence markedly the spectral sensitivity of vision in aquatic animals.

1.4. PSYCHOPHYSICAL MEASUREMENT PROCEDURES

A large body of data on the sensory capabilities of aquatic mammals obtained by behavioral (psychophysical) methods is reviewed in this book for comparison with physiological data. Proper evaluation of these results is possible only if the experimental procedures and evaluation criteria used in those studies are taken into account. Therefore, a brief introductory description of experimental procedures used in psychophysical measurements seems to be reasonable. It is intended to help a reader who is not experienced in psychophysical measurements to understand some of data presented below.

1.4.1. The Operant Conditioning Method

With the use of this method, measurements of sensory capabilities of an animal are based on establishing and maintaining the stimulus control of the subject's behavior. It implies that presentation of a certain stimulus results in a confidently detectable change in the subject's behavior; each time the subject perceives the stimulus, it performs a characteristic movement and does not perform this movement in the absence of the stimulus. In humans, the stimulus control can be easily achieved by a verbal instruction ("Push the button each time you hear a sound"). In animals, the stimulus control is achieved by training of the animal based on food reward or sometimes on other kinds of rewards. Such mode of training is known as operant conditioning (Skinner, 1961).

Once stimulus control has been established, the presence of the behavioral response can be used as an indicator of the capability of the subject to detect the stimulus. To maintain the established stimulus control, each correct response (the presence of the proper behavioral response in a stimulus trial or the absence of the response in a nonstimulus trial) is reinforced. Incorrect responses (the absence of the response in a stimulus trial, a wrong response, or a false alarm) are not rewarded. Often an intermediate reinforcement (a bridge signal, such as a certain sound) is used before the food reward to let the subject know as soon as possible that its response was correct. Thus, perception capabilities can be measured by varying any stimulus parameter and observing which stimulus does evoke the behavioral response and which does not.

However, both perceiving a stimulus and performing the behavioral response are not deterministic but probabilistic processes. One and the same stimulus may be detected by the animal in one trial and not detected in another similar trial. Similarly, even if the stimulus is detectable, the behavioral response may be absent in some trials, as well as in the absence of the stimulus, the behavioral response may appear (the false alarm). Therefore, procedures of statistical analysis of psychophysical data were elaborated. These procedures impose some requirements on the collection of experimental data. In well-designed psychophysical studies, the data acquisition is performed according to rules dictated by the need of subsequent quantitative data processing.

1.4.2. Conditioned Reflex

Apart from the operant conditioning method, the so-called Pavlovian conditioning (conditioned-reflex) method was used in some studies of sensory capabilities of aquatic mammals (e.g., Supin and Sukhoruchenko, 1974). In such studies, a stimulus is accompanied by a reinforcing stimulus that evokes a certain involuntary reflective response. For example, it may be a weak electric stimulation of the skin that evokes a complex of a few responses: the electrodermal (galvanic-skin) response (a slow change of the skin electric potential), change of respiration rate (in dolphins it manifests itself as an extra expiration-inspiration act), and change of the heart-beat rate. After a few coincidences of a sensory (acoustic, visual) and electric stimuli, the former becomes capable of evoking the same complex of responses. Thus, this response complex can be used as an indication of stimulus perception by the animal. An advantage of this approach is that it does not require long and careful training of the animal; as a rule, the conditioned reflex appears after only a few coincidences of the sound stimulus and the reinforcement. However, the conditions of measurements in this case are less natural than in operant-conditioning experiments.

1.4.3. The Statistical Basis for Threshold Evaluation

Most psychophysical studies of sensory abilities of animals are based on threshold determination. The purpose of such measurements is to find the lowest perceivable stimulus intensity (the absolute threshold) or the minimum detectable difference between stimuli (differential threshold). A psychophysically searched threshold is not a deterministic value: a near-threshold signal may evoke the response in one trial and may be ineffective in another trial. Therefore, a psychophysical threshold is a matter of statistical analysis. This analysis bases on a signal-detection theory, which is outlined in a few books, particularly as applied to psychophysical studies (Swets, 1964; Green and Swets, 1966).

According to the theory, there is no true sensory threshold since at a certain stimulus value, sensation randomly varies to a certain extent. This random variation can be presented as a sum of the signal and noise. Thus, a certain stimulus value is characterized not by a precise sensation value but by the probability density of sensation values (Fig. 1.1A). Different values (S_0, S_1, and S_2) of the varied stimulus parameter (say, intensity), result in different probability-density functions. As a particular case, S_0 may present the distribution of sensation levels at zero stimulus intensity (noise only, n), whereas S_1 and S_2 present distributions at the presence of nonzero stimuli (signal + noise, sn). To make a decision as to which stimulus (S_0 or S_1) was presented, a certain criterion β should be set assuming that at sensation levels above this criterion, the stimulus is assessed as S_1; at sensation levels below the criterion, the stimulus is assessed as S_0 (again, a particular case of S_0 may be zero value, the absence of a stimulus).

If the difference between the stimuli is large enough so that their probability-density functions virtually do not overlap (S_0 and S_2 in Fig. 1.1A), the decision criterion β_2 can be positioned at such a point that almost all sensation levels associated with S_2 are above the criterion value whereas all levels

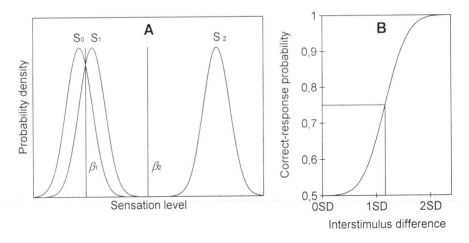

Figure 1.1. **A**. Probability density of sensation levels at different stimulus levels (S_0, S_1, S_2). β_1 – the best decision criterion to discriminate the stimuli S_0 and S_1; β_2 – the same for the stimuli S_0 and S_2. **B**. Correct-response probability as a function of interstimulus difference, at the best position of the decision criterion, assuming both probability-density functions being equal normal distribution; SD – standard deviation of the distributions.

associated with S_0 are below it. It means that the two stimuli can be discriminated with almost 100% probability. This high degree of performance becomes impossible, however, when two probability-density functions overlap because of small difference between the stimuli (S_0 and S_1). In this case, at any position of the decision criterion (β_1), some proportion of errors is inevitable since a part of S_1 distribution is below the criterion and a part of S_0 distribution is above it. It means that in some trials, the S_0 stimulus evokes sensation above the criterion and is wrongly assessed as S_1; in some other trials, the S_1 stimulus evokes sensation below the criterion and is wrongly assessed as S_0. Nevertheless, the number of correct responses (the area below the curve S_1 to the right of β_1 and the area below S_0 to the left of β_1) exceeds the number of errors (remaining parts of areas below these curves). It can be shown that the best ratio of correct to wrong responses is achieved when the decision criterion is at the interception point of the two distributions(β_1 in Fig. 1.1A). If the stimuli do not differ at all, the numbers of correct and wrong responses are equal at any decision criterion.

Thus, the ratio or correct to wrong responses depends on differences between the stimuli. The dependence of correct-response proportion on the difference between the discriminated stimuli is referred to as the psychometric curve (Fig. 1.1B). It is the integral of the probability-density function. At the best position of the decision criterion, proportion of correct responses varies within a range from 50% (indistinguishable stimuli) to 100% (completely distinguishable stimuli).

Estimation of a threshold is a matter of an arbitrarily chosen percentage of correct responses. In many studies, a 75% criterion is used as an middle point of the correct-response percentage range from 50 to 100%. If the probability-density distributions are equal normal distributions, this correct-response proportion appears at their shift relative to one another by 1.35 SD. In some studies, the difference of distributions equal to SD is taken as a threshold criterion; it gives 69% of correct responses. In some adaptive measurement procedures (see below), 71 or 79% of the criteria are usable.

An additional complication arises since the decision criterion is not always established at the best position for providing the highest correct-to-wrong response ratio. Because of motivation or other reasons, the subject may prefer to answer in one or another manner. For example, when the animal is trained to perform a certain movement in response to the stimulus (yes-response) and stay quiet when the stimulus is absent (no-response), it may prefer either to move or to stay. It corresponds to a shift of the decision criterion β_1 upward or downward. This is referred to as the subject's response bias. If the decision criterion is shifted upward, the proportion of no-responses (both correct no-responses to S_0 and wrong no-response to S_1) increases, whereas proportion of yes-response (both correct yes-responses to

S_1 and wrong yes-responses to S_0) decreases. Shifting the criterion down-ward yields the opposite result.

The signal-detection theory gives a basis to separate the influence of the discrimination capability and response bias on the subject's performance. For doing so, four probability values – those of both correct and wrong response to both S_0 and S_1 stimuli – have to be measured. In a particular case, when S_0 is the stimulus absence (noise only, n) and S_1 is the stimulus presence (signal + noise, sn), and two possible responses are yes (Y) and no (N), four possible response types are correct detection of the signal (hit, Y/sn), correct rejection (N/n), wrong rejection (miss, N/sn), and wrong detection (false alarm, Y/n). Probabilities of these response types are, respectively, $P(Y/sn)$, $P(N/n)$, $P(N/sn)$, and $P(Y/n)$ (Fig. 1.2).

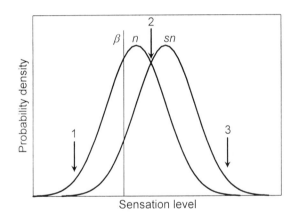

Figure 1.2. Probabilities of different response types at an arbitrary decision criterion. n – probability density of sensation level in the absence of the stimulus (noise), sn – that in the presence of the stimulus (signal + noise); β – a decision criterion. Probabilities $P(Y/sn)$ and $P(Y/n)$ are represented by the areas below the curves sn and n, respectively, to the right of the criterion; probabilities $P(N/sn)$ and $P(N/n)$ are represented by the remaining areas under the curves. Arrows 1–3 indicate a few decision-criterion positions.

It is axiomatic that

$$P(Y / sn) + P(N / sn) = 1 \tag{1.6}$$

and

$$P(Y / n) + P(N / n) = 1. \tag{1.7}$$

A standard presentation of these data is on a graph with $P(Y/sn)$ as the ordinate and $P(Y/n)$ as the abscissa (Fig. 1.3). It is referred to as the receiver operating characteristic (ROC) format. Consider presentation in this format of data obtained at various decision criteria as shown in Fig. 1.2.

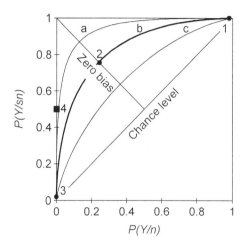

Figure 1.3. ROC-format graph: hit probability versus false-alarm probability. Three ROC-curves (a, b, c) represent different degrees of detectability; solid line (b) – the curve meeting the point of $P(Y/n) = 0.25$ and $P(Y/sn) = 0.75$. Points 1–3 correspond to decision-criteria 1–3 as shown in Fig. 1.2; point 4 represents $P(Y/sn) = 0.5$ at negligible $P(Y/n)$.

Criterion 1 represents extremely liberal strategy; both *n* and *sn* distributions are almost completely above the criterion, so both the false-alarm probability $P(Y/n)$ and hit probability $P(Y/sn)$ are close to 1. Although $P(Y/sn)$ is higher than $P(Y/n)$, their ratio is close to 1. It gives a point 1 at the graph of Fig. 1.3. Shifting the criterion to the right diminishes both the hit and false-alarm probabilities, but the ratio of hit to false-alarm probability increase. This ratio reaches its maximum at the criterion 2 (point 2 in Fig. 1.3). Further shift of the criterion to the right corresponds to more conservative strategies. Criterion 3 results in very low both hit and false-alarm probabilities (point 3 in Fig. 1.3). Thus, data obtained at a variety of decision criteria fall on a certain curve (the ROC-curve). The curve position depends on signal detectability. In Fig. 1.3, the curve *a* represents rather high detectability, whereas the curve *c* represents rather low one; the curve *b* represents intermediate degree of detectability. For a completely nondetectable signal, $P(Y/sn) = (Y/n)$, so the ROC-curve coincides with the positive diagonal (0, 0)–(1, 1) (chance-level line in Fig. 1.3). For a completely detectable stimulus, it passes on the left at upper borders of the graph (0, 0)–(0, 1)–(1, 1). The higher detectability, the higher position of the ROC-curve.

Thus, data presentation in the ROC-format makes it possible to estimate separately the signal detectability and response bias. The presence of an experimental point on a certain ROC-curve indicates the degree of detectability; one of these curves (such as that meeting a point of 75% hits and 25% false alarms) can be taken as an arbitrary threshold criterion. The position of a point on the curve indicates the bias: a shift to the upper-right corner indicates a more liberal strategy (preference to a yes-response); a shift to the lower-left corner indicates a more conservative strategy (preference to no-response). The position of a point on the negative half-diagonal corresponds to the neutral bias (zero bias line in Fig. 1.3).

If necessary, the bias can be manipulated purposefully by the reinforcement matrix and (or) probability: as a rule, the animal is capable of modifying its strategy in such a way as to get the maximum reinforcement. In aquatic mammals, it was demonstrated by Schusterman et al. (1975). In that study, the amount of reinforcement (number of fish) for hits and correct rejections was varied as 1:1, 1:4, and 4:1. The behavior of the animal closely followed the prediction of the theory. At the reinforcement ratio of 1:1 the bias was close to zero. At the ratio of 4:1 for hits and correct rejections, respectively (more reinforcement for hits), a bias to go-responses appeared: the number of both hits and false alarms increased. At the ratio of 1:4 (more reinforcement for correct rejections), a bias to no-go responses appeared: the number of both correct rejections and missing increased.

The procedure of threshold computation can be markedly simplified if the experimental procedure provides zero bias by definition. It is possible with the use of the forced-choice procedure. In this procedure, each trial contains two (sometimes more) temporal intervals or sites of stimulus presentation. One of these intervals (or positions) does contain a stimulus, whereas the other does not (absolute threshold determination), or they contain different stimuli (differential threshold determination). The subject has to detect which of the two intervals (positions) contains the stimulus or which of the stimuli is higher in intensity, pitch (acoustic), brightness (visual), and so on. The experimental situation is completely symmetrical both in the sensory respect (equal probability of the stimulus appearing in the first or in the second interval, in the right or left position) and in the effectory respect (say, depending on stimulus, the subject has to touch the right or left paddle). In this procedure, there are only two possible response variances: either true (hit) or false. The probability of hits in this paradigm varies from 1.0 (a completely detectable stimulus) to 0.5 (a random choice when a stimulus is not detectable). A certain arbitrarily chosen probability between these two levels can be adopted as a conventional threshold estimate: either 0.75 (as a middle point between the values of 1.0 and 0.5) or 0.71 or 0.79 (in staircase adaptive procedures, see below). The forced-choice paradigm is very convenient for psychophysical measurements since it gives easily interpretable results independent of the subject's bias to give positive or negative responses.

The forced-choice procedure is widely used in human psychophysics. Unfortunately, in many case it is difficult to train an animal for measurements in this paradigm, particularly for hearing measurements. The training may be successful if the required response is naturally associated with the stimulus; for example, the animal is required to touch the right paddle if a sound is emitted by the right speaker and to touch the left paddle if it is emitted by the left speaker. Such tasks are typical of spatial-selectivity measurements (e.g., Renaud and Popper, 1975). However, the task becomes very difficult in the absence of such natural association. For example, the animal is required to touch the right or left paddle depending on intensity or frequency of a sound emitted by *one and the same* speaker.

In some studies, spatial association of sound quality with a position of the paddle to be touched is established artificially by presenting sounds through two speakers: if the sound of a certain quality is emitted by the right speaker, the animal has to touch the right paddle; if it is emitted by the left speaker, the left paddle is to be touched (e.g., Vel'min and Dubrovskiy, 1975). In this case, animals can be trained successfully. However, a disadvantage of this approach is that measurement of a nonspatial sound parameter is mediated through the animal's ability to perform spatial discrimination.

Therefore, many psychophysical studies of hearing in aquatic mammals were carried out using the so-called go/no-go paradigm. In such experiments, each trial either contains or does not contain the stimulus. The animal is trained to perform a certain behavioral response (e.g., going to the paddle and touching it) to the stimulus (the go-response) and not to perform this response (remain at the start position) in the absence of the stimulus (the no-go response). In this paradigm, four response variances are possible: hit, miss, correct rejection, and false alarm. In this situation, a bias of the animal to go- or no-go response is important. So the threshold should be assessed using the ROC-format analysis.

Not all of hearing-threshold studies in aquatic mammals carried out in the go/no-go paradigm used the precise ROC-analysis of experimental data. It does not mean, however, that threshold estimates obtained in those studies were wrong. In most of the studies, the animal was trained carefully to reach a very low, almost negligible proportion of false alarms. One of the ways to reach a very low false-alarm probability was to punish the animal for each false alarm by including of 1 to 2 minute "time out" in the experiment. In this situation, it is reasonable to suppose that almost all go-responses indicate true detections. As a rule, a 50% go-response probability was used as an arbitrary threshold criterion, as a middle point between 100% go-response probability for completely detectable signals and near-zero go-response probability for nondetectable signals.

It should be noted that according to the signal-detection theory, a very low percentage of hits can be taken as a threshold criterion when the number of false alarms is negligible (for example, point 3 in Fig. 1.3). However, such low threshold criteria were never used since threshold estimates become very error-sensitive if they are based on only a few hits. As a rule, at least 50% performance was used as a threshold criterion. The point corresponding to 0.5 probability of hits and near-zero probability of false alarms belongs to a very high ROC-curve indicative of very high detectability (point 4 in Fig. 1.3). So from the point of view of the signal-detection theory, this threshold criterion is very conservative and yields some overestimation of the threshold. However, if the psychometric curve (see Fig. 1.1B) is steep enough, this overestimation is not very large and can be neglected.

In some investigations, experimental paradigms were used which combined some features of both two-alternative forced choice and go/no-go paradigms. For example, stimulus trials were either stimulus or nonstimulus (a feature of the go/no-go paradigm), and the animal was trained to touch one paddle when the stimulus was present and another paddle when the stimulus was absent (a feature of the two-alternative paradigm) (e.g., Møhl, 1968a; Terhune and Ronald, 1971; Terhune, 1988, 1989, 1991). Animals could be trained to perform this task, though it is more difficult than the go/no-go task. In these cases, a response bias was expected to be less than in the standard go/no-go paradigm because of equality of two required responses. However, even in this paradigm, some bias could not be excluded because of possible preference to one or another paddle.

A much more favorable situation appeared in many psychophysical measurements of visual capabilities of aquatic mammals. In these measurements, the true two-alternative forced-choice paradigm could be used because it was very easy to exploit the natural association between the location of the visual stimulus and location of the response paddle. For example, a simple experimental design implies two equal windows for presenting visual stimuli and two equal response paddles near these windows. The animal is required to push the paddle near the window where the stimulus appears. This task is rather easy for many animals. If the stimulus is presented randomly in one or another window, bias is absent or cannot influence the measurement results.

1.4.4. Data-Collection Procedures

To collect data appropriate for statistical threshold evaluation, a number of stimulus should be presented to the subject with variation of the investigated parameter (intensity, frequency, and so on) around the threshold. Then correct and wrong responses to each of the parameter values should be counted to calculate the correct-response probabilities and to use these data for threshold evaluation. The two most frequently used procedures of variation of the parameter are (1) the method of constant stimuli and (2) the up-down (adaptive, staircase) procedure.

With the use of the constant-stimuli method, the range of variation is selected prior to the testing. A number of values of the stimulus parameter within the selected range are presented randomly or in an ordered manner. Then percentage of hits (and percentage of false alarms if the ROC-analysis is to be applied) to each of the stimulus values is counted, and psychometric curves (see Fig. 1.1B) can be drawn. The threshold is usually found by interpolation between stimulus values that result in a hit percentage just above and just below the chosen threshold criterion, say, 75%, or the hit-to-false alarm ratio that corresponds to a certain ROC-curve. A more precise method is to approximate the experimental psychometric curve by a certain analytical function, such as the integral of Gaussian function, which is fitted to the data according to the least-square criterion. Interception of the approximating function with the threshold criterion gives the searched threshold estimate. The advantage of the last technique is that it uses more than two experimental points thus making the result more confident.

The constant-stimuli method requires anticipation of a certain threshold value in order to use stimulus values just above and just below the threshold. This is a disadvantage of the method: after completing the measurement session, it may occur that most of the presented stimuli were either far suprathreshold or far subthreshold, so the threshold cannot be evaluated properly. The adaptive (up-down, or staircase) procedure is free of this disadvantage because it implies choosing the stimulus values during the measurement session according to the subject's responses. In a simple version, the up-down procedure implies making the stimulus by a step more difficult for detection (for example, lower in intensity) after each hit and more easy for detection (higher in intensity) after each miss. It can be shown that such a procedure results in stimulus' level fluctuations around a value providing 50% correct response probability – that is, around the threshold criterion. As a rule, a measurement begins from an easily detectable stimulus value ("warming-up" trials), and the adaptive procedure soon brings the stimulus values close to the threshold. An estimate of the threshold is usually obtained by averaging the stimulus values at all reversal points. This estimate is close to that obtained by 50% criterion.

This version of the adaptive procedure is proper only when the false-alarm probability is negligible. It cannot be used in the two-alternative forced-choice paradigm when the theoretical hit probability is not less than 50%. For such cases, modifications of the adaptive procedure were elaborated known as one-up-two-down or one-up-three-down procedures (Levitt, 1971). In these versions, the stimulus is made one step easier to detect or discriminate after each incorrect response (just as in the original up-down procedure), but it is made one step more difficult to detect (discriminate) after *two* (one-up-two-down) or *three* (one-up-three-down) successive correct responses. These procedures result in stimulus' value fluctuation around a level of $0.5^{1/2} \approx 0.71$ or $0.5^{1/3} \approx 0.79$, respectively. These values are rather close to a standard 75% threshold criterion and thus can be used as arbitrary criteria. As in the original procedure, the threshold can be estimated by averaging stimulus values at the reversal points.

Chapter 2
HEARING IN CETACEANS

The auditory system of cetaceans, since they are capable of underwater hearing and adapted for echolocation, has attracted major interest for many years. More precisely speaking, one of the two cetacean suborders, *Odontoceti* (toothed whales, dolphins, and porpoises) was a subject of a particular interest. Probably all of them (at least, all species investigated to date) are capable of active echolocation. For echolocation, they use ultrasonic signals ranging to higher than 100 kHz (Kellogg, 1959; Norris et al., 1961; Norris, 1969; Au, 1993). Some information on auditory perception of odontocetes obtained in behavioral conditioning studies is presented in reviews by Popper (1980), Fobes and Smock (1981), Watkins and Wartzok (1985), and Au (1993).

In another cetacean suborder, *Mysticeti*, hearing is poorly investigated because they all are animals of a very large body size, which makes it difficult to use them in experimental studies. Their hearing capabilities can be only indirectly estimated basing on their vocalization, assuming that a vocalization frequency range is in accordance with the hearing sensitivity range. Mysticete vocalizations are significantly lower in frequency than those of odontocetes: from a few tens Hz to a few kHz (Weston and Black, 1965; Payne and Webb, 1971; Watkins, 1981; Clark, 1982, 1990; Watkins et al., 1987; Edds, 1982, 1988). So it is reasonable to suppose that their hearing is also low-frequency, contrary to the high-frequency hearing of odontocetes.

Below only the hearing physiology of odontocetes (mostly dolphins and porpoises) is considered since data on mysticetes are not available.

2.1. EAR MORPHOLOGY

2.1.1. Outer Ear and Middle Ear

Understanding of the many aspects of hearing physiology in cetaceans is not possible without taking into account peculiarities of the anatomy and biomechanics of their ear. A few reviews (Ketten, 1990, 1992a,b, 1997) consider this topic in detail; so only a brief overview is presented herein. All parts of the cetacean outer, middle, and inner ears are modified markedly as compared to the ear of terrestrial mammals.

Pinnae are absent in cetaceans, and external auditory canals are present but reduced. There is a small external meatus connected with the canal, but the canal itself is very narrow and filled with cells and cerumen; it does not connect with the tympanic membrane. Although initially some authors suggested that the external auditory canal might be the primary pathway for sound transmission in cetaceans (Fraser and Purves, 1954, 1959, 1960), it is commonly accepted now that it is not capable of being an acoustic pathway to the middle ear. It is not commonly agreed on how sounds are transmitted to the middle ear; this topic is discussed in detail below (see Section 2.10). In any case, it follows from the inability of the external auditory canal to function as a sound-conducting pathway that the eardrum hardly functions in the manner that it does in terrestrial mammals. Therefore, other ways should exist to transfer acoustical vibration to the inner ear.

The reduction of the external auditory canal is obviously an adaptation to the acoustic properties of water. In air, sounds cannot be transmitted directly into the body because of large difference in impedances of air and body tissues. Therefore, an air-filled canal is necessary to conduct airborne sounds to the sound-receiving membrane, the eardrum. In water, there are other ways do deliver sounds to the middle ear, so the external auditory canal was reduced.

The middle ear and inner ear of odontocetes are housed in two bulbous bones: tympanic bulla and periotic bulla, respectively. Very important, this tympano-periotic bullar complex in odontocetes is not firmly attached to the brain case; it is suspended in the peribullar cavity by a few ligaments and surrounded by a spongy mucosa. Thus, the bullar complex is isolated from the bony sound conduction (McCormick et al., 1970; Oelschläger, 1986). The isolation of the ears from the skull may be an important adaptation for underwater binaural hearing; under water, it prevents direct bony conduction of sounds to both ears equally. In mysticetes, the tympano-periotic complex is in direct contact with the skull through the mastoid process.

The middle ear exhibits a number of specific features resulting from its adaptation to underwater functioning. *In vivo* computer-tomography and magnetic-resonance imaging investigations suggested the intratympanic space is air-filled (Ketten, 1997). Thus, the ear must be adapted to prevent barotrauma from large pressure change in diving. One of the adaptations is tough and broad Eustachian tubes, which reduces the probability of tube closure and large external versus intratympanic pressure difference. Another adaptation is a thick, vascularized fibrous sheet, the *corpus cavernosum*, lining the middle-ear cavity, which can provide compensating changes of the cavity volume. Therefore, a question arises whether the change of the middle-ear cavity volume alters its resonance properties and frequency sensitivity. It should be noted that all measurements of hearing characteristics in cetaceans were carried out in experiments when the animal was positioned not far below the water surface; it remains still unknown whether all these characteristics are the same at a large depth underwater.

In accordance with the reduced external auditory canal, the tympanic membrane is deeply modified into an elongated, conical structure, the tympanic conus. It hardly plays the same role in sound reception as in terrestrial mammals. The lateral wall of the tympanic bulla has discrete areas of thin bone that align with the pan bone of the mandible that may function as a sound-conducting pathway (see below, Section 2.10).

All middle-ear ossicles (malleus, incus, and stapes) persist in cetaceans, but they are of specific shape (Reysenbah de Haan, 1956; McCormick et al., 1970; Fleischer, 1978; Nummela et al., 1999a, 1999b). There is no direct connection between the tympanic membrane (conus) and the malleus. Thus, contrary to terrestrial mammals, vibration of the ossicles are hardly transmitted through the tympanic membrane. Instead of the tympanic membrane, the malleus is fastened to the wall of the tympanic cavity by a bony ridge, the *processus gracilis*. The incus is tightly connected to the malleus thus forming a union. There is, however, an articulation between the stapes head and the incus.

A remarkable feature of all ossicles in odontocetes is their high density: mean 2.64 g/cm^3, which is markedly higher than in terrestrial mammals (2–2.2 g/cm^3) and even higher than in mysticetes (mean 2.35 g/cm^3); the high density probably indicates high stiffness (Nummela et al., 1999b). The whole ossicle complex becomes a stiff system capable of transmitting high-frequency vibrations. These features are not characteristic of mysticetes, which have more massive and not stiffened ossicles. The stapes footplate lies on the oval window being attached to the bone by a narrow annular ligament.

The mode of action of middle-ear ossicles in cetaceans is not completely understood. The stapes positioned in a usual manner on the oval window indicates the same mode of vibration transmission to the inner ear as in terrestrial mammals. But it was debatable how sound energy converts to stapes movements. McCormick et al. (1970, 1980) have shown that immobilizing the ossicular chain decreased cochlear potentials, whereas disrupting the external canal and tympanic cone had no effect. Basing on these data, they supposed that motion of the stapes relative to the cochlear capsule resulted from differential motion of the tympanic and periotic parts of the tympanoperiotic complex. It was suggested that the ossicular chain functions as a rigid spring mounted inside the tympanic cavity; due to this position, vibration of the bulla causes the stapes to move relative to the oval window in a "piston-cylinder" manner.

Another realistic model was suggested recently by Nummela et al. (1999a), Hemilä et al., (1999). This model bases on an assumption that in cetaceans, the reduced tympanic membrane is functionally substituted by the thin wall of the tympanic bone, the tympanic plate (Fig. 2.1). This thin bony membrane is much more stiff than the elastic tympanic membrane of terres-

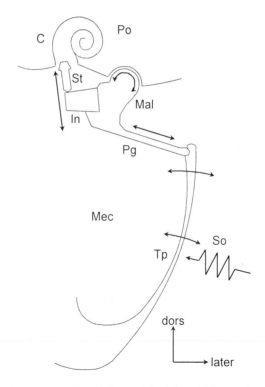

Figure 2.1. Schematic representation of the model of the middle-ear function in odontocetes (combined drawing from Nummela et al., 1999a, and Hemilä et al., 1999). Po – periotic bone, Mec – middle ear cavity, C – cochlea, St – stapes, In – incus, Mal – malleus, Pg – *processus gracilis*, Tp – tympanic plate, So – incident sound. Double-headed arrows show directions of vibration of the tympanic plate and ossicles; increasing lengths of the arrows from the tympanic plate to stapes show increasing vibration amplitude.

trial mammals, thus it is more appropriate to serve as a sound-receiving membrane in the high-impedance liquid media. The ventro-medial side of the tympanic plate is thick and massive; it vibrates very little at high frequencies, so the tympanic plate functions as a lever: vibration amplitude of its dorso-lateral edge is larger than the average vibration amplitude of the plate. The *processus gracilis* of the malleus forms a bone bridge from the rim of the tympanic plate to the malleus-incus union. It transmits vibrations to the malleus-incus. Since the head of the malleus rests against the massive periotic bone, it performs rotational vibration around this articulation. The opposite end of this lever, the incus, pushes the stapes to move relative to the oval window.

Thus, the system consisting of the tympanic plate and ossicles effectively transmits vibration to the oval window of the cochlea. Additionally, this system increases the velocity of vibration due to action of two levers: the tympanic plate, which provides higher amplitude of vibration of its dorso-lateral rim as compared to the average amplitude, and the malleus-incus union, with its incus arm longer than the malleus arm. This two-step increase of vibration velocity serves to match the high impedance of water with lower input impedance of the cochlea – the task just opposite that in terrestrial mammals, where the middle-ear ossicles serve to match the low impedance of air to higher input impedance of the cochlea.

At the moment, the model of Nummela et al. (1999a, 1999b), Hemilä et al., (1999) seems the most elaborated. Although not supported yet by direct physiological data, it nevertheless is based on commonly adopted anatomical findings. Computation based on middle-ear parameters of a few odontocete species (Hemilä et al., 1999) has shown a good coincidence of predictions of the model with hearing sensitivity and frequency range of all these species.

2.1.2. Inner Ear and Peripheral Neurons

The inner ear of cetaceans, as of all other mammals, consists of the auditory and vestibular labyrinths. However, the auditory labyrinth is much larger than the vestibular one – a feature that is not characteristic of many other mammals.

Detailed investigation of the cetacean cochlea was carried out by Wever et al. (1971a, 1971b, 1971c, 1972), Ketten (1992a, 1992b), Solntseva (1990). Qualitatively, the cetacean cochlea exhibits all principal features of the mammalian cochlea. It is organized as a spiral tube inside the periotic bone with a hollow bony axis, the modiolus. In the bottlenose dolphin *Tursiops truncatus*, which was studied in the most detail, the cochlea is of slightly more than two turns; in the Pacific white-sided dolphin *Lagenorhynchus obliquidens* it is of about 1.75 turns (for comparison: in humans, the cochlea is of nearly three turns). The tube is separated by the Reissner's and basilar membranes longitudinally into three ducts: the cochlear duct *scala media* filled with endolymph, tympanic duct *scala tympani*, and vestibular duct *scala vestibuli* filled with perilymph.

Stapes movement relative the oval window produce compressive waves deforming the basilar membrane. These waves spread along the membrane, and their amplitude vary along the membrane length depending on frequency, thus resulting in frequency selectivity of hair cells located on the basilar membrane. Thus, the basilar membrane dimensions and stiffness play a primary role in adjusting the dolphin's hearing to high sound frequencies.

As in all mammals, the basilar membrane in cetaceans is narrow and thick at the base and gradually thins and broadens to the apex. However, its quantitative characteristics are very specific in odontocetes. From base to apex, the basilar membrane increases in width 10 to 14 times and decreases in thickness five to six times. For comparison, in humans these ranges are five to six and two times, respectively. At the base, the basilar membrane of odontocetes actually looks like not a membrane but a bar: it is about 30 µm wide and 25 µm thick; at the apex, it is 300 to 400 µm wide and around 5 µm thick. Additionally, the basal region of the basilar membrane is supported by the ossified outer lamina, which further increases the stiffness of the construction. Due to such organization, the basilar membrane is able to resonate in a very wide frequency range. Its base resonates at extremely high frequencies, more than 100 kHz. At the same time, the thin and wide apical region is sensitive to relatively low frequencies of an order of tens of Hz.

In mysticetes, dimensions of basilar membrane are quite different from those in odontocetes: it is wide (about 100 µm at the base and more than 1000 µm at the apex) and thin (about 7 µm at the base and 2 µm at the apex). Thus, it is obviously tuned to low frequencies.

The organ of Corti containing the hair cells (acoustic receptors) and their supporting structures is located along all the basilar membrane in a manner common for all mammals. As in all mammals, the organ of Corti contains two types of hair cells: the inner (arranged in one row at the edge closer to the modiolus) and outer cells (arranged in a few rows at the opposite edge of the organ of Corti). The number of inner hair cells in *Tursiops* and *Lagenorhynchus* is near 3500, and the number of the outer hair cells is around 13,000. This ratio is very close to that in other mammals (in humans, near 3500 of the inner hair cells and 11,500 of the outer hair cells). This coincidence is remarkable, particularly, taking into account different roles of the inner and outer hair cells in sound perception. The inner hair cells are the acoustic receptors properly (they have synaptic contacts with afferent acoustic fibers), whereas the outer hair cells combine sensory and kinetic properties. Being excited by the basilar membrane vibration, they are capable of applying forces back to the basilar membrane, thus performing a positive feedback, which results in both high hearing sensitivity and sharp frequency tuning of the basilar membrane. Similar ratio of inner-to-outer hair cells in cetaceans and in other mammals suggests that this active mechanism of frequency tuning is able to operate, in particular, in the wide frequency range available for odontocetes.

The next level of signal transmission within the auditory system is the ganglion cell population. In all mammals, the number of ganglion cells is several times more than that of the inner hair cells (in humans, there are around 3500 of inner hair cells and around 30,000 of ganglion cells; the mean convergention ratio is around 8.5). Thus at least a few afferent fibers converge at each inner hair cell. In odontocetes, this ratio is markedly higher than in other mammals: according to Wever et al. (1972), there are from 60,000 to 70,000 ganglion cells in *Lagenorhynchus* and 95,000 cells in *Tursiops*. Thus, the mean convergention ratio is from 20 to 28.

The auditory nerve in cetaceans also demonstrates high degree of development. The number of auditory fibers (which are axons of ganglion cells) is rather high. In most dolphins, it is within a range of 70,000 to 100,000; in large whales, it exceeds 150,000 (contrary to about 30,000 in humans). In odontocetes, auditory fibers are thick (mean 12 μm in diameter), in mysticetes they are of smaller diameter (mean 3 μm) but still thicker than in most terrestrial mammals (Morgane and Jacobs, 1972; Gao and Zhou, 1991, 1992; Ketten, 1997)

2.2. AUDITORY EVOKED POTENTIALS IN CETACEANS

2.2.1. Intracranial Evoked Potentials

Since 1968, the auditory abilities of dolphins have been investigated using electrophysiological methods: recording of evoked potentials (EP) with implanted electrodes in the auditory centers of the brain. A pioneering study by Bullock et al. (1968) presented a first successful attempt to record EP in the dolphin brainstem using intracranial electrodes. Many findings of that study will be referred to below in corresponding sections of this chapter. Herein, it is noteworthy that the main properties of EP in the dolphin's brain stem were at first described in detail in that study. The recorded EP were of short latency and duration. They consisted of a series of short waves, each lasting 0.5 to 1 ms – that is, their duration was of an order characteristic of neuronal spikes rather than of longer synaptic potentials. EP amplitude was as large as tens of microvolts, so they were easily recordable and consistent.

Main properties of brainstem EP were confirmed in studies by Voronov and Stosman (1977), who used intracranial electrodes to record EP in a number of brainsem auditory nuclei of the harbor porpoise *Phocoena pho-coena*. They used intrabrain electrodes, which were stereotaxically inserted into the brain along the main brainstem axis. Thus, during one electrode penetration, they in consecutive order recorded EPs from the inferior collicu-lus, lateral lemnisc, superior olive, trapezoidal body, and cochlear nuclei. EP waveform varied depending on the recording point; nevertheless, in all the points the recorded EP was a complex of a few short waves, each about 0.5 ms long, with a common duration of several milliseconds. No special precau-tion was taken in that study to distinguish local evoked-potential activity (generated by neurons just around the recording point) and far-field activity (generated in other nuclei and recorded at a distance). A complicated wave-form of EP is an indication that recording of far-field potentials generated by a few sources was very probable. The weight of activity of different genera-tors depended on the electrode position; this resulted in different potential waveforms in different recording sites.

Frequency-threshold curves obtained by Voronov and Stosman (1977) us-ing EP recording in different sites within the brainstem did not exhibit any systematic trend though varied markedly. Using long tone bursts, the authors described evoked responses to both burst onset and offset. However, there were not enough data to conclude that the latter was the real off-response – that is, the response to sound cessation properly rather than to spectrum splatter at an instant of the sound offset.

Another series of evoked-potential studies in dolphins concerned re-sponses of the auditory cortex, also using intracranial recordings (Popov and Supin 1976a, 1976b; Popov et al., 1986). These studies were carried out us-ing electrodes implanted into the auditory cerebral cortex. Being inserted into the cortex up to 15 to 20 mm deep, the electrodes recorded pronounced EPs of more than 100 µV amplitude. Their waveform greatly differed from those recorded in the brainstem. Cortical EP were simple-shaped positive or negative (depending on the recording site) waves 15 to 20 ms long (Fig. 2.2). Variable polarity was not surprising taking into account uncontrolled elec-trode position relative to cortical layers. Responses were evoked effectively by both wide-band stimuli (clicks or noise bursts) and tone bursts as well as by level or frequency shifts. Response amplitude depended on the stimulus level within a range of 40 to 60 dB, and response threshold could be esti-mated.

Evoked potentials recorded by Ridgway (1980) through implanted epidu-ral electrodes probably are also of cortical origin. These EPs were longer

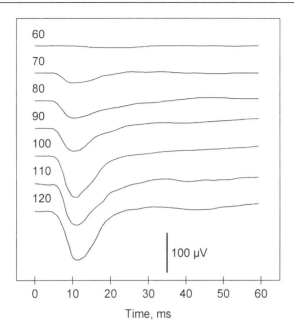

Figure 2.2. Intracranially recorded EP in a harbor porpoise *Phocoena phocoena*. Each record is an average of 30 sweeps. Stimulus is wide-band noise onset at the instant of record beginning. Stimulus intensity (dB re 1 μPa sound pressure) is indicated near records.

than those described above; their first-wave peak latency was about 30 ms. The difference in the temporal characteristics may be explained by different electrode positions. In the study by Ridgway, electrodes were positioned over the temporal cortex; as shown below (Section 2.11), this cortical area is not a primary auditory cortex.

2.2.2. Auditory Brainstem Responses (ABR)

After a period of invasive evoked-response investigations, it was reasonably admitted that the use of invasive, particularly intracranial, recordings in dolphins contradicts to ethic rules. In the last years, practical and legal reasons also prevented invasive investigations in cetaceans. Therefore, noninvasive techniques of evoked-potential-studies attracted the major attention. Among several types of noninvasively recorded EP, the so-called auditory brainstem responses (ABR) proved to be particularly valuable for investigations of hearing in cetaceans. ABR is widely used in studies on auditory perception of humans and is well detectable in many animals: the cat (Achor and Starr, 1980), rat (Shaw, 1990), hamster (Chen and Chen, 1990), monkeys (Legatt et al., 1986), and other mammals (Corwin et al., 1982); see also review by More (1983). It was found that ABR can be recorded in cetaceans (dolphins) as well.

Preliminary indications of possibility to record brain-stem evoked responses distantly were obtained in dolphins using electrodes positioned intracranially but not in the brain stem (Ladygina and Supin, 1970; Bullock and Ridgway, 1972; Supin et al., 1978). EP similar to those described by Bullock et al. (1968) were recorded through intracranial electrodes located in other brain structures, particularly, in the cerebral cortex.

The next step in this direction was the use of invasive but less traumatic extracranial recordings instead of intra-cranial ones. This technique allowed the recording of ABR-type EP from or near the skull surface (Ridgway et al., 1981). In that study performed on a bottlenose dolphin *Tursiops truncatus* and a common dolphin *Delphinus delphis*, the electrode position was used that as far as possible reproduced the standard recording position in humans; specifically, two recording electrodes were positioned at the vertex and the occipital bone. An averaging technique was used to extract the response from noise. In such conditions, well defined multipeak EP to short click was recorded, which was interpreted as ABR. Its overall duration was several ms, each wave lasting about 0.5 ms, and amplitude was as high as 10 μV. The ABR waves were numbered as I to VII in order of the positive peaks on the vertex (negative at the occipital bone) by analogy with the commonly adopted designation of ABR waves in humans.

Both intra- and extracranial ABR records were used in studies of Bibikov (1992) in harbor porpoises *Phocoena phocoena*. He picked up intracranial ABR of amplitude as large as 30 μV and extracranial responses up to 10 μV. Similarly to data of Ridgway et al. (1981), the responses consisted of a series of alternating positive and negative waves, each lasting 0.5 to 1 ms.

Further advance in evoked-potential studies of hearing in cetaceans was associated with really noninvasive recording of ABR in dolphins (Popov and Supin, 1985, 1990a, 1990b). In these studies, recording electrodes were secured at the body surface. Various types of noninvasive electrodes were used: either thin needle-shaped electrodes (diameter 0.3 mm) inserted 2 to 3 mm into the skin (no signs indicative of painfulness of such a procedure for the animal were observed), or electrodes mounted in suction cups fastened at the skin, or small plates fastened by adhesive tape or adhesive electro-conductive gel. All these electrode types proved to be effective. No attempt to mimic the electrode position standard for humans was made in these studies, and the so-called monopolar recording was used when one electrode (active) is placed near the potential source and the other one (referent) far from the source. The active electrode was placed at the dorsal head surface; the reference one could be placed at any part of the body far from the head, at the back or dorsal fin. When marine dolphins kept in sea water were studied, the electrodes (both the active and the reference one) were placed dorsally, and the animal was placed in the experimental tank in such a way as to keep the electrodes above the water surface. It was done to avoid shunting the electrodes by low-impedance sea water. However, a major part of the body remained in water to make experimental conditions comfortable enough for the animal. As a rule, to keep the animal in such a position, it was supported by a stretcher. When the animal was kept in high-impedance fresh water (when river dolphins were studied), both above-water and under-water electrode positions were effective. The experiments required neither anesthesia nor curarization. The animal could remain quiet in such a position for a few hours, allowing the recording of a large body of information during each recording session. It is noteworthy that no surgical procedures were used in such experiments.

A typical ABR recorded from the dolphin's head surface is demonstrated in Fig. 2.3 A. It consists of a sequence of waves, each lasting under 1 ms. The polarity of the ABR waves recorded from the head surface corresponds to that of the occipital electrode in recordings of Ridgway et al. (1981). The response onset latency without the acoustic delay (since the moment an acoustic pulse reached the animal's head) was 0.9 to 1 ms for clicks of high intensity.

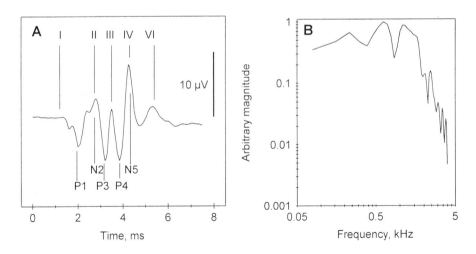

Figure 2.3. **A**. Waveform of ABR recorded from the dorsal head surface of a bottlenose dolphin *Tursiops truncatus*. Negativity of the active electrode is upward in this and all subsequent records. Zero point of the time scale corresponds to the instant of pulse excitation of the transducer; acoustic delay is 0.7 ms. Above the curve: designation of ABR waves by Ridgway et al. (1981); below the curve: designation by Popov and Supin (1990a). **B**. Frequency spectrum of the waveform presented in (A).

Since a correspondence between ABR waves in dolphins and humans is debatable, we did not use the numbering of ABR waves adopted for humans; instead of that, main both positive and negative waves were enumerated in succession as P1 to N5 (P – positive, N – negative). Wave peak latencies at clicks of high intensity were as follows (not including an acoustic delay): P1 – 1.5 ms, N2 – 2.1 ms, P3 – 2.6 ms, P4 – 3.2 ms, and N5 – 3.6 ms (Fig. 2.3A). Some additional ripples could be observed at the main waves.

Waves P3, P4, and N5 were the most prominent. In the bottlenose dolphin, the maximum ABR amplitude between the peaks P4–N5 constituted a few μV for stimuli of high intensity and exceeded 10 to 15 μV when the stimulation conditions were optimal.

This ABR waveform has a frequency spectrum that extends mostly up to 2000 Hz (Fig. 2.3B). Within this range, the spectrum exhibits a few peaks and valleys. The most prominent peaks are at 600 to 650, and 1000 to 1400 Hz. This peak-valley structure appears because of characteristics delays between ABR waves. It is well known that overlapping of equal or similar signals with a certain delay δt results in a ripple pattern of the frequency spectrum, the ripple frequency spacing $\delta f = 1/\delta t$. The ABR interwave delays from 1.5 to 1.6 ms (between P1 and P4, N2 and N5) to 0.6 ms (between P3 and P4) result just in frequency peaks from 600 to 1400 Hz.

Of course, as every noninvasively recorded EP of rather low amplitude, ABR requires to use a coherent averaging technique for its extraction from noise. The number of averaged sweeps depend on ABR amplitude, which, in turn, depends on stimulus parameters. As a rule, ABR becomes well detectable at averaging of only a few tens of sweeps; however, for precise measurements, it is desirable to average at least a few hundreds of sweeps, as a rule, 500 to 1000 sweeps, sometimes more. In this case, a very high response-to-noise ratio is achieved (Fig. 2.4). It should be noted that even at such number of averaged sweeps, ABR collection does not consume a very long time since these responses can be recorded at rather high rates of stimulus presentation, 10 to 30/s; the collection of 1000 sweeps takes 100 to 30 s, respectively.

The origin of ABR waves in dolphins was not purposefully investigated; nevertheless, some suggestions can be made. At least the first wave (P1) can be considered as a manifestation of activity of the acoustic nerve since its onset latency (less than 1 ms) is too short to suggest a synaptic delay. The latest waves (P4–N5) seem rather similar to the EP recorded by Bullock et al. (1968) and Voronov and Stosman (1977) directly from the inferior colliculus in dolphins; so it is reasonably to suggest that these ABR waves are of midbrain origin. This is a significant difference of ABR in dolphins from that in humans: the latter is generated mainly by auditory structures peripheral to the inferior colliculus.

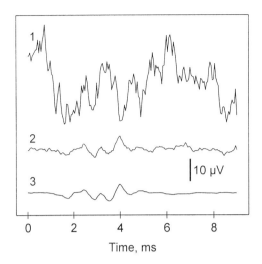

Figure 2.4. Extraction of ABR from noise at various number of averaged sweeps. 1 – single sweep, 2 – averaging of 64 sweeps, 3 – averaging of 1024 sweeps.

For successful use of ABR in acoustic measurements, the optimal position of electrodes should be used that provides the highest response amplitude and thus the best response-to-noise ratio. To find this position, ABRs were recorded from various sites on the dorso-lateral surface of the head. ABR amplitude dependence on electrode position along the middorsal line is shown in Fig. 2.5. The amplitude is maximal in a position of 6 to 9 cm caudal to the blowhole. In this region, all ABR waves are well pronounced. Moving off from this region, the response amplitude decreases at relatively low rate, so responses of considerable amplitude are recordable even at a distance of 15 to 20 cm from the focus. At a distance of 50 to 70 cm caudal from the blowhole, ABR amplitude falls to a negligible value; so this region can be taken as a reference electrode position. Taking into account these observations, the active electrode in most of experiments in bottlenose dolphins was located 6 to 9 cm caudal to the blowhole, while the reference electrode was placed near or at the dorsal fin.

To date, ABR were studied in a few species of odontocetes. Apart of the bottlenose dolphin *Tursiops truncatus*, which was investigated in the most detail, some data are available for the harbor porpoise *Phocoena phocoena* (Bibikov, 1992; Klishin and Popov, 2000), the common dolphin *Delphinus delphis* (Popov and Klishin, 1998), the Amazon river dolphin *Inia geoffrensis* and tucuxi dolphin *Sotalia fluviatilis* (Popov and Supin, 1990b), the beluga whale *Delphinapterus leucas* (Popov and Supin, 1987; Klishin et al., 2000), and the killer whale *Orcinus orca* (Szymanski et al., 1998, 1989).

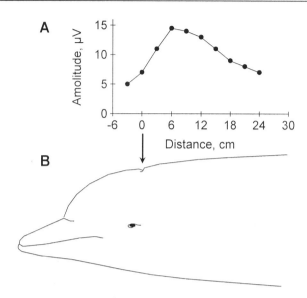

Figure 2.5. ABR amplitude dependence on the active electrode position along the longitudinal axis. Distances are indicated from the blowhole, which is taken as an arbitrary zero (arrow). Position of the abscissa axis in (**A**) corresponds to the dolphin's head profile shown in (**B**).

Examples of records obtained at similar stimulation conditions (a wide-band click of high intensity – such stimulus evokes a response of the maximal amplitude that is possible in the given species) are presented in Fig. 2.6. They demonstrate an obvious similarity of ABR waveforms in all the investigated species.

Apart of that, the examples show a noteworthy regularity: the larger body size of the animal, the less response amplitude. This regularity has an obvious physical basis. Indeed, far-field potentials of a dipole source are inversely proportional to cubed distance:

$$V = k_1 d / D^3 , \tag{2.1}$$

where V is the response voltage, d is the dipole moment, D is the distance, and k_1 is an adjusting factor. Supposing that d is proportional to the brain volume and hence to the brain mass ($d = k_2 m$), and D is proportional to the cube root of the body volume and hence of the body mass [$D = (k_3 M)^{1/3}$], it follows that

$$V = k_1 k_2 k_3 m / M , \tag{2.2}$$

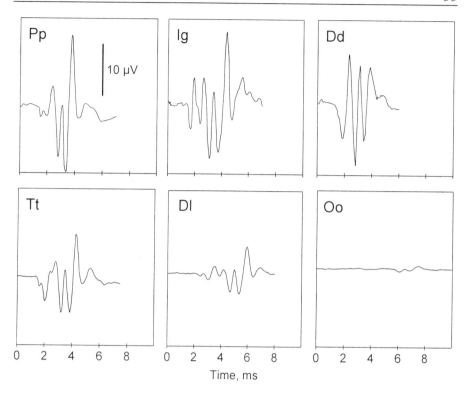

Figure 2.6. ABR examples of a few odontocete species. Pp – *Phocoena phocoena*, Ig – *Inia geoffrensis*, Dd – *Delphinus delphis*, Tt – *Tursiops truncatus*, Dl – *Delphinapterus leucas*, Oo – *Orcinus orca*. All records are presented at equal time and voltage scales.

where *m* is the brain mass, *M* is the body mass, and k_2, k_3 are adjusting factors. That is, other factors being equal, the response voltage is expected to be proportional to the brain/body mass ratio.

Of course, it is a very rough approximation that does not take into account a possible difference of body and brain shape, the degree of development of acoustic brain nuclei, and so on. Nevertheless, being applied to animals of similar constitution, for example, to odontocetes, it gives a satisfactory explanation of different evoked-potential amplitudes. Indeed, it is commonly known that the brain/body mass ratio decreases with increasing the body size. In particular, in the harbor porpoise, the brain/body mass is around 1:50 (0.8–1 kg brain mass, 40–50 kg body mass); in the bottlenose dolphin, this ratio is around 1:100 (1.5–1.8 kg brain mass, 150–200 kg body mass); in the beluga whale, the ratio is around 1:200 (about 2.5 kg brain mass, 500 kg body mass); in the killer whale, this ratio is around 1:1000 (3.5–4 kg brain mass, 3000–4000 kg body mass). This predicts a ratio of EP amplitudes in these four species as 20:10:5:1, which is rather close to the real ratio of ABR amplitudes (about 20 µV, more than 10 µV, less than 10 µV and about 1 µV, respectively). Thus, based on the highest ABR amplitude (which is important for measurement precision since higher amplitude provides better response-to-noise ratio), small cetaceans seem to provide better opportunities for evoked-potential studies of hearing.

2.2.3. Noninvasively Recorded Cortical Evoked Responses

Apart from ABR described above, another type of auditory evoked responses can be recorded noninvasively in dolphins (Popov and Supin, 1986; Supin and Popov, 1990). Temporal characteristics of this EP type are quite different from that of ABR: its onset latency is 8 to 10 ms, peak latency of the first wave about 15 ms, overall duration about 30 ms, and it looks like a simple positive-negative waveform (Fig. 2.7A).

The maximum response amplitude of this response type was observed in a restricted region about 20 cm behind the blowhole – that is, much more

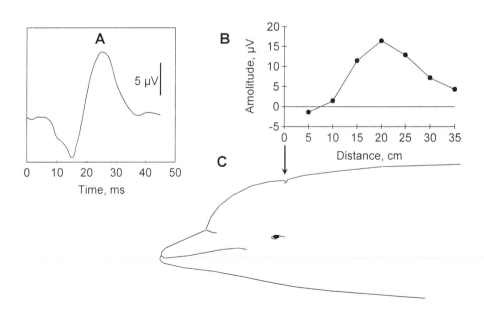

Figure 2.7. ACR properties. **A**. ACR waveform (recording from a point 20 cm behind the blowhole). **B**. ACR peak-to-peak amplitude dependence on the active electrode position along the longitudinal axis; positive amplitude values correspond to the waveform exemplified in (**A**), negative values correspond to the inverted waveform. Distances are indicated from the blowhole, which is taken as an arbitrary zero (arrow). Position of the abscissa axis in (**B**) corresponds to the dolphin's head profile shown in (**C**).

caudally than the area of the best ABR recording. Note that a position so caudal corresponds to the caudal margin of the skull or even behind the caudal margin. When the recording position shifted from the focus point, response amplitude fell steeply, and in rostral sites, response reversed in polarity (Fig. 2.7B).

Comparison of this responses with EP recorded from implanted intracortical electrodes (see Fig. 2.2) reveals an obvious similarity in their waveforms and temporal characteristics. Therefore, this response type can be interpreted as the far-field EP of the auditory cortex (the auditory cortical response, ACR). The position of the best-recording site (far caudal) indicates that the ACR-generating dipole is tilted to the rostroventral-dorsocaudal direction.

ACR is not as consistent and easily recordable as ABR. It exhibits a marked interindividual variability, and its recording requires careful positioning of the active electrode. So ACR was not used for investigation of the cetacean auditory system as widely as ABR. Nevertheless, the use of this response type can be helpful for comparison of functional characteristics of different levels of the auditory system.

2.2.4. Rhythmic Evoked Potentials

A specific version of noninvasively recorded EPs is their rhythmic sequence evoked by amplitude-modulated sounds or short-pulse trains. In such cases, evoked responses to successive stimulation cycles markedly overlap and fuse into a complex rhythmic response which may be composed of several types of individual evoked potentials. Origin of such rhythmic responses requires a special analysis.

2.2.4.1. Envelope-Following Response.

A widely used stimulus type to produce rhythmic EP is sinusoidally amplitude-modulated (SAM) sound (tone or noise carrier). The data obtained with the use of this stimulus type can be a subject of frequency-spectrum analysis. In humans, potentials recorded from the head surface were shown to follow both the low frequencies of pure tones (Moushegian et al., 1973; Batra et al., 1986) and the envelopes of amplitude-modulated carriers (Rodenburg et al., 1972; Hall, 1979; Galambos et al., 1981; Rickards and Clark, 1984; Kuwada et al., 1986; Rees et al., 1986). The former is known as the frequency following response (FFR), while the latter is termed the amplitude-modulation following response (AMFR) or envelope following response (EFR). EFR is also well detectable in some other mammals (Dolphin and Mountain, 1992).

In dolphins, FFR was not studied in detail since it appears at low tone frequencies that are little effective for the dolphin's hearing. However, EFR can be easily obtained and were studied in dolphins. Two modes to elicit EFR were used in these studies: steady-state stimulation (Dolphin, 1995; Dolphin et al., 1995) and stimulation by short amplitude-modulated sound bursts (Supin and Popov, 1995a).

In studies of Dolphin (1995), Dolphin et al. (1995) carried out on the false killer whale *Pseudorca crassidens*, beluga whale *Delphinapterus leucas*, and bottlenose dolphin *Tursiops truncatus* with the use of steady-state stimulation, mainly low carrier frequencies were tested, from 500 to 10,000 Hz, with wide variety of modulation rates and high level (120–130 dB *re* 1 μPa). Magnitude of the recorded responses was evaluated by their frequency spectra. It was found that evoked potentials recorded from the head surface are capable of to follow rather high modulation rate, up to a few kHz.

Steady-state stimulation and recording are very appropriate for frequency analysis, but these conditions provide less information concerning the response origin since the latency of individual response waves cannot be detected directly. Nevertheless, the latency can be estimated indirectly basing on the response-phase dependence on stimulation rate. If a rhythmic response has a certain group delay relative to the stimulus, its phase-versus-frequency function is of a constant slope. This slope has the dimensionality of time (cycle/kHz = ms) and quantitatively is equal to the delay. However, a constant phase-vs-frequency slope was not found in the studies of Dolphin et al.; the slope varied from 20 ms at low modulation rates (tens Hz) to 0.3 to 0.4 ms at high rates (more than 1000 Hz). This inconstancy may be interpreted either as a result of a whole-cycle ambiguity of the phase shift or as an indication of multisource origin of EFR. Long delay (around 20 ms) at low stimulation rates may indicate a contribution of evoked responses of cortical origin. Middle-range delays (5 to 6 ms) correspond to main waves of ABR. Delays as short as 0.3 to 0.4 ms (even not including the acoustic delay) hardly can indicate contribution of any physiological source, thus contamination by stimulus artifacts can be suspected.

In the study of Supin and Popov (1995a), EFR was evoked by short (12 ms) SAM tone bursts. Amplitude-modulated tone bursts evoked robust rhythmic EFR that followed the modulation rate. Figure 2.8 demonstrates representative EFRs at various modulation rates. In order to obtain EFR in more pure form, the tone burst had shallow rise. Such stimulus evoked a small transient on-response, which after a few milliseconds was replaced by a quasi-sustained EFR. Both the start and the end of the response appeared with a lag of a few milliseconds relative to the stimulus. The response-free

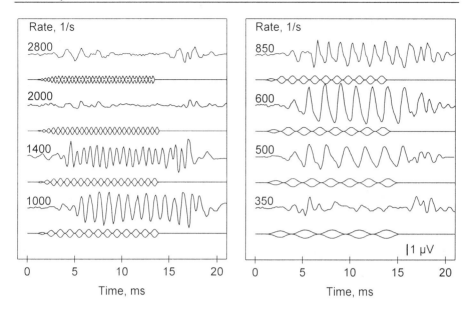

Figure 2.8. Envelope-following responses to various modulation rates of pure tone (64 kHz, 120 dB re 1 µPa, 60 dB above response threshold) in a bottlenose dolphin *Tursiops truncatus.* In each pair of records, the upper one is EFR and the lower one is the stimulus envelope. Modulation rate (s⁻¹) is indicated near records.

initial part of the records indicate that the records were not contaminated by stimulation artifacts.

A characteristic feature of EFR in dolphins is its ability to follow stimulation rate within a wide rate range. Figure 2.8 demonstrates high-amplitude EFR at stimulation rates more than 1000 Hz, although low-amplitude responses were recorded at stimulation rates near 3000 Hz. In spite of the sinusoidal stimulus envelope, rhythmic responses to modulation rates below 1000 Hz had complicated, nonsinusoidal waveform. At higher modulation rates, the response waveform became simpler and closer to a sinusoid.

To investigate EFR origin, its delay relative to the stimulus was estimated basing on its phase shift at various stimulation rates. Figure 2.9 shows phase shifts for EFR to the amplitude-modulating sinusoid as a function of modulation rate. At modulation rates below 2000 Hz, phase shifts fall on a straight line with a slope of 4.26 cycle/kHz. This slope corresponds to a group delay of 4.26 ms.

Another way to find out EFR delay is to measure its lag relative to the stimulating burst. As mentioned above, both the EFR start and end were delayed relative to the onset and offset of the stimulus burst. This lag may be used to estimate EFR latency. It was important to minimize the response to

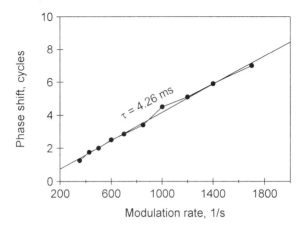

Figure 2.9. EFR phase shift dependence on modulation rate in a bottlenose dolphin *Tursiops truncatus*. Tone carrier of 64 kHz, 120 dB *re* 1 μPa (60 dB above the response threshold). The straight line is the regression line, its slope is 4.26 cycle/kHz.

the burst onset and offset in order to obtain EFR in as clear a form as possible. Therefore, in these experiments, EFR was evoked by a modulation burst imposed on a continuous carrier, using a modulation index of no more than 0.5. Figure 2.10 shows EFR obtained in such a way. The EFR end lag is clearly detectable. There was some uncertainty as to which point must be taken as the EFR end; however, this uncertainty did not exceed a fraction of a millisecond.

The lag was found to be about 4.5 ms at modulation rates below 2000 Hz (rates of 600 and 1000 Hz are exemplified in Fig. 2.10). This value agrees with that found by the phase-shift technique. At rates above 2000 Hz (2400 Hz in Fig. 2.10), the lag was markedly shorter, about 2.5 ms. These lags included the acoustic delay, which was 0.5 ms. Thus the true latencies were about 4 ms for stimulation rates below 2000 Hz and 2 ms for higher rates.

Comparison of the delays obtained by the phase-shift and end-lag measurements with ABR evoked by a short sound click shows (Fig. 2.10D) that the delay of 4.5 ms (including the acoustic delay) corresponds to the main positive-to-negative wave of ABR. The 2.5 ms delay corresponds exactly to the earliest small ABR wave.

These data indicate that EFR obtained in dolphins is a rhythmic ABR sequence. It appears also that at least two ABR generators contribute to EFR: one with a 4-ms latency responding to modulation rates below 2000 Hz, and another with a 2-ms latency responding to modulation rates up to 3400 Hz. As shown above (Section 2.2.3.2), the late ABR part probably reflects midbrain activity, whereas the earliest wave is probably the auditory nerve

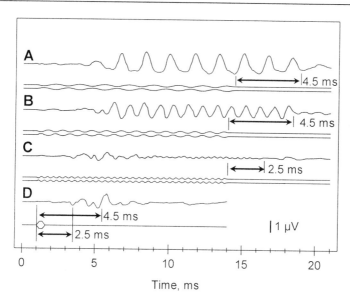

Figure 2.10. EFR end lag relative to the stimulus in a bottlenose dolphin *Tursiops truncatus.* In each pair of the records, the upper trace is the evoked-potential record and the lower trace is the stimulus envelope. Stimulus is the modulation burst in continuous tone carrier of 64 kHz, 60 dB above the response threshold. Modulation rates are 600 Hz (**A**), 1000 Hz (**B**), and 2400 Hz (**C**). Arrows show the end delay of 4.5 ms (**A** and **B**) and 2.5 ms (**C**). **D** ABR to a tone pip of 64 kHz, 60 dB. Arrows show delays of 2.5 and 4.5 ms relative to the pip onset.

response. Certainly, participation of intermediate auditory levels (for example, the cochlear nucleus and superior olivary complex) in generation of ABR and EFR can not be excluded.

A possible neuronal basis of different EFR origin at different modulation rates is that in progressively more central nuclei of the auditory system, neurons have transfer functions with progressively lower cut-off frequencies, from a few hundred Hz in the cochlear nucleus (Møller, 1972; Frisina et al., 1990) to several times lower cut-off frequencies in the inferior colliculus (Rees and Møller, 1983) and even lower rates in the auditory cortex (Schreiner and Urbas, 1986); see Langner (1992) for a review.

2.2.4.2. Rate-Following Response

A specific version of rhythmic evoked responses is that evoked by rhythmic click sequence. Contrary to sinusoidal amplitude envelopes, rhythmic click sequences have wide frequency spectrum at any rates and thus are capable of evoking responses at unlimitedly low rates.

Responses to rhythmic click trains of a limited duration were investigated in detail in the bottlenose dolphin *Tursiops truncatus* (Popov and Supin, 1998). They are demonstrated in Fig. 2.11. At a low rate (10 s^{-1}), each click evoked a typical ABR as it was described above: a few waves occurring mainly within the first 5 ms after the click; later waves of the highest amplitude. At rhythmic stimulation with rates up to 200 to 300 s^{-1}, successive clicks evoked distinct ABRs similar to those evoked by single clicks. At higher rates, ABRs merged into a complex quasi-sustained waveform. The term *quasi-sustained* is used since small but noticeable changes in the response waveform and amplitude were observed during the stimulus trains, so the response was not truly sustained. By analogy with the envelope-following response (EFR) to amplitude-modulated sounds, this response type was designated as a rate-following response, RFR.

At click rates up to 800 s^{-1}, RFR was of a complex nonsinusoidal waveform. At higher rates, RFR waveform became simpler and closer to a sinu-

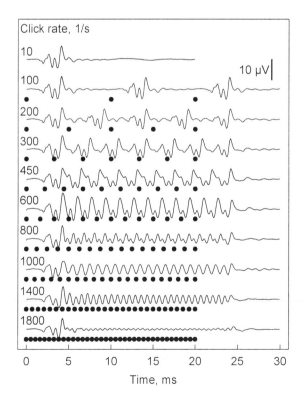

Figure 2.11. RFR of a bottlenose dolphin *Tursiops truncatus* to rhythmic clicks at rates from 10 to 1800 s^{-1}, as indicated. Stimulus level 50 dB above ABR threshold. Dots show instants of click presentations.

soid. At rates up to 1000 s^{-1}, RFR amplitude was close to that of the single ABR. Significant decrease in the RFR amplitude appeared at rates above 1400 s^{-1}. At rates higher than 600 s^{-1}, the response may be separated into an initial transient response (ABR) to the train beginning and the subsequent quasi-sustained RFR of a smaller amplitude. An end lag of RFR (the ongoing response after the last pulse of a series) 4 to 4.5 ms long was always observed.

These data indicate that RFR is a rhythmic sequence of ABRs. Transition from individual ABRs to RFR can be traced at stimulation rates below 450 to 600 s^{-1} when RFR waves closely resemble the ABR waveform. The end lag of the response indicates the response group delay of 4 to 4.5 ms. This delay coincides with latencies of the highest ABR waves. This is also an indication of the RFR origin as an ABR sequence.

Similar results were obtained in killer whales *Orcinus orca* (Szymanski et al., 1998) except a little lower RFR capability to follow high click rates (less than 1000 s^{-1}).

2.2.5. Contribution of Various Frequency Bands to ABR

When EP are used in experimental studies of hearing, one of problems is to determine which parts of the cochlear partition are involved in their generation; that is, which frequency band of a complex stimulus contributes to the recorded EP. It is a common problem for evoked-potential studies in both humans and animals. A natural direct approach seems to be the use of narrow-band stimuli (tone bursts or filtered clicks) in order to find out which frequencies are effective to provoke EP, in particular, ABR. However, in many animals and particularly in humans this approach was difficult to realize since sound effective to provoke EP, such as clicks or short bursts, have broadband frequency spectra. This makes it difficult to assess the real contribution of a certain frequency band to ABR generation. Masking of the spectrum splatter (Picton et al., 1979) improved the frequency specificity of ABR, but did not solve the problem completely.

Therefore, an indirect approach based on the use of wide-band stimuli and high-pass masking noise was widely adopted. High-pass masking noise eliminates basal (high-frequency) cochlear contribution thus revealing the apical (low-frequency) contribution. This technique was initially developed for electrocochleography (Elberling, 1974; Eggermont, 1976) and then was successfully applied to ABR analysis (Davis and Hirsh, 1976; Don and Eggermont, 1978). It was suggested that the contribution of specific portions of the basilar membrane to the ABR can be revealed by a derived response technique: successive subtraction of response waveforms obtained with successive high-pass cut-off masking noise conditions. This technique yields narrow-band (between the cut-off frequencies) contributions. The derived-response technique was applied for both electrocochleography (Teas *et al.*, 1962; Elberling, 1974) and ABR analysis (Don and Eggermont, 1978; Parker and Thornton, 1978a,b; Eggermont, 1979; Burkard and Hecox, 1983; Gould and Sobhy, 1992; Donaldson and Ruth, 1993). These studies have shown increased contribution (as estimated by ABR amplitude) of high-frequency parts relative to low-frequency ones.

Dolphins provide good opportunity to approach this problem. ABR amplitude in dolphins is an order of magnitude higher than in humans, which makes possible precise measurements even in cases of low-effective (narrow-band and low-level) stimuli. The wide frequency range of the dolphin's hearing allows one to use high-frequency stimuli, which can combine a short duration (which makes them effective to evoke ABR) with a strictly limited frequency band. Therefore, it was possible to use a direct approach to evaluate the contribution of different frequency bands to ABR by recording ABR to filtered clicks and noise bursts (Popov and Supin, 2001).

A feature of that study was the use of digitally generated stimuli – clicks

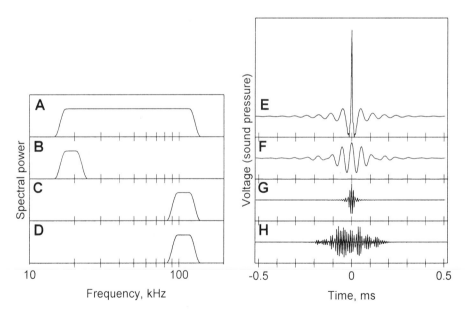

Figure 2.12. Examples of stimulus spectra (**A–D**) and waveforms (**E–H**) used to study the frequency-band contribution to ABR. **A** and **E**. Three-octave filtered click 16–128 kHz. **B** and **F**. Half-octave filtered click 16–22.5 kHz. **C** and **G**. Half-octave filtered click 90–128 kHz. **D** and **H**. Half-octave filtered noise burst 90–128 kHz. All the spectra are equalized by spectrum power per octave; thus, both of the half-octave clicks (16–22.5 and 90–128 kHz) are of equal overall energy, whereas the three-octave click (16–128 kHz) is of six-fold higher energy.

and noise bursts with strictly defined frequency bands. Some typical stimuli spectra and waveforms are exemplified in Fig. 2.12. The stimulus bands were restricted by ramps bounding a plateau of a constant level (as expressed in spectrum power per octave). To provide these spectrum features of acoustic stimuli, the spectra of the electrical signals were corrected for the frequency response of the transducer.

The use of the sound clicks with band-filtered spectra has shown that the ABR amplitude was dependent on both intensity and center frequency (Fig. 2.13). High-frequency stimuli evoke ABR of much higher amplitude than low-frequency stimuli of the same bandwidth (in octave measure), spectrum level, and energy (compare Fig. 2.13, A and B). The response amplitude at a certain stimulus level and also the amplitude limit achieved at high stimulus intensities (120–130 dB) are markedly greater at high-frequency rather than at low-frequency stimuli.

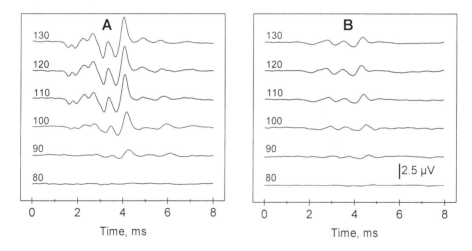

Figure 2.13. ABR to half-octave band-filtered clicks in a bottlenose dolphin *Tursiops truncatus.* **A.** 90 to 128 kHz stimulus band. **B.** 11.2 to 16 kHz stimulus band. Stimuli of different bands are equalized in their spectrum power density per octave and thus (since their bandwidths are equal in octave measure) in overall energy. Stimulus intensity is specified near records in dB re 1 μPa^2/oct.

ABR amplitude dependence on both stimulus frequency and intensity is plotted in Fig. 2.14 at two click bandwidths: 1 oct (A) and 0.25 oct (B). The plots show that (1) ABR amplitude increased with increasing the stimulus intensity, until it reached a limit at levels above 105–110 dB; (2) ABR amplitude increased with shifting the stimulus band upward on the frequency scale (the plots shift upward); and (3) the wider the stimulus bandwidth, the

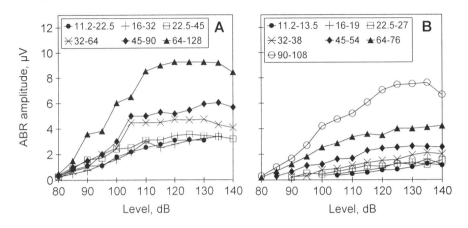

Figure 2.14. ABR amplitude dependence on stimulus intensity at different stimulus frequency bands in a bottlenose dolphin *Tursiops truncatus*. **A**. Octave bands. **B**. Quarter-octave bands. Stimulus intensity is specified near records in dB *re* 1 μPa^2/oct. Frequency bands are indicated in the legend (kHz).

higher the amplitude limit (in panel A, the curves reach higher levels than in panel B).

Comparison of contributions of different frequency bands to ABR comes across some problem since relation between physical and physiological intensities may be different for different frequencies. However, the asymptotic limits reached by ABR amplitude at high stimulus levels provide a reliable basis to compare contributions of different frequency bands. Amplitude limits are presented in Fig. 2.15 as a function of the frequency band position. ABR amplitude increased several times with shifting the stimulus band upward. Thus, higher frequencies (basal cochlear parts) much more contribute to generate ABR than lower frequencies (apical cochlear parts).

Besides, ABR amplitude limits increased with increasing the frequency band. In Fig. 2.15 this effect manifests itself in upper position of the plot for 1-oct stimulus than 0.5-oct and 0.25-oct plots. In more detail, ABR dependence on stimulus bandwidth was studied using low-pass and high-pass filtered stimuli. For low-pass filtered clicks, the low passband boundary was kept constant as 11.2 kHz, and the upper boundary varied from 16 kHz (0.5-oct bandwidth) to 128 kHz (3.5-oct bandwidth) by half-octave steps. For high-pass filtered clicks, the upper passband boundary was kept constant as 128 kHz and the lower boundary varied from 90 kHz (0.5-oct bandwidth) to 11.2 kHz (3.5-oct bandwidth), again by half-octave steps. Similarly to experiments with band-pass filtered stimuli, the amplitude limits reached at high stimulus levels were used to characterize ABR dependence on the stimulus bandwidth.

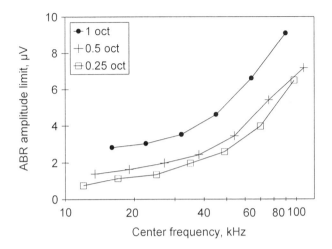

Figure 2.15. ABR amplitude limit dependence on stimulus center frequency in a bottlenose dolphin *Tursiops truncatus.* Data for 0.25, 0.5, and 1-oct stimulus bandwidth, as indicated in the legend.

Figure 2.16 presents ABR amplitude limits as a function of the stimulus bandwidth for low-pass and high-pass filtered clicks. In both cases, bandwidth increase resulted in amplitude-limit increase. In accordance with results obtained with narrow-band clicks, the amplitude-versus-bandwidth

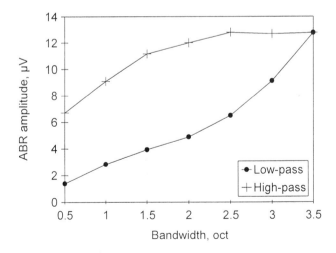

Figure 2.16. ABR amplitude limits as a function of the stimulus bandwidth. Low-pass filtered clicks from 11.2 to 16 kHz (0.5 oct) to 11.2 to 128 kHz (3.5 oct). High-pass filtered clicks from 90 to 128 kHz (0.5 oct) to 11.2 to 128 kHz (3.5 oct). Averaged data from two bottlenose dolphins *Tursiops truncatus.*

functions were different for low- and high-pass filtered clicks. Low-pass filtered clicks of a small (half-octave) bandwidth (11.2 to 16 kHz) evoked low-
amplitude ABR, thus the amplitude steeply increased when more effective
high-frequency regions were involved with the bandwidth increase up to 128
kHz. High-pass filtered clicks evoked ABR of a rather high amplitude even
at a half-octave bandwidth (90 to 128 kHz); involvement of low-frequency
bands (until 11.2 kHz, 3.5 oct) resulted in a less steep amplitude increase.

When clicks of various bandwidths are used as stimuli, it should be taken
into account that temporal pattern of filtered clicks depends on their spectrum band: the wider the band, the shorter the click (compare Fig. 2.12,A–C
and E–G). To eliminate possible effects of the temporal pattern, stimuli were
used that were constant in temporal pattern when their spectrum parameters
were varied. It is possible, particularly, with the use of filtered noise bursts
of various spectra but of constant duration (see Fig. 2.12,D, H).

Experiments with band-, low-, and high-pass filtered noise bursts were
carried out using the same filter forms as for the filtered clicks. In all these
series, noise bursts had a constant envelope form and duration (one 0.5-ms
long cycle of a cosine function). These experiments gave results very similar
to those obtained with correspondingly filtered clicks (Fig. 2.17). At band-
pass-filtered half-octave bursts, ABR amplitude increased with the upward
shift of the stimulus band: from about 2 μV at a band of 16 to 22 kHz to
about 5 μV at a band of 90 to 128 kHz (Fig. 2.17A). At low-pass filtered

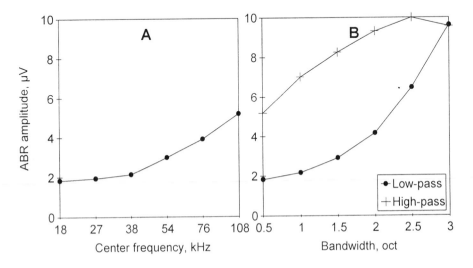

Figure 2.17. ABR amplitude limit dependence on filtered noise burst parameters. **A.** Dependence on the half-octave band position. **B.** Dependence on the bandwidth of low- and high-pass
filtered bursts. Averaged data from two bottlenose dolphins *Tursiops truncatus.*

bursts, ABR amplitude increased with widening the band more steeply: from about 2 µV at a band of 16 to 22 kHz to about 9 µV at a band of 16–128 kHz (Fig. 2.17B). At high-pass filtered bursts, ABR amplitude was rather high even at a narrow band (around 5 µV at a band of 90–128 kHz) and rose slowly with widening the band, up to about 9 µV at a band of 16–128 kHz (Fig. 2.17B). Thus, the found dependence of ABR amplitude on the stimulus center frequency and bandwidth really reflects the contribution of various frequency bands to ABR, not influence of the temporal pattern of the stimulus.

From these data, it may be concluded that contribution of various cochlear parts to ABR generation is unequal; higher frequencies are much more effective than lower ones. This suggests that ABR is the most appropriate for investigation of higher frequency range of hearing in cetaceans. On the other hand, at lower frequencies, ABR amplitude is also significant though lower than at higher frequencies. Therefore, ABR is indicative as well of hearing processes in low-frequency ranges if the stimulus spectrum is controlled precisely enough.

2.3. EVOKED-POTENTIAL PROCEDURES IN HEARING MEASUREMENTS

Studies of hearing capabilities using the evoked-potential technique are based first of all on evoked-potential amplitude measurements. For threshold measurements, it is usually assumed that a stimulus reaches its threshold level when the evoked-response amplitude falls down to zero. So it is important to measure properly the response amplitude when it becomes comparable with the record noise.

2.3.1. ABR Threshold Measurements

The simplest approach is to measure directly the peak-to-peak response amplitude until the peak becomes undetectable. Then two ways are possible to assess the response threshold. The first one is to find a stimulus level that evokes a response of a certain arbitrarily chosen low (near-zero) amplitude. This stimulus level can be found by interpolation between tested levels that evoke responses just above and just below the criterion amplitude. The other way is to approximate a lower part of the amplitude-versus-level dependence by a certain function – say, a straight regression line – and to extrapolate this line to the zero-amplitude point; the corresponding stimulus level is a good estimate of the threshold.

Direct peak-to-peak amplitude measurement is usable at a rather high response-to-noise ratio, which allows a criterion amplitude low enough. If high precision of measurement is desirable at comparable response and noise magnitudes, a more sophisticated approach is appropriate. It is based on match filtering of the records. Evaluation of small EP amplitude in noise actually means finding the weight of the characteristic EP waveform in a noisy record. This problem can be solved by passing the signal trough an optimal filter (a filter specifically tuned to the searched waveform). The temporal transfer function of this filter is the time-inverted searched-for waveform. Such filtering is just the same as calculation of the cross-correlation function (CCF) between the analyzed record and the standard EP waveform. This approach allows the quantitative evaluation of the evoked-response magnitude in cases when response peaks can not be identified with a satisfactory precision.

In experiments when low-amplitude EP should be extracted from noise, we used the measurement procedure as follows (Fig. 2.18). CCF was calcu-

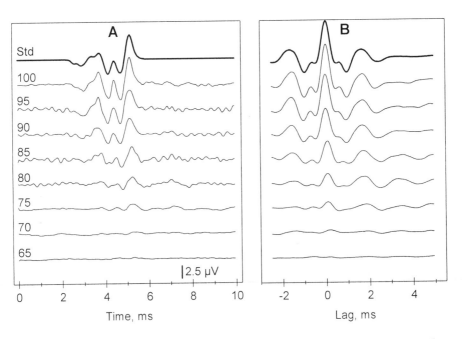

Figure 2.18. The use of match-filtering for measurement of low EP amplitudes. **A**. ABR recorded in a bottlenose dolphin *Tursiops truncatus* to clicks of various intensities. **B**. Cross-correlation functions between these records and a standard ABR waveform. Stimulus levels are indicated near records in dB *re* 1 µPa; Std – the standard ABR waveform used for match-filtering and its autocorrelation function.

lated between the measured record and a standard low-noise record containing high-amplitude EP (the standard EP waveform). Since EP of lower amplitude had longer latency, CCF peaked at a lag from zero to a few hundred µs. The peak value of this function was taken as a measure of the EP waveform weight in the analyzed record. Autocorrelation function (ACF) of the standard EP waveform was also calculated; it gave the ratio between the correlation-function peak value and the peak-to-peak response amplitude. Using this ratio, the found CCF values were converted to response amplitudes (peak-to peak voltage).

Cross-correlation functions of the type presented in Fig. 2.18 allowed a threshold amplitude criterion as low as 0.05 to 0.07 µV. Thresholds were found by interpolation between the stimulus levels giving response magnitude just above and just below the criterion value. In the case exemplified in Fig. 2.18, this threshold estimate was 60.0 dB.

Another way to estimate a threshold was as follows. ABR amplitude (measured directly or by the use of the match-filtering procedure) was plotted as a function of stimulus intensity (Fig. 2.19). As a rule, such plot consisted of two branches: the oblique one at lower stimulus intensities (up to 110 dB in Fig. 2.19) and constant-amplitude one at a high-intensity range where EP amplitude reached its maximum limit (above 120 dB in Fig. 2.19). The oblique branch could be approximated by a regression straight line which was extrapolated to zero response magnitude. This point was taken as a threshold estimate. As an example, the regression line drawn in Fig. 2.19 through points from 60 to 105 dB indicates a threshold of 61.0 dB.

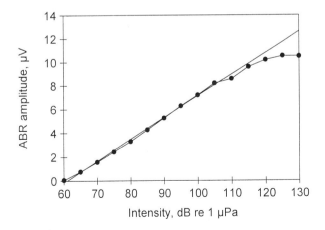

Figure 2.19. ABR amplitude dependence on stimulus intensity in a bottlenose dolphin *Tursiops truncatus.* Stimulus is a wide-band (8–128 kHz) click. The straight line is a regression line for points within a range of 60 to 105 dB.

Estimates obtained in two described ways differ by 1 to 2 dB. Both of these modes of threshold estimation are usable to assess thresholds, though the second one (regression line) seems preferable since it is free of arbitrary selection of the threshold criterion.

2.3.2. EFR and RFR Threshold Measurements

One more approach to precise measurement of response amplitude is based on the use of rhythmic sound modulations: sinusoidal amplitude modulation (SAM) of a carrier tone or sound pulse sequence. These stimuli evoke rhythmic responses, EFR or RFR, which can be the subject of Fourier analysis. Figure 2.20A exemplifies EFR records obtained at various stimulus lev-

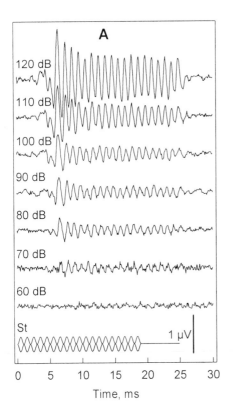
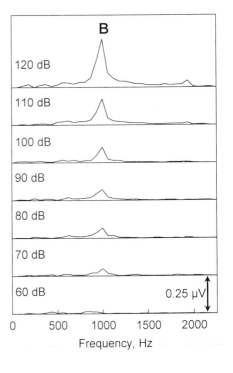

Figure 2.20. **A**. Examples of EFR in a beluga whale *Delphinapterus leucas* to SAM sound of various level. **B**. Frequency spectra of EFR. Carrier frequency 64 kHz, modulation rate 1000 Hz, stimulus levels are indicated near the records and spectra in dB *re* 1 μPa, St – stimulus envelope.

els. To measure hearing threshold, EFR to amplitude-modulated tone bursts of various levels was recorded. Modulation rate was 1000 Hz since rates of 600 to 1000 Hz are effective for evoking EFR. Sound level was varied in steps of 5 dB (to reduce the figure, only records separated by 10-dB steps are exemplified). With lowering the stimulus level, EFR amplitude diminished until the response disappeared in noise. In order to quantify EFR magnitude, a 16-ms long fragment (from 7.5 to 23.5 ms after the stimulus onset) of each record was Fourier transformed to obtain the frequency spectrum (Fig. 2.20B). The position of the transformed window was selected in such a way as to cover a major part of the rhythmic response but did not include the initial transient part of the response. It is noteworthy that the analyzed window (16-ms long) contained a whole number of stimulation/response cycles; this is a condition for appropriate Fourier analysis of a signal of a limited duration. The spectra reveal a well-defined peak at the modulation frequency of 1000 Hz. The magnitude of this peak was taken to express the response magnitude in terms of RMS voltage.

Using such evaluation, response magnitudes were plotted as a function of stimulus level (Fig. 2.21). A typical EFR-magnitude dependence on stimulus level markedly differs from that of a single ABR to a short stimulus. It consists of three branches. At middle levels (70 to 105 dB in Fig. 2.21), EFR magnitude was little dependent on stimulus level; it was shown (Supin, Popov, 1995a), that within this range, EFR more depended on modulation depth rather than on stimulus level. The little dependence of amplitude on stimulus level may be a result of short-term adaptation appearing during the prolonged rhythmic stimulation. At excessively high levels (higher than 110

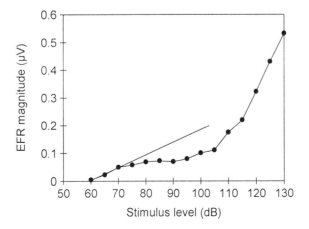

Figure 2.21. EFR magnitude as a function of amplitude-modulated tone level in a beluga whale *Delphinapterus leucas*. Tone frequency 64 kHz, modulation rate 1000 Hz. A straight line is the regression line approximating the plot within a region of 60 to 70 dB.

dB in Fig. 2.21), EFR magnitude rose steeply, which probably was a result of a spectrum splatter and involvement of wider frequency band to response generation. At low levels (below 70 dB in Fig. 2.21), EFR magnitude fell steeply with the level decrease. Apparently, within this range the short-term adaptation cannot compensate the intensity decrease, so EFR amplitude becomes intensity-dependent.

Using the magnitude-versus-level function, hearing thresholds can be estimated. Again, two threshold estimates are usable. The first one is a stimulus level at which the response magnitude reached a certain arbitrary criterion. Spectra of the type presented in Fig. 2.20B allowed a threshold criterion as low as 0.01 μV. Thresholds were found by interpolation between the stimulus levels giving response magnitude just above and just below the criterion value. In the case exemplified in Fig. 2.21, this threshold estimate was 61.4 dB.

Another way to estimate a threshold is to approximate the low-level branch of a plot like that in Fig. 2.21 by a regression line that is extrapolated to the zero response magnitude. This point can be taken as a threshold estimate. As an example, the regression line drawn in Fig. 2.21 through points at 60, 65, and 70 dB indicated a threshold of 59.3 dB. Estimates obtained in these two ways differ by 1 to 2 dB. Both of these modes of threshold estimation are usable to assess thresholds, though the second one (regression line) seems preferable as free of arbitrary selection of the threshold criterion.

The spectrum-peak magnitude is a convenient measure of response amplitude. It can be evaluated more confidently than EP amplitude in the original record. Correspondingly, threshold evaluation is also more confident and precise than using visual detection of EP in records.

2.4. HEARING SENSITIVITY AND FREQUENCY RANGE

2.4.1. Psychophysical Data

A commonly adopted way to investigate hearing sensitivity in cetaceans is a behavioral (psychophysic) experiment. Hearing sensitivity is expressed in terms of hearing thresholds – the lowest level of perceivable sound (the lower threshold, the better sensitivity). Threshold measurements at a variety of frequencies provide the audiogram, which shows threshold as a function of sound frequency. The audiogram characterize both sensitivity and frequency range of the investigated animal. As a rule, obtaining the audiogram is a very first step in the investigation of hearing characteristics of a certain species.

Long ago Kellogg et al. (1953) and Schevill and Lawrence (1953) noticed that dolphins are able to hear high-frequency sounds, above 100 kHz. Precise measurements, however, were performed later. Johnson (1967) was the first who measured in a bottlenose dolphin *Tursiops truncatus* hearing thresholds as a function of pure-tone frequency within a wide frequency range, from 75 Hz to 150 kHz. This study was carried out using a standard go/no-go paradigm with conditioning the animal against committing false alarms and in conjunction with the adaptive (up-down) testing procedure. This study revealed the highest sensitivity within a frequency range of 50 to 80 kHz; in this range, thresholds were as low as 40 to 50 dB *re* 1 µPa (10^{-13}–10^{-14} W/m^2). The presented audiogram exhibited good sensitivity (within 10 dB *re* the best threshold) within a frequency range from 15–20 to 100–110 kHz. With lowering the frequency beyond this region, thresholds rose gradually with a rate of about 10 dB/oct, whereas frequency increase above 110–120 kHz resulted in very steep threshold increase, near 500 dB/oct.

Until now, the audiogram presented by Johnson (1967) remains one of the most precise and detailed. Later, audiograms were obtained in psychophysical operant-conditioning experiments in a number of other species of odontocetes (Fig. 2.22): the harbor porpoise *Phocoena phocoena* (Andersen,

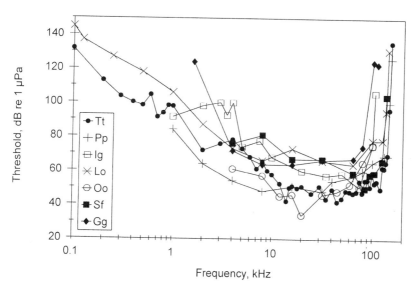

Figure 2.22. Representative psychophysical audiograms of a few odontocete species. Tt – *Tursiops truncatus* (Johnson, 1968), Pp – *Phocoena phocoena* (Andersen, 1970), Ig – *Inia geoffrensis* (Jacobs and Hall, 1972), Lo – *Lagenorhinchus obliquidens* (Tremel et al., 1998), Oo – *Orcinus orca* (Szymanski et al., 1999), Sf – *Sotalia fluviatilis* (Sauerland and Denhardt, 1998), Gg – *Grampus griseus* (Nachtigall et al., 1995). Plotted by table data presented in the papers referred to.

1970), killer whale *Orcinus orca* (Hall and Johnson, 1971; Bain and Dal-hiem, 1994; Szymanski et al., 1999), Amazon river dolphin *Inia geoffrensis* (Jacobs and Hall, 1972), beluga whale *Delphinapterus leucas* (White et al., 1978; Awbrey et al., 1988; Johnson, 1992), Pacific bottlenose dolphin *Tursiops gilli* (Ljungblad et al., 1982), false killer whale *Pseudorca crassidens* (Thomas et al., 1988), Chinese river dolphin *Lipotes vexillifer* (Wang et al., 1992), Risso's dolphin *Grampus griseus* (Nachtigall et al., 1995), Pacific white-sided dolphin *Lagenorhynchus obliquidens* (Tremel et al., 1998), and tucuxi dolphin *Sotalia fluviatilis* (Sauerland and Denhardt, 1998). Some additional threshold data were obtained in a very low frequency range, down to 75 Hz, for a false killer whale *Pseudorca crassidens* and a Risso's dolphin *Grampus griseus* (Au et al., 1997). All of them were obtained in various versions of the go/no-go paradigm. Thus, a substantial body of data was collected in rather similar experimental conditions, which made it possible to characterize common features of odontocete hearing.

All the audiograms of odontocetes demonstrate some common features. All of them feature the frequency range wider than 100 kHz with the lowest thresholds at frequencies higher than 20–30 kHz. The only exception was a study of Hall and Johnson (1971), who found in the killer whale a high frequency limit of 35 kHz; however, these findings were not confirmed in later studies. Above the best-sensitivity frequency region, thresholds rise very steeply at a rate of hundreds dB/oct, so frequencies above 150 kHz are almost ineffective. Below the best-sensitivity region, thresholds rise rather slowly at a rate of 10 to 15 dB/oct, so as thresholds are measurable at frequencies as low as hundreds Hz. The only exception is the audiogram of the Risso's dolphin by Nachtigall et al. (1995) which was almost flat at frequencies down to 4 kHz and rose steeply at 1.6 kHz; however, since this steep rise is defined by only one experimental point, this finding should be taken with caution. In the highest-sensitivity region, thresholds mostly are as low as 40 to 50 dB *re* 1 μPa; in some cases, higher thresholds were obtained (60–70 dB); however, there is some possibility that in these cases higher thresholds were a result of conservative behavior of the animal, which responded positively only to well detectable stimuli.

Similar results were obtained in the harbor porpoise *Phocoena phocoena* with the use of the Pavlovian-conditioning method (Supin and Sukhoruchenko, 1974) (Fig. 2.23). In that study, conditioned electrodermal response, changes of breathing and heart-beat rate were used as an indicator of sound (pure tone) perception. Both the upper cut-off frequency and the best threshold values were almost the same as in operant-conditioning experiments (Andersen, 1970). However, within a low-frequency range (below 30–40 kHz), thresholds were markedly higher. Perhaps the cause of the difference is that during the experiments of Supin and Sukhoruchenko, the animal

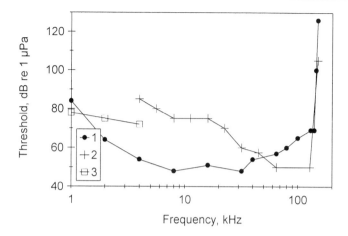

Figure 2.23. Psychophysical audiograms obtained in harbor porpoises *Phocoena phocoena* by various methods. 1 – underwater sound presentation, operant conditioning (Andersen, 1970); 2 – underwater sound presentation, Pavlovian conditioning (Supin and Sukhoruchenko, 1974); 3 – in-air sound presentation, operant conditioning (Kastelein et al., 1997).

was in a shallow bath near the water surface. It should be noted, however, that within a frequency range below 10–20 kHz, the audiogram of Andersen features much lower thresholds than in all other odontocetes, and these data have not been confirmed yet in other operant-conditioning experiments; thus, the source of the disagreement is debatable.

Apart of standard underwater sound presentation, a study was performed when the auditory sensitivity of a harbor porpoise *Phocoena phocoena* was measured in air using airborne sounds (Kastelein et al., 1997). The animal was trained to beach itself on a platform and then to touch a response target when he heard a sound from an in-air speaker (a version of the go/no-go paradigm was used). Thresholds were estimated by a criterion of 50% correct responses at low percentage of false alarms. Thresholds found in such a way varied from 66 dB *re* 20 µPa at 500 Hz to 46 dB at 4 kHz; higher frequencies were not tested. Being estimated in terms of aerial hearing sensitivity, these thresholds look much higher than those of most terrestrial mammals and humans. However, being expressed in levels relative 1 µPa, these thresholds appear to be close to normal underwater hearing thresholds (Fig. 2.23, only data at 1 to 4 kHz are presented).

Similarity of underwater and aerial thresholds when they are expressed in terms of sound pressure is a confirmation of that the cetacean ear functions as a sound-pressure receiver. Being adapted to perception of underwater sounds, it requires rather high sound pressure to perceive; nevertheless, in the high-impedance aquatic medium, this sound pressure corresponds to sound power low enough. In the low-impedance aerial medium, the same pressure carries much higher sound power.

All experimental data presented above were obtained in odontocete species. Precise experimental measurements with trained animals were never conducted in mysticetes. However, attempts were made to estimate hearing sensitivity in mysticetes basing on their responses to sounds in natural conditions. Dahlheim and Ljungblad (1990) observed responses of gray whales *Eschrichtius robustus* in a marine bay to projected tones of various frequencies and levels. Startle reactions, changes in course, transit speed, or respiration within 5 s after the tone projection were taken as indications of the sound perception. The lowest threshold of 95 dB *re* 1 µPa was found at 800 Hz; this threshold was close to the level of ambient noise in the bay. With frequency increase, thresholds increased up to 142 dB at 1800 Hz. Limited available data do not allow us to conclude whether these estimates represent the real hearing thresholds, but they clearly indicate low-frequency hearing in gray whales (mysticetes) contrary to high-frequency hearing of odontocetes.

2.4.2. Evoked-Potential Data

Apart from behavioral data, evoked-potential technique proved to be an effective way to study hearing sensitivity and frequency range of cetaceans. Since ABR amplitude is dependent on sound level, it is possible to find a level at which the response amplitude approaches or reaches the zero amplitude; this sound level can be adopted as a threshold estimate. Using this technique, audiograms were obtained in a number of species: the harbor porpoise *Phocoena phocoena* (Popov et al., 1986), bottlenose dolphin *Tursiops truncatus* (Popov and Supin, 1990a), beluga whale *Delphinapterus leucas* (Popov and Supin, 1987), common dolphin *Delphinus delphis* (Popov and Klishin, 1998), Amazon river dolphin *Inia geoffrensis* (Popov and Supin, 1990c), and killer whale *Orcinus orca* (Szymanski et al., 1999).

In early evoked-potential studies of hearing sensitivity cited above, short tone pips of varying frequency were used as test stimuli. Duration of the stimuli was short because just this stimulus type is effective to provoke ABR. As a rule, rise-fall time of the stimuli should be less than 1 ms. For example, it could be a pip enveloped by one cycle of a cosine function, 1-ms long or shorter. Such stimuli are characterized by significant spectrum splatter: the shorter the pip, the wider the spectrum, which makes less precise attribution of the found threshold to a certain frequency. This limitation, however, was not usually taken as very important because a spectrum splatter of an order of a few kHz is much narrower than the frequency range of the dolphin's hearing (more than 100 kHz).

Another disadvantage of short test stimuli is that their intensity depends on both sound power and duration. If a stimulus is so short that its duration is within the limit of complete temporal summation, the intensity depends on energy (the product of power and duration). If the temporal summation is incomplete, estimation of stimulus intensity becomes problematic. It leads to problems in comparison of results of evoked-potential measurements with those obtained by behavioral methods; the latter use rather long test stimuli (obviously longer than the temporal summation limit) that are characterized by sound power flux density (or corresponding sound pressure).

Therefore, in recent studies, SAM stimuli were used instead of short pips (e.g., Klishin et al., 2000). As mentioned above (Section 2.2.4), at modulation rates of an order of a few hundreds Hz, amplitude-modulated stimuli evoke rhythmic responses that follow the modulation rate (EFR). This stimulation mode and response type have a few advantages: (1) the level of the long tone burst can be specified unambiguously in terms of the root-mean-square (RMS) sound pressure; (2) low-amplitude responses to near-threshold stimuli can be measured precisely using Fourier transform and evaluation of the magnitude of the spectral peak at the modulation frequency.

To measure hearing threshold at a certain frequency, EFR were recorded to amplitude-modulated tone bursts of various sound levels. Modulation rates were from 600 to 1000 Hz since these rates are well effective to evoke EFR. Sound level was varied in steps of a few dB (as a rule, 5 dB) until the response disappeared in noise. The measurement procedure was illustrated above (see Figs. 2.20 and 2.21). The spectrum peak magnitude at the modulation frequency was taken to express the response magnitude in terms of RMS voltage and thus to find a threshold at a given frequency.

A few typical audiograms of several odontocete species obtained by evoked-potential method are presented in Fig. 2.24. They are mainly similar: U-shaped with a steeper high-frequency branch and a less steep low-frequency branch. The frequency bands of the audiograms were from 100 kHz (the killer whale) to 150 kHz (the tucuxi dolphin). In most of the studied species, the minimum thresholds were at frequencies from 60 to 80 kHz, the minimum thresholds were below 50–55 dB *re* 1 μPa. The audiogram of the killer whale is shifted to a little lower frequency as compared to the audiograms of smaller odontocetes.

When the audiograms of one and the same species were obtained by both psychophysical and evoked-potential methods, there was a good correspondence between them. In some species, however, audiograms of specific forms were found. In the Amazon river dolphin, the audiogram was of an unusual W-shape because of the presence of two low-threshold frequency

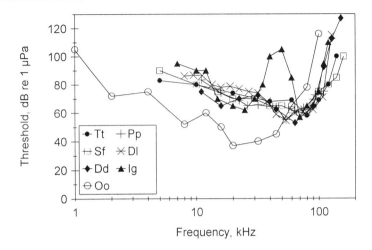

Figure 2.24. Representative evoked-potential audiograms of a few odontocete species. Tt – *Tursiops truncatus*, Pp – *Phocoena phocoena*, Sf – *Sotalia fluviatilis*, Dl – *Delphinapterus leucas*, Dd – *Delphinus delphis*, Ig – *Inia geoffrensis*, Oo – *Orcinus orca*. The audiogram of *Orcinus orca* is plotted by table data in: Szymanski et al., 1999; others are authors' data.

regions at frequencies of 20 to 25 and 70 to 80 kHz. The two low-thresholds regions were separated by a deep high-threshold hiatus.

Sometimes, evoked-potential audiograms show a little higher thresholds than the psychophysical ones; the difference probably results from insufficient temporal summation within short pip probes used in the evoked-potential measurements. But in general, the correspondence is good enough. This encourages us to use the evoked-potential method for the study of cetacean species that are not available for long keeping in captivity and for the long careful training that is necessary for psychophysical investigations. The evoked-potential technique is much less time consuming; thus it can be applied for a much larger number of cetacean species.

2.5. TEMPORAL RESOLUTION

Temporal resolution is the ability of the auditory system to discriminate successive stimuli following one after another. It determines the ability to perceive temporal patterns of acoustic stimuli, which is as important for stimulus recognition as the spectral pattern.

2.5.1. Psychophysical Studies

The first attempt to study temporal processing in the dolphin's auditory system was a study by Johnson (1968b) who measured auditory sensitivity to tone pulses of variable duration. This study revealed a typical process of the auditory temporal summation: at short pulse duration, the auditory thresholds (expressed in sound pressure) decreased as the duration of a tone pulse increased. At short stimulus duration, this threshold shift was close to 3 dB per duration doubling thus indicating a complete summation of the pulse energy; at a pulse duration long enough, the threshold reached an asymptotic limit. In experiments of Johnson, this limit was achieved at a rather long duration, hundreds of milliseconds.

To characterize the temporal summation more precisely, a simple relation is used to approximate the threshold-versus-duration function:

$$I(t) = I_\infty (1 + \tau / t), \tag{2.3}$$

where $I(t)$ is the threshold intensity at the pulse duration t, I_∞ is the threshold intensity at infinitely long duration, and τ is the temporal-summation constant; the more τ, the longer the summation. The temporal summation constant is a very important parameter of temporal processing of signals. Signals following one after another with a delay markedly shorter than the integration constant fuse into a single one and hardly can be perceived separately; signals with intervals longer than the integration constant can be discriminated in time. According to the data by Johnson (1968b), the constant τ in dolphins was as long as 100 to 200 ms at tone frequencies lower than 10 kHz; this is almost the same value as the temporal-summation constant in humans (Plomp and Bouman, 1959; also rev. Green, 1984; de Boer, 1984). At higher frequencies, the temporal summation constant diminished down to tens of milliseconds.

Indications of long temporal summation were also obtained in experiments when sensitivity to rhythmic clicks was measured as a function of the repetition rate (Dubrovskiy, 1990). Noticeable decrease of hearing thresholds was observed as soon as the repetition rate approached or exceeded approximately 10 s^{-1}, thus indicating temporal summation within a range of tens to hundreds of milliseconds. However, in this range, threshold dependence on click rate was much shallower than 3 dB/oct, as could be expected for the complete energy summation. The dependence as steep as 3 dB/oct occurred only at click rates above 3000–4000 s^{-1}, which corresponds to integration time of 250 to 300 µs. Thus, on the one hand, wide-band click tests demonstrated some degree of long temporal summation, similarly to data of Johnson (1968b) with pure tones; on the other hand, contrary to Johnson's data, wide-band click stimuli revealed the complete energy summation within a very short integration time.

Very short integration time was obtained also in a number of other psychophysical experiments. With the use of double clicks, it was shown (Au et al., 1988; Au, 1990) that when interclick intervals were shorter than 200–250 µs, detection threshold decreased approximately by 3 dB, thus indicating the complete energy summation; at longer intervals, the threshold immediately returned to the level characteristic of a single click. The integration time was assessed as 264 µs.

There were also obtained other indications that the integration time in the dolphin's auditory system is as short as 200 to 300 µs. In particular, this is a time limit of backward masking. Backward masking appears when the time delay between two signals, the first (probe) and the second (masker), is too short to perceive them separately, so the backward masking limit is an indicator of the integration time in the auditory system. Moore et al. (1984) have shown in go/no-go echo detection experiments with a bottlenose dolphin *Tursiops truncatus* that masking of an echo signal by a subsequent noise burst appears at intervals not longer that 500 µs; the calculated 70% detection threshold corresponded to a delay of 265 µs.

Forward masking efficiency also increased markedly when intervals between the masking and probe click shortened from 500 to 250 µs (Vel'min and Dubrovskiy, 1975). Discrimination of interclick intervals was found to be better when the intervals were within the same limit of 200 to 300 µs, rather than at longer intervals; it was suggested that at intervals shorter than 200–300 µs, clicks merged into an "acoustic whole" (Vel'min and Dubrovskiy, 1975; Dubrovskiy, 1990).

Thus, many data suggested a capability of the dolphin's auditory system to integrate acoustic energy within a very short time. This short integration time should be the basis of very high temporal resolution since it allows dolphins to perceive separately short sounds at intervals as short as fractions of a millisecond. However, at some conditions, the dolphin's auditory system featured a quite different integration time, as long as tens of milliseconds – two orders of magnitude longer. Thus, the problem of the temporal integration time in the dolphin's auditory system needed to be elucidated.

Apart from studies concerning the integration time, a capability of a bottlenose dolphin *Tursiops truncatus* to discriminate temporal intervals was investigated (Yunker and Herman, 1974). In that study, the dolphin was trained to respond by different movements (to the left and to the right) to sound stimuli of a "standard" and a longer duration. It should be stressed that the standard duration exceeded the temporal-integration limit; it was from

300 to 1200 ms. So this test revealed a temporal-processing capability dictated by a quite different mechanism than the temporal-integration limit. The dolphin was capable to discriminate signals differences from 18.3 ms at 300-ms standard (6.1%) to 49.2 ms at 1200-ms standard (4.1%). It is a little better than thresholds in humans at similar standard durations; for example, according to Abel (1972), thresholds were 35.3 ms at 320-ms standard (11.0%), 58 ms at 640 and 1000-ms standards (9.1% and 5.8%, respectively).

2.5.2. Dependence of ABR on Stimulus Duration.

An important indicator of temporal resolution in the auditory system is the EP dependence on the duration of sound. Such measurements were carried out in dolphins using wide-band noise bursts in conjunction with ABR recording (Popov and Supin, 1990a; Supin and Popov, 1995b). In these experiments, noise burst duration varied within a range from 20 µs to 5 ms. Signals shorter than 20 µs could not be reproduced at a steady sound pressure, and it was not sensible to use signals longer than a few ms since their duration would be beyond th`e duration of ABR.

The dependence of the ABR amplitude on the stimulus duration and level is shown in Fig. 2.25A. At low stimulus intensity (70 dB), ABR amplitude markedly depended on stimulus duration: it rose with the duration increase up to 0.5 ms (that is, obvious temporal summation occurred). Further pro-

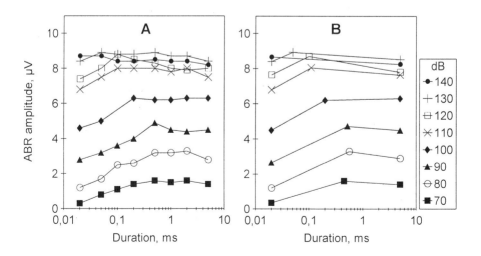

Figure 2.25. **A.** Dependence of ABR amplitude on noise burst duration at various noise intensities in a bottlenose dolphin *Tursiops truncatus*. **B.** Approximation of the plots by two-branch regression lines. Intensities are indicated in dB re 1 µPa.

longation of the stimulus produce no effect on the amplitude. At higher intensities (80 to 120 dB), ABR amplitude was rather high even at the shortest of the available durations (20 μs); it also increased with increasing the stimulus duration, but this increase was small. At the highest of intensities (140 dB), ABR amplitude reached its maximum level even at the shortest duration (20 μs) and did not increase with stimulus prolongation.

The plots in Fig. 2.25A show that the stimulus intensity influenced the integration time (the stimulus duration required to reach the amplitude maximum): it was longer at near-threshold levels (about 0.5 ms at 70–80 dB) and shortened at higher intensities. To estimate the integration time more precisely, each of the plots in Fig. 2.25A was arbitrarily subdivided into two parts: the oblique and horizontal ones. Each of the parts was approximated by a regression straight line (Fig. 2.25B). The interception pint of these lines was adopted as indicating the integration time. With the use of this evaluation, the integration time dependence on stimulus intensities looked like presented in Fig. 2.26. It accounted for about 0.5 ms at stimulus intensities of 70 to 90 dB re 1 μPa (which corresponded to 10–30 dB above the long-stimulus response threshold) and decreased to values below 0.1 ms at higher intensities.

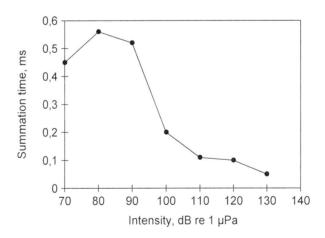

Figure 2.26. Temporal integration time dependence on noise intensity in a bottlenose dolphin Tursiops truncatus.

The integration time can also be estimated basing on threshold measurements. For this purpose, ABR thresholds were found at each of the stimulus durations using the regression-line technique. The result is presented in Fig. 2.27. With increasing the stimulus duration from 0.02 to 0.5 ms, thresholds

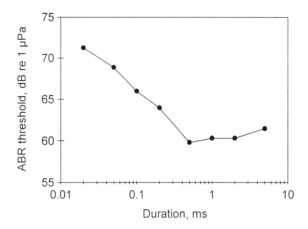

Figure 2.27. ABR threshold dependence on noise burst duration in a bottlenose dolphin *Tursiops truncatus*.

decreased by more than 20 dB. Stimulus prolongation above 0.5 ms did not result in further threshold decrease. Thus, both the response amplitude and threshold data indicate integration time not longer than 0.5 ms.

It should be noted that temporal summation was studied only using wide-band noise, not narrow-band tone pips. It was done because the short time of summation dictates to use very short stimuli with even shorter rise-fall times; the short rise-fall ramps inevitably would be wide-band. If the stimuli were tone bursts, the wide-band ramps would produce clicks making the stimulation incorrect. Therefore, only wide-band noise stimuli were appropriate for these measurements.

2.5.3. ABR Recovery at Double-Click Stimulation

The ability of ABR in dolphins to recover rapidly after the previous stimulus was noticed long ago for rhythmic (Ridgway et al. 1981) and double stimuli (Supin and Popov, 1985). A similar effect was reported earlier for EP recorded by intracranial electrodes (Bullock et al., 1968; Bullock and Ridgway, 1972). Further, this effect was investigated in much more detail.

A widely used method to assess the recovery of excitability is the double-click test. In this test, responses are recorded to two short acoustic stimuli (clicks, short noise bursts, or tone pips) separated by a certain interstimulus interval (ISI). The first stimulus of the pair is the conditioning one, and the second stimulus is the probe. At short ISI, the amplitude of the response to the probe stimulus is less than that of the control response to the similar single stimulus. The probe response amplitude recovers when ISI increases. The recovery function is the probe response amplitude as a function of ISI. This recovery function is a valuable indicator of the temporal resolution: the sorter the recovery function (faster recovery), the higher temporal resolution.

The double-click technique was used in conjunction with ABR recordings to assess the temporal resolution of the auditory system in dolphins (Supin and Popov, 1985; Popov and Supin, 1990a). In conditions of double-click stimulation, ABR showed high temporal resolution: complete recovery of the second (probe) response required ISI of only a few ms, and a just detectable response to the probe click was observed at intervals as short as 200–300 μs.

More detailed quantitative analysis of ABR recovery was carried using a variety of stimulus intensities (Supin and Popov, 1995c). Contrary to most of the previous studies in which clicks were produced by activation of a transducer by short pulses, in that study clicks were generated by activation of a transducer by short (30 μs) bursts of quasi-white noise (Fig. 2.28A). Short noise bursts instead of pulses were used for the reason that two identical signals (such as pulses) separated by a short delay τ result in a characteristic "rippled" frequency spectrum of the combined signal with a ripple spacing of $1/\tau$ (Fig. 2.28B). Therefore, variation of interpulse intervals is associated with spectrum variation, which would have a confounding effect. This effect could be neglected at long interpulse intervals τ, when the ripple spacing $1/\tau$ is too small to be resolved. However, it cannot be neglected at intervals as short as hundreds μs, which were also tested. The use of short noise bursts that randomly varied in waveform diminished this effect though did not ex-

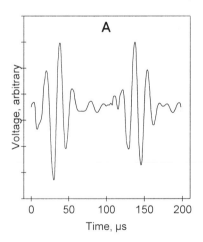
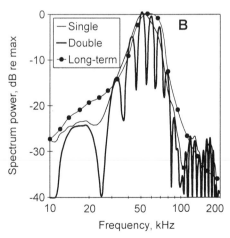

Figure 2.28. Quasi-random double-click waveform and spectra. **A**. Waveform example of a pair of click with a 100-μs interval. **B**. Spectra of the first click of the record, the pair of clicks, and a long sequence of click pairs, as the legend indicates.

clude it completely in each pair of clicks. However, ABR records are collected by presentation of a few hundreds of stimuli. The long-term spectrum of such continuous train of click pairs is free of the rippled-spectrum effect because of random variation of ripple positions in individual pairs (Fig. 2.28B). So the obtained effects of interclick interval variation can be attributed to a temporal rather than spectral pattern of the stimulus.

Apart from variation of ISI, stimulus intensities were also varied in these experiments; however, intensities of both clicks in a pair were always kept equal. Figure 2.29 presents several examples of ABR records obtained at two click intensities (80 and 120 dB re 1 μPa – that is, 20 and 60 dB above the threshold) and a few of the tested ISIs. At ISI shorter than 2 to 3 ms, the first and second responses were partially superimposed; therefore, the second (probe) response was obtained in the pure form by point-by-point subtraction of the ABR to the conditioning stimulus from the ABR to the paired stimuli. With the use of the subtraction technique, a just detectable response to the second click was observed at intervals as short as 200 to 300 μs.

The records show nearly complete recovery of the probe response at ISI of several (2 to 10) ms. However, some difference was found between the recovery courses at different click intensities. Shorter recovery time was observed at a lower stimulus level (80 dB in Fig. 2.29A): at ISI of 2 ms, the

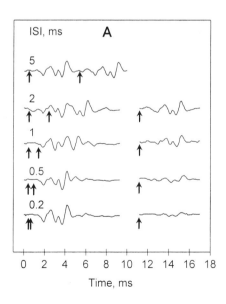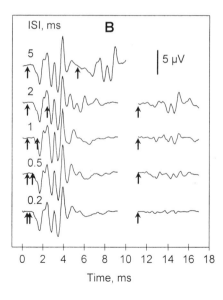

Figure 2.29. ABR evoked by double clicks of different intensities: 80 dB (**A**) and 120 dB re 1 μPa (**B**). In each panel, the left column – original records, right column – responses to the probe stimulus obtained by the subtraction procedure. ISI are indicated near records, and upward-headed arrows mark click instants (only the test click in the right columns).

second (probe) response was already equal to the first (conditioning) one. Longer recovery time was observed at a higher level (120 dB in Fig. 2.29B): at ISI of 5 ms, the probe response was still about 1.5-times less than the conditioning one.

However, the response recovery shows little dependence on stimulus intensity when the probe response amplitude is estimated in terms of its absolute voltage, not relative to the conditioning response amplitude. Figure 2.29 demonstrates that at ISI of 0.2 to 1 ms, the test responses to 80- and 120-dB stimuli little differed in amplitude, although the conditioning responses differed more than three times. At longer ISI, the probe response to the 120-dB double-click stimulus continued to grow in amplitude with ISI prolongation (2 and 5 ms), while the probe response to the 80-dB stimulus reached its upper amplitude limit.

Another characteristic feature of the recovery process was that the probe response obtained by the subtraction procedure was never suppressed completely. Even at the shortest delays (less than 0.3 ms) when the two clicks fused into a single one, the subtraction procedure resulted in a small remainder response (Fig. 2.29, ISI of 0.2 ms). This remainder response persisted at unlimitedly short (near-zero) ISI because two clicks fused into a single one had double intensity (3 dB higher level) than each of the clicks, thus evoking ABR of somewhat higher amplitude than the single conditioning click. This difference resulted in a small subtraction-derived response.

Figure 2.30 presents all the data on ABR recovery at various stimulus in-

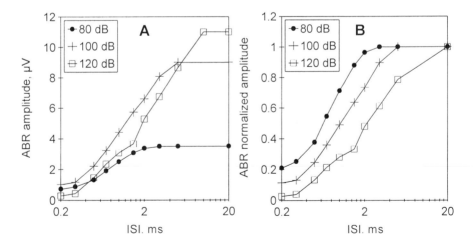

Figure 2.30. ABR recovery functions for click intensities of 80, 100, and 120 dB re 1 μPa, as indicated. Test response amplitude is presented in its absolute voltage (**A**) and as a fraction of response amplitude to the single click (**B**).

tensities. It shows the second (probe) response amplitude as a function of ISI, taking stimulus intensity as a parameter. The plots demonstrate in more detail the features of the recovery process which were illustrated by the original records in Fig. 2.29. ABR amplitude dependence on ISI appeared as soon as ISI exceeded 0.2–0.3 ms and lasted at least a few ms. When response amplitude was presented as absolute voltage (Fig. 2.30A), the plots were grouped together at short ISI, but at longer ISI they diverged: the higher stimulus intensity, the higher response amplitude limit. The shortest recovery time (about 2 ms for complete recovery) was observed at the lowest stimulus intensity used (80 dB re 1 µPa), the longest recovery time (more than 10 ms) at the highest intensity (120 dB), the intensity of 100 dB resulted in an intermediate recovery time of about 5 ms (Fig. 2.30B). The plots demonstrate also small responses that persisted when the delay between two clicks was as short as 0.2 to 0.3 ms.

Quick ABR recovery as presented in Fig. 2.30 was observed when both conditioning and probe clicks were of equal intensities. If the conditioning click exceeded the test one in intensity, the recovery time increased with increasing the difference. Figure 2.31 presents experiments when probe click levels varied within a range from 75 to 115 dB re 1 µPa, whereas conditioning click levels independently varies from 75 up to 135 dB re 1 µPa. Thus, the difference between the two clicks varied within a range from 0 dB (equal

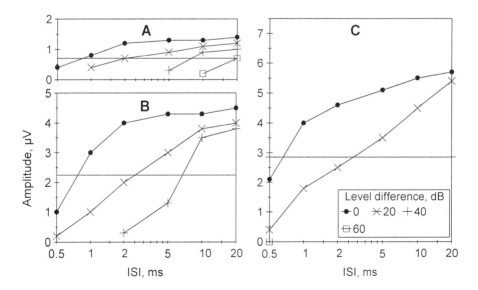

Figure 2.31. ABR recovery functions at unequal levels of the conditioning and probe clicks. Probe click levels: **A**. 75 dB, **B**. 95 dB, **C**. 115 dB. Excess of the conditioning click level above the probe one is indicated in the legend. Horizontal lines indicate 50% recovery level.

conditioning and probe levels) to as large as 60 dB (135-dB conditioning click and 75-dB probe click). The more the level difference, the longer ABR recovery.

The ABR recovery dynamics at various levels of conditioning and probe clicks can be summarized in a plot as presented in Fig. 2.32. The 50% recovery time of the probe response is plotted as a function of the level difference between two clicks. Being presented on a logarithmic ordinate scale, the dependence between the two variables was almost linear: an increase in the level difference by 40 dB led to a 10-fold prolongation of the recovery time.

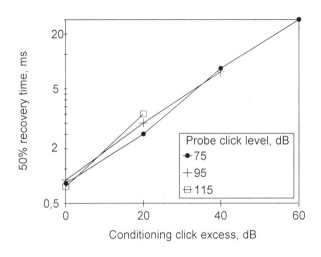

Figure 2.32. 50% recovery time of ABR as a function of level difference between the conditioning and probe clicks, taking the probe level as a parameter. Data from a bottlenose dolphin *Tursiops truncatus.*

Thus, on the one hand, similarly to the temporal-summation experiments, the double-click test revealed a rather high temporal resolution in the auditory system: ABR amplitude was sensitive to ISI as soon as it exceeded 0.2 to 0.3 ms. On the other hand, no signs of temporal summation was observed later than 0.3–0.5 ms, whereas the double-click test revealed much longer events: even at equal conditioning and test levels, ABR amplitude dependence on ISI lasted a few ms and could be longer than 10 ms; at unequal levels, it could be as long as a few tens ms. This contradiction calls for further study of temporal processing in the dolphin's auditory system.

2.5.4. Gap-in-Noise Detection Measurements

A widely adopted method to estimate the auditory temporal resolution is the temporal gap-detection technique: evaluation of the shortest detectable silent interval (gap) in an otherwise continuous sound. This method was very productive for estimating temporal resolution of the auditory system, particularly in humans (Plomp, 1964; Penner, 1977; Smiarowski and Carhart, 1975; Fitzgibbons and Wightman, 1982; Fitzgibbons, 1983; Moore et al., 1993). ABR-recording technique allowed to apply this method to study the temporal resolution in dolphins (Supin and Popov, 1995b; Popov and Supin, 1997).

In that study, stimuli were wide-band noises generated by activation of a transducer by quasi-random binary sequence (sampling rate of 500 kHz), which were gated in bursts (Fig. 2.33). The frequency band of the noise made it possible to use short rise-fall ramps of the burst, thus producing short gaps in noise, although because of transducer ringing and reverberation in the experimental bath, gaps could not be shorter than 0.1 ms. Gap-containing stimuli consisted of a pregap burst, a gap of varying duration, and a postgap burst. The gap duration varied within a range from 0.1 to 10 ms.

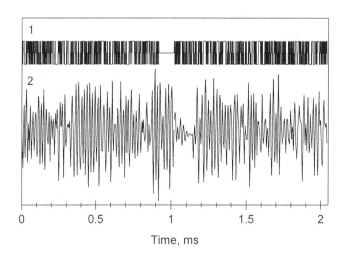

Figure 2.33. A fragment of electric (1) and acoustic (2) signals containing a 0.1-ms gap.

ABR to noise bursts with a gap are exemplified in Fig. 2.34. The stimulus record shows the pregap noise burst, which was 5 ms long; then it was followed by a gap of varying duration and the 5-ms long postgap burst. All the records exhibit the on-response to the first burst onset and a response to the gap.

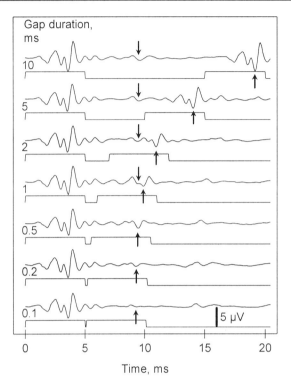

Figure 2.34. ABR evoked by two noise bursts with a gap of various duration between them. In each pair of records, the upper one is the EP record, and the lower one is the stimulus record (noise onset upward). Sound intensity 100 dB *re* 1 µPa. Gap duration (ms) is indicated near the curves. At long gaps (1–10 ms), downward headed arrows show off-responses (to the gap beginning), and upward headed arrows show on-responses (to the gap end); at short gaps (0.1–0.5 ms), gap-responses (fused off- and on-responses) are shown by arrows.

The first on-response was identical in all the records: it was a typical ABR as described above. Its amplitude was as large as a few microvolts. The response to a gap depended on the gap duration. At long duration (2–10 ms), it contained the separated off-response to the burst offset (gap beginning) and on-response to the following burst onset (gap end). When gap duration was of 10 ms or longer, the on-response reached its maximal amplitude; by shortening the gap, the response amplitude diminished. The off-response was of lower amplitude and simpler configuration than the on-response; the latencies of the main waves of the off-response were 0.4 to 0.5 ms longer than those of the on-response.

With shortening the gap, the delay between the off-response to the gap beginning and on-response to the gap end became respectively shorter. At gap duration shorter than 1 ms, the two responses fused into a single one, which is reasonable to designate as the gap-response. Since the off-response had a longer latency than the on-response, at the 0.5-ms gap duration they overlapped almost completely. This combined gap-response to the 0.5-ms gap was of a higher amplitude than the on-response (the more so, higher than the off-response) at longer gap durations. Further shortening of the gap resulted in diminishing of the response. However, a small but detectable response was observable at the shortest available gap duration of 0.1 ms. Thus below we use the term *on-response* for gap duration of 1 ms and longer when this response can be distinguished from the off-response, and the term *gap-response* for gap duration of 0.5 ms and shorter.

The obtained data are presented in Fig. 2.35 as plots showing the dependence of the gap- and on-response amplitude on gap duration. The functions were obtained at several noise intensities, from 80 to 120 dB re 1 µPa (20 to 60 dB above the response threshold). At a gap duration from 0.1 to 0.5 ms, amplitude of the combined gap-response was plotted; at longer durations, the on-response amplitude was plotted. The plots show the main effects as were exemplified by original records in Fig. 2.34: response amplitude increased with gap prolongation, except for the gap-response to the 0.5-ms gap higher than the on-response to 0.7–1 ms gaps.

Another feature of the presented plots is that they group densely within a gap-duration range from 0.1 to 0.5 ms, although they become widely separated at longer gaps. In other words, responses to short gaps were almost independent of the noise level, whereas responses to the noise onset (after a long silence) did depend, in a large measure, on level. This effect looks very

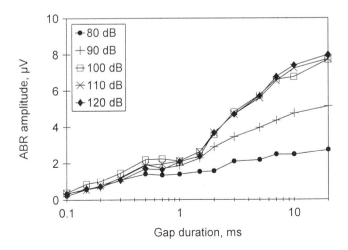

Figure 2.35. ABR amplitude dependence on gap duration at sound intensities of 80 to 120 dB *re* 1 µPa, as indicated.

similar to that obtained at double-click stimulation (see Section 2.5.3). ABR amplitude became independent of level when the response was evoked by both a click and a noise onset after a short delay.

The data show that the dolphin's hearing is sensitive to much shorter gaps than other animals and humans. Since detectable ABR was evoked by a gap as short as 0.1 ms, it is reasonable to assume that the gap detection threshold in dolphins is no more than 0.1 ms. For comparison, in humans, gap-in-noise detection thresholds revealed by psychophysical measurements were 3 to 5 ms, almost two orders of magnitude longer than in dolphins (Plomp, 1964; Penner, 1977; Smiarowski and Carhart, 1975; Fitzgibbons and Wightman, 1982; Fitzgibbons, 1983). At the best combinations of the noise bandwidth and upper cutoff frequency, gap thresholds in humans may be as short as 2.2 ms (Snell et al., 1994); however, this value is still more than an order of magnitude longer than in dolphins. Gap-detection thresholds in sinusoids are 6 to 8 ms at frequencies of 400 Hz and above (Moore et al., 1993). Experiments with detection of short stimulus in a gap of a masker ("temporal window" measurements) have also shown the temporal window duration of several milliseconds (Moore et al., 1988; Plack and Moore, 1990). In several animal species, gap thresholds are of the same order – that is, several milliseconds (rev. Fay, 1988, 1992).

Very probably, the extremely low gap-detection thresholds in dolphins are associated with the wide (more than 100 kHz) frequency range of their hearing. Indeed, many studies pointed out that gap-detection thresholds decrease as the signal frequency increases (Viemeister, 1979; Fitzgibbons, 1983; Shailer and Moore, 1985; Formby and Muir, 1988). It was expected since at higher frequencies, peripheral auditory filters have wider passbands, thus transferring more rapid temporal modulations. Later these data have been reevaluated because in those experiments the noise bandwidth was co-varied with center frequency, and this was a confounding factor since fluctuations inherent in noise depend on the bandwidth, thus influencing the gap detection efficiency. Indeed, gap sensitivity is markedly affected by the inherent noise fluctuations (Glasberg and More, 1992; Snell, 1995). Keeping the bandwidth constant, temporal resolution was less dependent on sound frequency (Shailer and Moore, 1985, 1987; Moore and Glasberg, 1988; Grose et al., 1989; Plack and Moore, 1990; Eddins et al., 1992; Moore et al., 1993). These findings suggested the integration time is dictated mostly by a central integrator rather than the width of the auditory filters. Nevertheless, both the sound frequency and bandwidth influence the gap detection (Snell et al., 1994). Thus, upper cutoff frequency determines gap sensitivity to a large extent because of both widening of peripheral filters and decreasing intrinsic noise fluctuations. The dolphin's auditory system operates with frequencies an order of magnitude higher than those for human's hearing. At these high frequencies, peripheral auditory filters should have rather wide passbands. Even if a central integrator plays a primary role in temporal resolution, its integration time may be evolutionary adjusted to the temporal resolution of peripheral filters. Thus, the gap-detection threshold becomes very low in dolphins.

2.5.5. Derivation of the Temporal Transfer Function of the Auditory System

Temporal resolution of the auditory system may be expressed in terms of its temporal transfer function which, has integration properties: the shorter the integration time, the better the temporal resolution. The data presented above (Sections 2.5.1–2.5.4) did indicate rather high temporal resolution in the dolphin's auditory system. However, they did not provide precise evaluations of its temporal transfer function and integration time. For better understanding the temporal processing in the auditory system, it was important to find the integration time course in dolphins. Apart from these reasons, some experimental finding needed explanation. In particular, it was the fact that response recovery after a conditioning click or a gap in noise lasted up to tens of milliseconds, whereas temporal summation lasted not longer than hundreds of microseconds. Probe-response amplitude independence of level at short ISI or gap durations also at a first sight seemed paradoxical and needed to be explained.

To describe integration processes in the auditory system, a widely adopted model is commonly used that consists of successive stages: a bank of bandpass peripheral filters, each followed by a square-law rectifier, an integrator itself (low-pass filter), and an output device that produces the measured response (Viemeister, 1979; Moore et al., 1988; Plack and Moore, 1990). The integrator is assumed to play a primary role in limiting the temporal resolution. As to the output device, in psychophysical studies it was assumed to be a decision device. For evoked-response studies, the output device must be the evoked-response generator.

Suppose that a constant input signal appears at the integrator input at an instant t_0 (the burst onset in Fig. 2.36A). Then the output signal $R(t)$ is

$$R(t) = I \int \phi(t - t_0) dt , \qquad (2.4)$$

where, $\phi(t)$ is the temporal transfer function. If the stimulus duration is $t-t_0$, then the output signal reaches a certain value $R(t-t_0)$ and then decays. Thus,

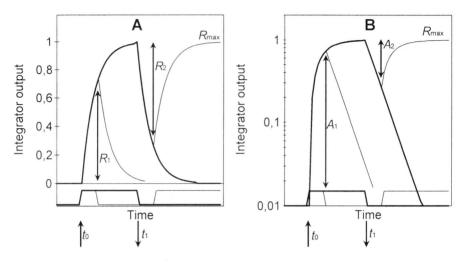

Figure 2.36. The model of temporal-summation and gap-resolution events with nonlinear transform of the integrator output. **A**. Linear presentation of the integrator output. **B**. Nonlinear presentation on the log ordinate scale. Lower solid line presents a long burst; lower thin line presents short-burst (at an instant t_0) and short-gap (at an instant t_1) stimuli. Upper solid and thin lines present the integrator output response to these stimuli. R_1 and R_2 – magnitudes of the linear integrator output responses to the short-burst and short-gap stimuli; A_1 and A_2 – the same responses after the non-linear transform.

the function $R(t)$ in this case simulates the temporal summation course. If the input burst is long enough, the integrator output reaches an asymptote:

$$R_{max} = I \int_0^\infty \phi(t)dt .$$ (2.5)

.

Similarly, if a constant signal disappears at the integrator input at an instant t_1 (a gap beginning), then the output signal $R'(t)$ is

$$R'(t) = R_{max} - \int \phi(t - t_1)dt ,$$ (2.6)

where R_{max} is an output value before the gap, and $\phi(t)$ is the temporal transfer function. The end of the gap cancels the decay and restores the integrator output up to the previous value R_{max}. Thus, the response to the gap is evoked by a shift of the integrator output from R_{max} to $R(t)$ (off-response), or from $R(t)$ to R_{max} (on-response), or both (gap-response). The function $R(t)$ simulates the response recovery with increasing the gap duration.

As follows from these equations and Fig. 2.36A illustrates, the integrator outputs are mirror-symmetrical in conditions of temporal summation and gap detection. However, the results presented above showed that temporal summation and gap detection time courses differed by more than an order of magnitude: in temporal-summation experiments, ABR reached its maximum within less than 0.5 ms, whereas in gap detection conditions, complete ABR recovery required more than 10 ms.

As to the double-pulse stimulation, it may be simulated as if short pulses (each much shorter than the integrator transfer function) appear at the integrator input. Each of the pulses evokes an integrator output response that reproduces the transfer function $\phi(t)$. The second response adds to the first one, and if the integrator acts linearly, this addition does not depend on ISI and is always of the same amplitude as the first response (Fig. 2.37A). Actually, the second response was shown to be strongly suppressed at short ISI and recovered at long ISI.

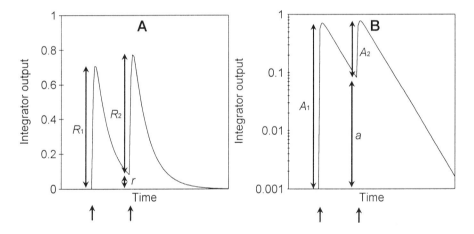

Figure 2.37. Model of ABR recovery in conditions of double clicks. **A.** Linear presentation of integrator output dependence on time. R_1 and R_2 – output responses to two successive input pulses (indicated by arrows), r – remainder of the first response at an instant of the second stimulus. **B.** The same function presented on logarithmic ordinate scale as a simulation of nonlinear transform; A_1 and A_2 – integrator output response amplitudes to the first and second pulses respectively after the nonlinear transform, a – remainder of the first response after the nonlinear transform.

The data obtained in temporal summation, gap detection, and double-click stimulation experiments become explainable when taken into account the nonlinear dependence of ABR amplitude on signal intensity. As shown above (Section 2.3.1), ABR amplitude almost linearly depends on dB-measure of sound level – that is, its dependence on sound power is essentially nonlinear. Figures 2.36B and 2.37B show schematically the appearance of the integrator output after nonlinear amplitude transform. In these schemes, the temporal transfer function of the integrator is assumed to be exponential, and the nonlinear transform is assumed to be logarithmic; it does not mean, however, that the real temporal and input-output transfer functions in the auditory system are precisely the same.

To simulate the effect of a nonlinear transform on burst- and gap-evoked responses, the very same function as in Fig. 2.36A is presented on a log ordinate scale in Fig. 2.36B. The logarithmic transform extends lower range of the signal and compresses the higher range. As a result of such transform, the integrator output rapidly reaches values close to the maximum after the signal onset and decays slowly after the stimulus offset. Thus, one and the same temporal transfer function results in rather short temporal-summation time and long-lasting dependence on the gap duration.

The same model explains test response dependence on ISI in double-click stimulation conditions. When two clicks are presented, the second response of the integrator is partially superimposed on the first response (Fig. 2.37A) – that is, the second response appears at a "pedestal" r remaining of the first one. In a linear system, both of these responses are equal (R). When transformed nonlinearly, the integrator output takes the form shown in Fig. 2.37B. Due to compression of higher signal values and because of the "pedestal" under the second response, the first and second responses A_1 and A_2 become unequal. The shorter ISI, the higher "pedestal" under the second response, and the lower its amplitude.

To describe these events quantitatively, suppose that the output value $R(t)$ evokes a response of amplitude A, according to a nonlinear input-output transfer function of the evoked-response generator:

$$A(R,t) = \Phi\big[R(t)\big] \tag{2.7}$$

in conditions of temporal summation, or

$$A(R,t) = \Phi\big(R_{\max}\big) - \Phi\big[R(t)\big] \tag{2.8}$$

in conditions of gap resolution, where $\Phi(R)$ is the intensity-to-amplitude transfer function.

In conditions of double-click stimulation, the second response amplitude $A(t)$ depends on remainder r of the first response as

$$A(t) = \Phi[R + r(t)] - \Phi[r(t)], \tag{2.9}$$

(Fig. 2.37B), where $\Phi(R)$ is a nonlinear transfer function, R is the integrator response amplitude, and $r(t)$ is the remainder of the first response at an instant of the second pulse (pedestal).

Suppose the function $\Phi(R)$ reproduces the ABR amplitude dependence on stimulus intensity. Using this function, the integrator output $R(t)$ can be obtained from ABR amplitude data $A(t)$. If the $R(t)$ function was obtained in conditions of double-click stimulation (that is, the time is ISI), it directly reproduces the integrator temporal transfer function. If the $R(t)$ function was obtained in conditions of temporal-summation or gap-resolution experiments (that us, the time is stimulus or gap duration), this function should be differentiated to obtain the ultimate temporal transfer function.

It is noteworthy that a few predictions follow from this relation. All of them agree well with experimental data.

(i) If the assumed nonlinear amplitude-versus-intensity function is logarithmic – that is, $\Phi(R) = k \cdot \log(R)$ – and the integrator output after a short pulse decays as $r = R \cdot \varphi(t)$, where k is an adjusting factor, t is time after the pulse, and $\varphi(t)$ is the integrator transfer function, then equation (2.9) takes a form

$$\begin{aligned} A(t) &= k \log[R + R\varphi(t)] - k \log[R\varphi(t)] = \\ &= k \log[1 / \varphi(t) + 1] \end{aligned} \tag{2.10}$$

The final form of the expression does not contain the term R. It means that the second evoked-response amplitude $A(t)$ depends only on ISI t; at a given t, the second response amplitude is independent of the integrator output amplitude R and hence independent of stimulus intensity.

This mechanism is illustrated in Fig. 2.38. Two curves show integrator output signals at two input signal levels. Being presented in the linear scale (A), the curves differ in their amplitudes (R_1 and R_2) by several times. However, after the logarithmic transform (B), the same curves differ only in their shift along the ordinate scale, while their probe-responses amplitudes (A_1 and A_2) become equal. It was indeed observed, as grouping of plots in Fig. 2.30A shows. When ISI is long enough so as the integrator output reaches the response threshold level, the probe response becomes intensity-dependent.

The very same is the situation for gap-responses. Again, assuming $\Phi(R) = k \log(R)$, equation (2.8) takes a form

$$\begin{aligned} A[R(t)] &= k[\log R_0 - \log R_0 \varphi(t)] = \\ &= k \log[1/\varphi(t)] \end{aligned} \tag{2.11}$$

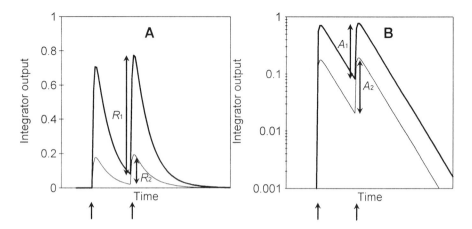

Figure 2.38. Mechanisms of probe-response amplitude independence on stimulus level in conditions of double-click stimulation. **A.** Linear presentation of the integrator output response to two pulses (indicated by upward-headed arrows). **B.** The same after the nonlinear transform simulated by presenting the function on the log ordinate scale. Solid and thin lines present responses to two different stimulus intensities. R_1 and R_2 – amplitude of linear probe responses; A_1 and A_2 – amplitudes of nonlinearly transformed probe responses.

– that is, the gap-response amplitude is independent of R_0 and thus of stimulus level. This feature of the gap-response was really observed.

(ii) At very short ISI, when no significant decay of the integrator output is expected, it can be adopted: $r = R$. At this condition, equation (2.9) takes a form

$$A = \Phi(2R) - \Phi(R) \tag{2.12}$$

– that is, the test response amplitude is equal to a difference between response amplitudes to stimuli differing twice in intensity (i.e., by 3 dB). It was really observed at ISI of 0.2 to 0.3 ms (see Fig. 2.30).

(iii) If the conditioning stimulus increases in level relative to the test one, the decay of the integrator output begins from a higher value; thus, it takes longer time to reach a certain value of the "pedestal" under the test response. Since the pedestal value dictates the probe response amplitude, the probe response recovery takes longer time. The recovery time dependence on conditioning stimulus level should reproduce the integrator temporal transfer function. Indeed, ABR recovery time increases with increasing the conditioning stimulus level relative to the probe one (see Figs. 2.31 and 2.32). The rate of the recovery time prolongation (ten-times prolongation per 40-dB level difference) is very close to the decay rate of the integrator temporal transfer function as it is computed below.

The good fit of the predictions to the data confirms the applicability of the model to temporal-processing mechanisms in the dolphin's auditory system.

For quantitative data analysis according to the presented model, ABR amplitude dependence on intensity (that is, the function $\Phi(R)$) should be known. Measurements made for the short-burst clicks have shown that ABR amplitude was intensity-dependent until approximately 120–130 dB re 1 μPa (60–70 dB above the response threshold), beyond which it reached asymptote. Within this range, the dependence was described as roughly close to a straight line when the intensity was expressed on a dB (logarithmic) scale and the amplitude on a linear scale (Popov and Supin, 1990a, 1990b). More detailed measurements in a low-intensity range (Supin and Popov, 1995c) have shown that the deviation of the inclined part of the function from a straight line was not negligible (Fig. 2.39A). In the low-intensity range, the slope of the function was shallower than at middle intensities; at higher intensities, the slope reduced to zero when the response amplitude reached asymptote. Nevertheless, the inclined part of the function was much closer to a straight line when the intensity was expressed on a dB (logarithmic) scale than on a linear scale. In linear scales, both in the sound pressure and sound power domains, the slope was maximal near the response threshold and was markedly diminished with intensity increase (Fig. 2.39B and C).

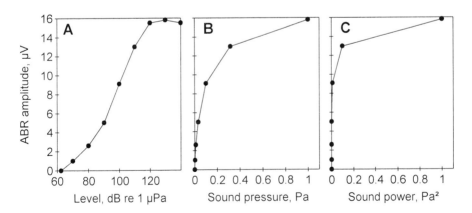

Figure 2.39. ABR amplitude dependence on click intensity expressed in various measures: dB-level (**A**) , sound pressure (**B**) and sound power flux density (**C**).

Using the ABR amplitude dependence on sound power, double-click data (Supin and Popov, 1995c) were analyzed to find the integrator temporal transfer function. Since the temporal transfer function of any system may be defined as its output response to a short input signal, it was assumed that the response of the integrator to a click reproduces the temporal transfer function. Thus, the task to define the temporal transfer function reduced to the task to find the output response of the integrator to a short click. The computation is based on the equation. (2.9) which relates the probe response amplitude A to the integrator output values R and r. However, actual values of the integrator output are not known. Therefore, sound intensity was used as an indirect measure of the integrator output supposing that it is directly proportional to intensity. Thus, equation (2.9) relating the test response amplitude to integrator output, can by substituted by

$$A = \Phi(I + i) - \Phi(i),\tag{2.13}$$

where I and i are sound power in dB-measure corresponding to the integrator output amplitude R and remainder r, respectively. Taking the experimentally found function relating ABR amplitude to stimulus intensity (Fig. 2.39) as $\Phi(i)$, the functions $\Phi(I+i) - \Phi(i)$ were computed at three stimulus intensities I values, 80, 100, and 120 dB; they are presented in Fig. 2.40A.

Plots in Fig. 2.40B reproduce ABR recovery functions (see Fig. 2.30) – that is, test ABR amplitudes depending on ISI. Since the plots in Fig. 2.40A relate the response amplitudes to the integrator output, this output r can be derived, projecting plots of Fig. 2.40B via the corresponding functions in Fig. 2.40A into the sound intensity domain. The result is plotted in Fig. 2.40C. All three stimulus intensities revealed similar time course of the integrator output. The derived values were virtually constant for about 200 μs after the stimulus. Then a decay in slope of 10 to 11 dB per time doubling (about 35 dB/decade) occurred, such that the data form a roughly straight line on the dB-intensity versus logarithmic time coordinates. The point of −3 dB decay was around 0.3 ms.

Taking that after a short input pulse, the integrator output reproduces its temporal transfer function, the curves in Fig. 2.40C represent the required temporal transfer function.

An approximate analytical description of this transfer function may be made as follows. Integrator output is presented herein in the sound intensity domain on the dB scale. Assuming that afferent signal is converted via a square-law rectifier, a 10-dB change of sound intensity leads to the 10-fold change of the integrator input and output – that is, the −35-dB change corresponds to a $10^{-3.5}$ change on the linear scale. Hence the function of 35 dB/decade decay may be defined as:

$$r(t) = R(t\ \tau)^{k},\tag{2.14}$$

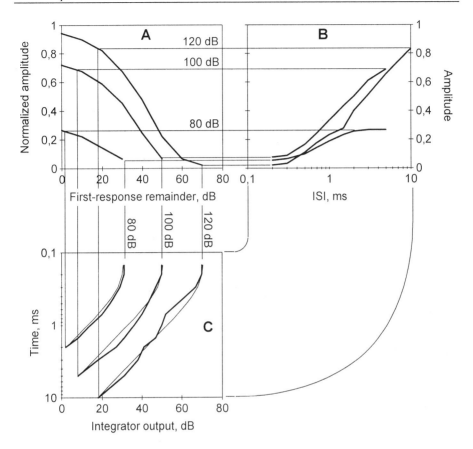

Figure 2.40. Computation of the temporal transfer function from ABR recovery data. **A.** Functions relating the conditioning-response remainder r to the probe-response amplitude $A(t)$, derived from the plot of Fig. 2.39 according to equation (2.13) for click intensities of 80, 100, and 120 dB re 1 µPa, as indicated. **B.** ABR recovery functions for click intensities of 80, 100, and 120 dB, as indicated. **C.** Integrator output derived at these three intensities; the panel is turned side to make the integrator output scale of this panel coinciding with the sound intensity scale in the panel (**A**); thin lines present analytical approximations according to equation (2.15) at the three intensities. Straight lines between the panels show projections of key points of plots (**B**) through the functions of the panel (**A**) to the panel (**C**).

where $k = -3.5$, R and τ are adjustable parameters.

This equation approximates the function decay only at large t values. To approximate both the initial flat part and subsequent decay of the function, we modify Eq. (21) to another simple relation:

$$r(t) = \frac{R}{1 + (\tau/t)^k},\qquad\qquad (2.15)$$

where R is the initial (immediately after a short input pulse) output level, and k and τ are parameters determining the time course. The parameter k dictates the function slope at large t: equation (2.15) reduces to equation (2.14) when $t/\tau \gg 1$, and decay of 35 dB/decade corresponds to $k = -3.5$. However, $r(t)$ cannot exceed R: equation (2.15) reduces to $r(t) = R$ if $t/\tau \ll 1$. The parameter τ determines the transition point from the flat to oblique branch of the plot: $r(t) = 0.5R$ when $t = \tau$. Since the obtained curves (Fig. 40C) decline to the 0.5 level (−3 dB) at around 0.3 ms, it was reasonable to assume $\tau = 0.3$ ms. Functions computed according to equation (2.15) assuming $k = -3.5$ and $\tau = 0.3$ ms (thin lines in Fig. 2.40C) agree well with the experimental data. Figure 2.41 shows the averaged functions, both derived from experimental data and computed according to equation (2.15), presented on linear (A) and log (B) scales. This result is an estimation of the temporal transfer function of the integrator.

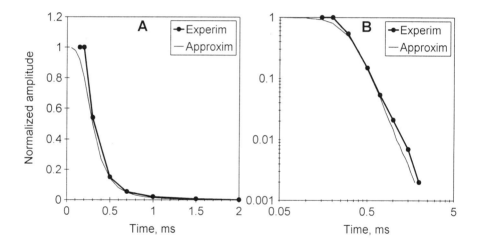

Figure 2.41. Temporal transfer function of the dolphin's hearing as derived from the double-click data. **A.** The function presented on linear scales. **B.** The same function presented on log scales. Solid line – averaged experimental data, thin line – approximation by equation (2.15).

The temporal transfer function was actually obtained in the sound intensity domain since the computation was based on the function relating ABR amplitude to sound intensity. It was supposed that the integrator input is proportional to the sound power because of the action of a square-law rectifier. However, other relations cannot be excluded. If the integrator input is proportional to the sound pressure, then the −3 dB decay (at the 0.3 ms delay) indicates the transfer function level of 0.71 instead of 0.5, and the subsequent decay of 35 dB/decade corresponds to the slope of $t^{-1.75}$ instead of $t^{-3.5}$. However, the shape of a transfer function just in the sound intensity domain is important for understanding many aspects of sound processing, and this form does not depend on properties of the supposed rectifier.

Does the suggested function agree with other widely adopted analytical descriptions of temporal transfer functions in the auditory system? In many studies, the temporal integration in the auditory system is described by a model with an exponential decay. The results presented above show another type of the decay – namely, inversely proportional to a power of time. However, this result is not very surprising since several psychophysical studies in humans have also shown a similar type of decay. A pioneering study of Plomp (1964) has shown that the decay of sensation occurs in a straight line on log-log coordinates (sensation level in dB versus time plotted logarithmically), whereas the exponential function looks like a straight line when the time scale is linear. Nevertheless, in recent studies, a "rounded exponential" (*roex*) function was used to describe the shape of the temporal window (Plack and Moore, 1990). However, that study has also shown the temporal decay to be shallower than an exponential one; therefore, at least two exponents with different time constants were necessary to fit experimental data. The temporal transfer function obtained in dolphins could also be approximated by a combination of several exponents with different time constants. However, description by a single simple equation like equation (2.15) seems preferable.

A temporal transfer function provides a detailed description of temporal processing in the auditory system. Nevertheless, for many purposes, temporal resolution needs to be characterized in a simpler manner, by a single characteristic value. A commonly adopted metric of temporal resolution is the equivalent rectangular duration (ERD) (Plack and Moore, 1990). It is the duration of an idealized rectangular function of the same integral value as the real transfer function. For the function described by equation (2.15), ERD is

$$D_{ER} = \int_{0}^{\infty} dt \left[1 + (\tau / t)^k \right] =$$

$$= \pi \tau\, k \sin(\pi / k)$$

(2.16)

Assuming $k = -3.5$ and $\tau = 0.3$ ms, $D_{ER} \approx 1.15\tau \approx 0.35$ ms.

Similar analysis was performed for gap-detection data. Using sound power as a relative measure of R, equation (2.8) can be substituted by

$$A(t) = \Phi(I) - \Phi(i),$$ (2.17)

where I and i are the sound powers corresponding to R_{max} and $R(t)$, respectively, and $\Phi(i)$ is the intensity-to-amplitude transfer function. Since $\Phi(I) = A_{max}$, it follows

$$i = \Phi'[A_{max} - A(t)],$$ (2.18)

where Φ' is a function inverse to Φ. Then the sought-for temporal transfer function $\varphi(t)$ can be found as a derivative of $i(t)$.

Successive stages of the analysis are illustrated in Fig. 2.42. Panel A shows $A_{max} - A(t)$ values, found from experimental data presented in Fig. 2.35. Panel B shows $\Phi'[A_{max} - A(t)]$ values calculated using the ABR amplitude dependence on intensity as the $\Phi(I)$ function. In so doing, an additional assumption should be made to solve a problem as follows. As was pointed out, the gap-response combining both on- and off-components is of higher amplitude than the on-response only – that is, intensity-to-amplitude transfer functions $\Phi(I)$ are different for these two response types. It results in breaks of plots between 0.5 and 1 ms in Fig. 2.42A. Therefore, at short gaps, $\Phi'[A_{max} - A(t)]$ values cannot be calculate correctly, since A_{max} is the on-response amplitude and $A(t)$ is the gap-response amplitude. In order to solve the problem, it was supposed that $\Phi(I)$ function for gap-responses is approximately 1.5 times steeper than that for on-responses. This eliminates the breaks in plots of Fig. 2.42B. Figure 2.42C and D shows $\varphi(t)$ as a result of differentiation of functions presented in Fig. 2.42B. Note that the functions were calculated from the time of 0.05 ms, although experimental data were obtained only for gap duration of 0.1 ms and longer. The values between $t = 0$ and $t = 0.1$ ms were calculated supposing $A(0) = 0$.

The temporal transfer function $\varphi(t)$ found in such a way demonstrates a decline to the 0.5 level at a time of about 0.18 ms. Then a further long decay appears with a slope of 35 to 40 dB/decade – that is, the function is inversely proportional to the 3.5th to fourth power of time. The initial part of these functions is better demonstrated when presented on linear scales (Fig. 2.42C), whereas its later part is better demonstrated on the double logarithmic scale (Fig. 2.42D). ERD of the temporal transfer function was found to be 0.27 ms. It is rather close to that found as a result of analysis of the double-click data.

The integration time in the dolphin's auditory system (ERD of around 0.3 ms) is an order of magnitude shorter than in many other animals and humans. In humans, no method showed an integration time shorter than a few milliseconds. Careful "temporal window" measurements have shown ERD

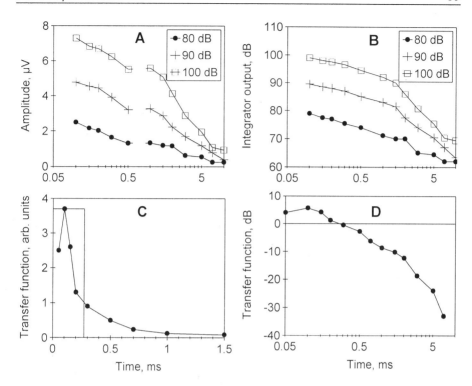

Figure 2.42. Successive stage of calculation of the integrator transfer function basing on gap-detection data. **A.** $A_{max} - A(t)$ values versus gap duration for sound intensities of 80 to 100 dB re 1 µPa, as indicated. At gap durations of 1 ms and longer, on-response amplitude is used as $A(t)$; at gap durations of 0.5 ms and shorter, gap-response amplitude is used; on-response amplitude after a long silence is used as A_{max}. **B.** $\Phi'[A_{max} - A(t)]$ values versus gap duration obtained by transform of the plots (**A**). **C.** Integrator transfer function obtained by differentiation and dividing of plots (**B**) by intensity, linear presentation. **D.** The same presented on log-log scales.

of the temporal window to be of several milliseconds (Moore et al., 1988; Plack and Moore, 1990). Amplitude modulation detection also indicated the integration time constant of an order of several milliseconds, with the shortest values of about 2 ms (Eddins, 1993). In several animal species, the temporal resolution was found to be of the same order – that is, of several ms (rev. Fay, 1988, 1992).

It may be problematic to compare directly behavioral data with those defined by the evoked-potential method. Nevertheless, such comparison is possible as a preliminary approach, taking into account that behavioral data on integration time in dolphins (Au et al., 1988; Au, 1990; Dubrovskiy, 1990) agree well with the evoked-potential results. Thus, the integration time of the dolphin's auditory system seems really to be extremely short in comparison with other animals and humans.

There is a question, however, of how the data indicating very short integration time reconcile with the data mentioned above (Section 2.5.1) indicating much longer integration time (Johnson, 1968, 1991, 1992; Vel'min and Dubrovskiy, 1975; Dubrovskiy, 1990). Those studies indicated the integration time an order of magnitude longer than the integration-time estimates presented above. However, there is no real contradiction between these data. The integration time found by ABR recording probably reflects properties of peripheral parts of the auditory system. It does not exclude the existence of other integrators with longer integration time at higher levels of the auditory system. The action of these integrators may manifest itself in temporal-summation experiments allowing the accumulation of the stimulus energy over a much longer time than the short peripheral integration time. A similar relation is known in humans: the integration time found in temporal-summation experiments (200 ms or more) is one to two orders of magnitude longer than the values (of several milliseconds) found in other temporal-resolution experiments (rev. Green, 1984; de Boer, 1984). Similarly, it can be supposed that dolphins have a very short integration time in the auditory periphery and a longer integration time at higher levels.

2.5.6. Rhythmic Amplitude-Modulation Test and Modulation Transfer Function

In psychophysical and physiological studies, sinusoidally amplitude-modulated (SAM) sounds are widely used to estimate the temporal resolution in the auditory system. Such studies determine the ability to transfer various modulation rates, which is characterized by the modulation rate transfer function (MTF). As well as the temporal transfer function, MTF is a quantitative characteristics of the temporal resolution but is presented in frequency instead of time domain.

In human psychophysical experiments, MTF was obtained as the threshold modulation depth depending on modulation rate (Viemeister, 1979). In physiological studies, either neuronal responses or evoked potentials can be used to obtain MTF. Among evoked potentials, the envelope-following responses (see Section 2.2.4) is the most appropriate for MTF measurements. The dependence of EFR magnitude on modulation rate can be taken as MTF.

In odontocetes, EFR was studied by Dolphin (1995), Dolphin et al. (1995), and Supin and Popov (1995a). In the study of Dolphin (1995) and Dolphin at al. (1995), three species were investigated: the false killer whale *Pseudorca crassidens*, the beluga whale *Delphinapterus leucas*, and the bottlenose dolphin *Tursiops truncatus*. A feature of that study was the use of steady-state stimulation by SAM stimuli of rather low carrier frequency –

from 0.5 to 10 kHz. In conditions of steady-state stimulation, EFR magnitude was estimated only by frequency spectra of long-lasting records, so only indirect-control experiments (placing the recording electrodes into sea water in the position of the animal) were used as an evidence that records were not contaminated by artifacts. It was found that in all the investigated species, EFR can follow rather high modulation frequencies, so MTFs were low-pass in shape with cut-off frequencies of 1–2 kHz.

Another type of SAM stimulus was used in the study of Supin and Popov (1995a) in bottlenose dolphins *Tursiops truncatus*. It was a burst of 12 ms duration with a linear rise-fall time of 2 ms. This stimulus type evokes a robust EFR as described above (Section 2.2.4). The bursts were presented at a rate of 10/s; thus the burst duration was markedly shorter that the interburst silent interval. With this stimulus type, the response (EFR) temporal pattern could be traced, and the absence of envelope-following potentials during a few millisecond latency provided a direct evidence of artifact absence. The SAM signals were produced by multiplying a tone carrier by a sum of a cosinusoid and a DC signal, so that the sound amplitude was modulated as $1+m\cos2\pi ft$, where t is time, m is the modulation index ($0<m<1$), and f is the modulation rate. Modulation rates from 250 to 4000 Hz in approximately ¼-octave increments were tested in detail.

To evaluate EFR magnitude, a part of the record containing a whole number of response cycles were Fourier transformed to find the weight of the modulation-rate fundamental which was taken as the response amplitude. Figure 2.43 shows EFR amplitude defined in such a way, as a function of modulation rate; three plots represent the functions obtained in three experimental animals. The plots show a typical form of this function. It featured a few peaks, the highest of them at rates of 600 and 1000 Hz, a less prominent peak at 1400 Hz, and troughs at 700–850, 1200, and 2000 Hz. The function declined sharply at rates above 1700 Hz, and was hardly detectable at 2000 Hz. However, above 2000 Hz, one more small peak appeared again at a rate of 2400–2800 Hz, and the response was detectable at rates as high as 3400 Hz.

These functions may be adopted as a first approximation of the MTF. For more precise MTF estimation, it was necessary to ascertain whether EFR-generating mechanisms were linear. Strictly speaking, the term MTF can be applied only to a linear system since otherwise response dependence on modulation rate is not constant at various stimulus parameters. Linearity of the system implies, in particular, that EFR amplitude is linearly dependent on modulation depth.

To evaluate EFR linearity, its amplitude was measured as a function of modulation depth at sound intensities from 80 to 140 dB re 1 μPa (20 to 80 dB above response threshold). The results of measurements at an intensity of

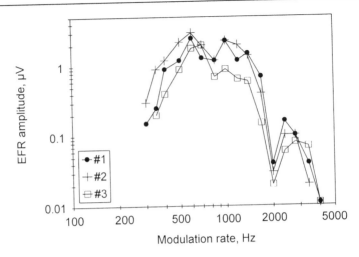

Figure 2.43. EFR amplitude dependence on modulation rate in bottlenose dolphins, *Tursiops truncatus.* Tone carrier of 64 kHz, 60 dB above threshold, modulation depth 1. Data for three animals.

120 dB are presented in Fig. 2.44. The plots show that at all modulation rates and within the major part of the modulation depth range, EFR amplitude was linearly dependent on modulation index: on double log scales, the plots appear to be straight lines with a slope close to 1. Other stimulus intensities gave similar results.

The linear relationships between modulation depth and EFR amplitude made it possible to apply the term MTF to this response and obtain MTF in a more precise manner than with the use of EFR amplitude. To obtain MTF values, a ratio A/m (EFR amplitude to modulation depth, µV) was used. This ratio as a function of modulation rate was taken to plot MTF.

Figure 2.45 presents a family of MTF-plots obtained in such a way at a variety of carrier frequencies. All the curves are of a similar form. They have several peaks: the most prominent peaks at 600 and 1000 Hz and a less prominent one at 1400 Hz. MTFs declined sharply at modulation rates higher than 1700 Hz. However, an additional peak was still visible at a frequency of 2400 Hz, and the functions extended up to 3400 Hz.

To characterize quantitatively MTFs at various carrier frequencies, their bandwidths at a level of 0.1 (–20 dB) relative to the maximal value were used. This level was chosen since it was below the peaks and troughs within a range of up to 1400 Hz, so the function bandwidth at this level was determined by the steep decrease at rates above 1400–1700 Hz. The MTF bandwidth defined in such a way was virtually constant throughout the entire frequency range, from 4 to 110 kHz, and comprised 1700 to 1800 Hz.

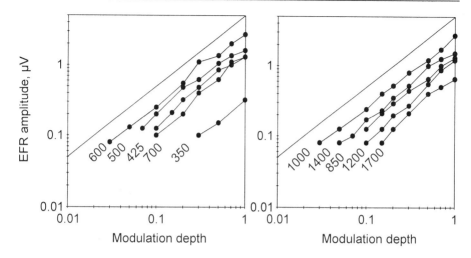

Figure 2.44. EFR amplitude dependence on modulation depth at various modulation rates in the bottlenose dolphin *Tursiops truncatus*. Tone carrier of 64 kHz, 120 dB re 1 μPa. To make the plots better readable, they are separated into two groups: lower rates (350 to 600 Hz, left panel) and higher rates (1000 to 1700 Hz, right panel). Oblique straight lines show the slope of 1 at amplitude/modulation-depth ratio of 5 μV.

It was shown above (Section 2.2.4) that EFR in dolphins can be interpreted as a rhythmic sequence of ABR waves. Thus, MTF obtained based on EFR records reflects mainly the properties of those parts of the auditory system that generate ABR (presumably, from the auditory nerve to the inferior colliculus). This interpretation explains some properties of EFR-based MTF, particularly, the appearance of several alternating peaks and troughs in this function. The same peaks and troughs appear in the ABR spectrum. Their origin was discussed above (Section 2.2.4). Since ABR reflects responses of several generators (different auditory centers), both its waveform and spectrum are dependent on delays between the responses of these generators. Overlapping of potentials produced by several generators with different delays results in peaks and troughs in the ABR spectrum, and these spectrum features manifest themselves in MTF. This spectrum pattern has little to do with the real temporal resolution of the auditory system. MTF decay at rates below 500 Hz can also be attributed to the ABR waveform since the ABR spectrum has the same decay.

However, independently of evoked potential waveform and spectrum, the appearance of potentials at a certain rate indicates a capability of some neuronal structures to follow the rate. So the actual temporal resolution of the dolphin's auditory periphery is expected to be not less than the obtained MTF bandwidth (1700 Hz).

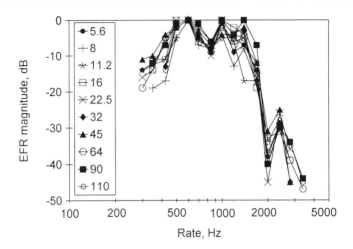

Figure 2.45. MTF at various tone frequencies (4 to 110 kHz, as indicated) in the bottlenose dolphin *Tursiops truncatus.* Tone level 120 dB re 1 µPa.

Thus, neglecting the alternating peaks and troughs of the obtained MTF, its main part can be assessed as a low-pass filtering function with a cut-off point (as estimated by 20-dB decay) at around 1700 Hz. It is noteworthy that this bandwidth is almost independent of the carrier frequency (see Fig. 2.45). The same is true in humans: temporal resolution of their hearing is little dependent on sound frequency (Shailer and Moore, 1985; Moore and Glasberg, 1988; Grose et al., 1989; Plack and Moore, 1990; Eddins et al., 1992; Moore et al., 1993). The interpretation of these data is that peripheral bandpass auditory filters play a minor role in the temporal resolution, because the filter passbands enlarge with frequency. The temporal resolution, therefore, is limited mostly by a more central integration process. The same interpretation is suitable for dolphins. MTF cut-off at about 1700 Hz may reflect the action of a low-pass (integrating) process that is little dependent on sound frequency. Data on the origin of ABR and EFR show that this process may be attributed to the brainstem or lower levels.

Since MTF is a widely adopted way to characterize temporal resolution, it is possible to compare the MTF in dolphins with those in other animals and humans, as well as to compare MTFs obtained in electrophysiological and psychophysical experiments. In humans, MTF obtained by EFR recording had a cut-off of 50 to 70 Hz (Rees et al., 1986; Kuwada et al., 1986). It agrees well with psychophysical studies that have also shown the MTF cut-off frequency of about 50 Hz (Zwicker, 1952; Viemeister, 1979; Eddins, 1993). Qualitatively similar functions were obtained in other mammals behaviorally (Salvi et al., 1982) and electrophysiologically (Dolphin and

Mountain, 1992). Thus, psychophysical and electrophysiological data are in good agreement.

In dolphins, MTF has a much wider bandwidth (about 1700 Hz) than in humans. This difference may be partially due to another EFR origin (brainstem, not cortical). Psychophysical MTF was not obtained in dolphins, and the temporal resolution of their hearing was psychophysically measured in terms of integration time only. As shown above, both psychophysical and evoked-potential studies revealed the integration time in dolphins of about 300 μs. MTF bandwidth and integration time can be compared using Fourier transform. A first approximation may be to substitute the temporal integration function by its equivalent rectangular duration (ERD) and MTF by its equivalent rectangular bandwidth (ERB). The frequency spectrum of a rectangular temporal function is

$$W(f) = (\sin \pi f \tau \, / \, \pi f \tau)^2, \tag{2.19}$$

where $W(f)$ is the spectral power, f is frequency, and τ is ERD. ERB of this spectrum is $\varphi = 1/2\tau$ (φ is ERB). This simple relation, in particular, describes the relationship between ERD and ERB in humans: MTF bandwidth of 50 to 70 Hz in humans corresponds to ERD of 7 to 10 ms which agrees with the results of "temporal window" measurements (Moore et al., 1988; Plack and Moore, 1990). MTF bandwidth of 1700 Hz in dolphins corresponds to ERD of 300 μs, which agrees with both the behavioral and evoked-potential data described above.

Generation of EFR by lower auditory centers is not a general rule for all species. In humans, EFR appearing at modulation rates up to 50–70 Hz exhibits latency as high as 30 ms suggesting the cortical origin of the response (Kuwada et al., 1986; Picton et al., 1987). At modulation rates of 100 to 400 Hz, low-amplitude EFR was observed in humans with a latency as short as 7–9 ms. This points to the midbrain origin (Kuwada et al., 1986). In the gerbil, EFR at modulation rates of up to 50 Hz had also a delay suggesting cortical generation, while EFR at higher rates was supposed to be of peripheral origin (Dolphin and Mountain, 1992).

A possibility must not be ruled out is that the brainstem origin of EFR in dolphins was a consequence of the used electrode position. The recording electrode was located in the point where ABR amplitude was maximal; correspondingly, ABR had a dominant role in forming the EFR. Another electrode position makes it possible to record evoked potentials of a presumably cortical origin (Supin and Popov, 1990). We can concede that this electrode position might give EFR of a cortical origin and different MTF.

2.5.7. Rhythmic Click Test

Rhythmic sequence of short clicks was the first test used in measurements of temporal resolution of hearing in dolphins since it was very easy to present such stimulus type. In those studies, the sound-emitting transducer was activated by short uniform pulses, and only the rate of pulse repetition was the variable parameter. Clicks evoked ABRs, so the ability of ABR to follow the stimulation rate provided a convenient estimate of the temporal resolution. An important difference between SAM sounds and rhythmic clicks as test stimuli is that clicks are able to evoke pronounced ABRs at unlimitedly low rates, whereas low-rate SAM sounds are less effective.

Several studies (Ridgway et al., 1981; Popov and Supin, 1990b; Popov and Klishin, 1998) used long (steady-state) sequences of rhythmic clicks that were generated continuously throughout all the time of evoked-response collection (Fig. 2.46). This test was applied to a number of dolphin species: the bottlenose dolphin *Tursiops truncatus*, common dolphin *Delphinus delphis*, harbor porpoise *Phocoena phocoena*, beluga whale *Delphinapterus leucas*, tucuxi dolphin *Sotalia fluviatilis*, and Amazon river dolphin *Inia geoffrensis*.

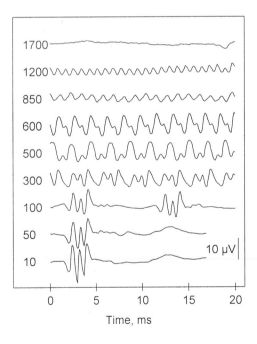

Figure 2.46. ABR to rhythmic steady-state click stimulation of various rates in a common dolphin *Delphinus delphis*. Click level 120 dB re 1 µPa. Stimulation rate (s^{-1}) is indicated near records.

All of them demonstrated a similar manner of ABR amplitude dependence on click rate (Figs. 2.46 and 2.47). Amplitude of click-evoked ABR remained almost constant until the click rate reached a few tens s^{-1} and decreased with the further rate increase.

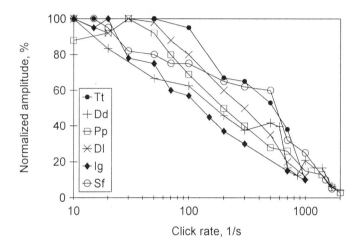

Figure 2.47. ABR amplitude dependence on click rate (steady-state stimulation) in several dolphin species. Tt – *Tursiops truncatus*, Dd – *Delphinus delphis*, Pp – *Phocoena phocoena*, Dl – *Delphinapterus leucas*, Ig – *Inia geoffrensis*, Sf – *Sotalia fluviatilis*.

The ability to follow stimulation rates of a few tens s^{-1} can be assessed as an indication of rather high temporal resolution. Nevertheless, it is obviously less than the ability revealed by the SAM test, which resulted in the highest EFR amplitude at rates around 1000 Hz, whereas responses to steady-state rhythmic pulses decreased at rates above a few tens s^{-1}.

Very probably, the source of disagreement between the data obtained with rhythmic sound pulses and SAM stimuli may be a long-term adaptation during the steady-state rhythmic-click stimulation. It should be stressed that in the above-mentioned experiments, ABRs were collected with the use of long steady-state pulse sequences. With the pulse amplitude being constant, the mean power of the pulse sequence is proportional to the pulse rate (Fig. 2.48A). Hence, the long-term adaptation may increase with rate thus resulting in the ABR amplitude decrease. This effect cannot appear in the case of SAM stimuli since their mean level is independent of modulation rate.

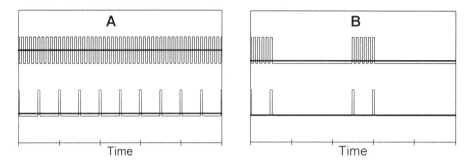

Figure 2.48. Mean power dependence on rate of standard pulses. **A**. Steady-state pulse sequence. **B**. short pulse trains. Upper trace – high pulse rate, lower trace – low rate. Solid horizontal lines are mean power values.

A way to avoid the confounding effect of long-term adaptation is to use short pulse trains separated by long intertrain intervals. Such stimuli have low mean-power value (Fig. 2.48B), so the adaptation effect is much less than with the use of steady-state sequences. Even if the adaptation exists, it can be evaluated by the amplitude of a response to the first pulse of the train. Therefore, a study was undertaken to measure the capability of ABR in bottlenose dolphins to follow rhythmic sound pulses (clicks and pips) presented as short trains separated by longer intervals (Popov and Supin, 1998).

In that study, stimuli were rhythmic trains of wide-band clicks or short pips; the latter had envelope of one 0.5-ms cycle of a cosine function. Sound pulses (clicks or pips) were presented in trains of 20 ms duration and were repeated at a rate of 10 s^{-1}. Thus, intertrain silence intervals were much longer than the trains.

Evoked potentials to rhythmic click trains (the rate-following response, RFR) were described and exemplified above (see Fig. 2.11). RFR amplitude as a function of stimulation rate is shown in Fig. 2.49. In these experiments, RFR was recorded at stimulus intensities from 70 to 130 dB re 1 μPa peak-to-peak sound pressure, which corresponded to 10 to 70 dB above the ABR threshold. As Fig. 2.49 shows, RFR amplitude was almost constant at rates up to 200 s^{-1}. Further rate increase resulted in a few small peaks and valleys in the function: peaks at 500–600, 1000–1200, and around 2400 s^{-1}; valleys at 400–450, 700–800, and around 2000 s^{-1}. Superimposed on these peaks and valleys, a general trend was the amplitude decrease at rates above 1000 s^{-1}, so that 10-fold amplitude decrease occurred at a rate of 1700 s^{-1}. Small but detectable responses appeared at much higher rates (100-fold amplitude decrease at 2800–3200 s^{-1}). With variation of the stimulus level, the function shifted along the ordinate axis (the higher level, the higher amplitude) but its form was little affected.

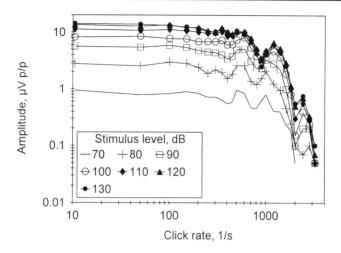

Figure 2.49. RFR peak-to-peak amplitude dependence on click rate in a bottlenose dolphin *Tursiops truncatus*. Short-train stimulation. Click intensities are indicated in dB re 1 μPa.

Comparison of the amplitude-vs-rate function with the ABR spectrum (see above, Fig. 2.3) shows that the peaks of the function at 500–600, 1000–1200, and 2400 s^{-1} corresponded fairly to those of the ABR spectrum. However, the function did not reproduce the ABR spectrum exactly: at low rates (below 200 s^{-1}), the amplitude-versus-rate function was constant, whereas the ABR spectrum magnitude decreased at low frequencies; at high rates (above 1700 s^{-1}), the amplitude-versus-rate function passed below the ABR spectrum.

To characterize RFR properties in terms of a frequency transfer function, frequency spectra of the rhythmic responses were computed (Fig. 2.50). All the spectra had prominent peaks at the fundamental and harmonics of the stimulus rate. At low rates (up to 200 s^{-1}), the magnitude of the fundamental was much lower than those of harmonics; the highest harmonic magnitudes were at 500–600 and 1000–1200 Hz – that is, at frequencies corresponding to the maxima of the ABR spectrum. At higher rates, magnitudes of the fundamental increased being the highest at 600 and 1000 Hz and rather high at 450 to 1400 Hz – that is, when the stimulation rate coincided with high-magnitude components of the ABR spectrum. However, at stimulation rates above 1400 s^{-1}, the fundamental magnitude decreased markedly: at a rate of 1800 s^{-1}, the fundamental peak was very small. It should be noted that RFR to lower rates (200 to 600 s^{-1}) contained harmonics of 1800 Hz that were markedly higher than the fundamental at 1800 s^{-1} rate.

Fundamental magnitude dependence on frequency (that is, on stimulation rate) is presented in Fig. 2.51. Similarly to peak-to-peak amplitude, this func-

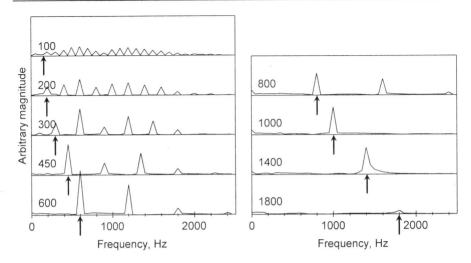

Figure 2.50. Frequency spectra of RFR in a bottlenose dolphin *Tursiops truncatus*. Click rates from 100 to 1800 s^{-1}, as indicated. Linear time and magnitude scales. Stimulus level 110 dB re 1 μPa. Arrows mark fundamentals of the stimulation rates.

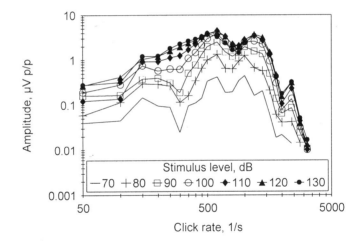

Figure 2.51. RFR fundamental magnitude as a function of rhythmic click rate (fundamental frequency) in a bottlenose dolphin *Tursiops truncatus*. Click levels are indicated in dB re 1 μPa.

tion contained peaks at 500–600, 1000–1200, and around 2400 Hz with valleys between them. However, contrary to the peak-to-peak amplitude, the fundamental magnitude decreased at low rates. It means that at low rates, the response was composed mainly of higher harmonics.

In the functions describing RFR both peak-to-peak and fundamental amplitude dependence on frequency, there were several peaks and valleys, just as they were in the EFR-derived MTF. Similarly, this peak-valley pattern should be explained by the ABR waveform, which consists of a few waves with different delays reflecting successive excitation of several evoked-response brainstem generators. Overlapping of potentials produced by several generators with different delays results in peaks and valleys in both the ABR spectrum and RFR transfer function. This pattern does not reflect the real temporal resolution of the auditory system. Neglecting the peak-valley pattern, the RFR transfer function looks like a low-pass filter. Thus, the temporal resolution of the dolphin's auditory brainstem is expected to be not less than the obtained bandwidth of the RFR transfer function, which has a cut-off frequency of about 1700 Hz at a 0.1 amplitude level.

The RFR transfer function obtained with the use of short pulse trains was very similar to that obtained with SAM sounds: in both cases, the cut-off frequency was around 1700 Hz. These data agree with the integration time of the dolphins hearing of around 300 μs: the low-pass cut-off frequency of 1700 Hz corresponds to the integration time of 300 μs. As to lower rate-following limits obtained in experiments with steady-state pulse trains, it really can be explained by a long-term adaptation that becomes deeper with increasing the click rate. In short-train stimulation conditions, the adaptation was insignificant; it was indicated by the fact that transient on-responses to the burst onset were not dependent on stimulation rate.

The RFR characteristics described above were obtained in the bottlenose dolphin. In other odontocete species, they may differ to some extent. In par-

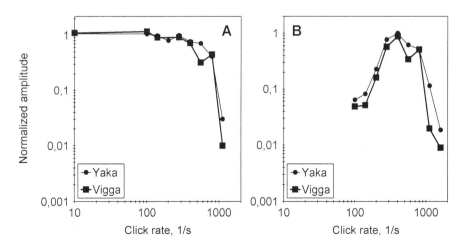

Figure 2.52. **A**. RFR peak-to-peak amplitude dependence on click rate for two killer whales *Orcinus orca*. **B**. RFR fundamental magnitude dependence on click rate for the same animals. Plotted by data in Szymanski et al. (1998).

ticular, a study of RFR evoked by rhythmic click trains was performed in killer whales *Orcinus orca* (Szymanski et al., 1998). This study yielded results similar to those obtained in the bottlenose dolphin, except somewhat lower cut-off frequency as revealed by both peak-to-peak amplitude and fundamental magnitude of RFR (Fig. 2.52); the cut-off frequency was around $1000 \, \mathrm{s}^{-1}$.

2.6. FREQUENCY TUNING

Frequency-selective properties of the inner ear can be simulated by a bank of overlapping bandpass filters. The tuning acuteness (quality) of these equivalent filters determines the ability of the auditory system to perform frequency analysis of sounds. A number of attempts were made to quantify the frequency tuning in dolphins.

2.6.1. Critical Ratios and Critical Bands

Attempts to estimate the frequency tuning in dolphins began long ago. These studies were based on psychophysical measurements and used standard paradigms – first of all, the critical band and critical ratio paradigms. The critical ratio technique suggested by Fletcher (1940) is the simplest way to estimate the frequency tuning. This technique is based on the assumption that when a pure tone is masked by noise, the masking is caused by only a narrow band of the noise surrounding the tone frequency – that part of the noise spectrum that is transferred by the same bandpass filter as transfers the tone. It was supposed also that at the masked threshold, the noise power is equal to the tone power. From these assumptions, it follows

$$I_{th} = I_n = N_0 \Delta f \,, \tag{2.20}$$

where I_{th} is intensity of the tone at threshold (μPa^2), I_n is intensity of the masking noise (μPa^2), N_0 is noise spectrum density ($\mu Pa^2/Hz$), and Δf is the passband of the filter centered at the tone frequency. Using this equation, the passband of the auditory filter can be estimated basing of the data of masked threshold measurements. Being expressed in dB-measure, this value is termed as the critical ratio:

$$CR = 10 \log(\Delta f) \,. \tag{2.21}$$

The first attempt to estimate the critical ratio in the bottlenose dolphin *Tursiops truncatus* was made by Johnson (1968a). He measured masked threshold for a variety of tone frequencies, from 5 to 100 kHz. According to his data, the critical ratio increased almost proportionally with frequency suggesting a constant quality of the equivalent auditory filters (Fig. 2.53). The filter quality Q is defined as

$$Q = f_0 / \Delta f ,$$

(29)

where f_0 is the center frequency and Δf is the bandwidth of the filter. The critical-ratio data could be approximated by a constant-Q line of $Q = 14.4$.

Later similar measurements were performed in a beluga whale *Delphinapterus leucas* (Johnson et al., 1989) and in a false killer whale *Pseudorca crassidens* (Thomas et al., 1990). In the beluga whale, within a frequency region above 10 kHz, Q estimates of the same order as in the bottlenose dolphin (higher than 10) were obtained (Fig. 2.53); at lower frequencies, critical ratio was 16 to 20 dB (40 to 100 Hz). In the false killer whale, critical ratios were from 17 dB at 8 kHz to 42 dB at 115 kHz (50 Hz to 16 kHz, respectively). The latter critical ratio values correspond to Q from 160 (at 8 kHz) to 7.2 (at 115 kHz). Thus, there was a large scatter of tuning estimates obtained by the critical-ratio method.

The problem is, however, that the critical ratio in principle is a poor estimate of the frequency tuning. It depends much more on efficiency of signal

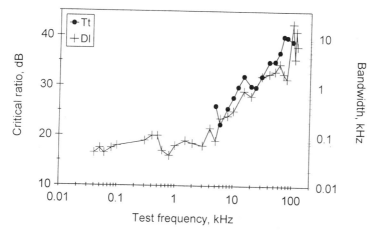

Figure 2.53. Psychophysically measured critical ratios as a function of frequency. Tt – *Tursiops truncatus* (Johnson, 1968a); Dl – *Delphinapterus leucas* (Johnson et al., 1989). The left ordinate scale – critical ratio, the right scale – corresponding filter passband. Plotted by table data in the papers referred to.

detection in noise rather than on filter quality. Indeed, the assumption that I_{th} = I_n is voluntary and has no evidence. The exact value of the ratio of I_{th} to I_n should be known with high precision taking into account that change of this ratio by a few dB results in dramatic change of the filter passband estimate. For example, a change of the signal detection efficiency by 3 dB results in a twofold change of the estimate of the filter passband; change of the efficiency by 5 dB corresponds to 3.16 times change of the passband estimate, and so on. Actually, however, this ratio is not known, and it is quite probable that its difference from unit is not negligible. Moreover, the same order of uncertainty arises because of inevitable data scatter. An error of 3 to 5 dB that is tolerable in routine threshold measurements is unacceptable in critical-ratio measurements as resulting in two to three times error in the filter passband estimate.

Another more direct technique to measure the auditory filter quality is the critical-band method (Fletcher, 1940). In this technique, the masker thresholds are determined at various masking noise bandwidths. It is assumed that if the noise bandwidth is within the filter passband, the masked threshold should increase proportionally to the noise bandwidth, keeping noise spectrum density constant. When the noise width becomes equal or greater than the filter width, the masked threshold should no longer increase as the noise width increases since the filter does not transfer more noise. Thus, if the masked threshold is plotted as a function of the noise bandwidth, the inflection point of the plot indicates the filter width. The filter bandwidth determined in such a way is referred to as critical bandwidth.

Contrary to the critical-ratio technique, the critical-band estimates are directly based on a frequency parameter (noise width), not on the threshold value. It makes this technique less sensitive to the unknown efficiency of signal detection in noise, which is obviously an advantage of this method. On the other hand, the critical band is a multipoint measure – that is, to find one filter width value, not one but a number of threshold values should be determined. This limits the available data since psychophysical measurements are time-consuming, and identification of the inflection point on a threshold-versus-noise width plot sometimes is problematic.

An attempt to measure critical bands (as well as critical ratios) in a bottlenose dolphin *Tursiops truncatus* was made by Au and Moore (1990). Their critical-ratio estimates differed little from those of Johnson (1968a). As to the critical bands, measurements at three frequencies (30, 60, and 120 kHz) yielded an incredibly low filter quality, as low as 2.2. Taking into account a possible uncertainty of finding the inflection point among scattered threshold values, this estimate should be taken with caution.

2.6.2. Tuning Curves

Among a variety of masking techniques used for frequency-tuning measurements, the tone-tone masking is the most demonstrative. In this paradigm,
one pure tone (the probe) is masked by another pure tone (the masker). The
method is based on assumptions as follows. Suppose a probe tone of frequency f_0 is transferred by a filter centered on this frequency. If a masker
frequency f_m differs from f_0, it is transferred through this filter with some
attenuation $a(f)$. Assumption that at the masked threshold, there is a constant
ratio c between the transferred probe intensity I_p and masker intensity $I_m/a(f)$,
results in the relation:

$$I_m(f) = cI_p a(f),\qquad\qquad\qquad (2.23)$$

– that is, keeping the probe intensity I_p constant, the masker intensity at the
masked threshold is proportional to the filter attenuation. Thus, the masked
threshold dependence on frequency directly reproduces the filter form. The
masker level as a function of frequency (keeping probe frequency and level
constant) is referred to as the tuning curve. In some studies, masking curves
were obtained in an opposite manner, keeping masker parameters constant
and varying probe frequency and level. Of course, this mode results in an
inverted tuning curve. In all other respects, both techniques should yield
identical results for a linear filter. Actually auditory filters are not precisely
linear, so different modes of measurement yield somewhat different tuning
curves. This difference, however, is not very large and can be left out of
scope.

2.6.2.1. Psychophysical Tuning Curves

In psychophysical studies of cetaceans, the tuning-curve paradigm was used
in the only study by Johnson (1971). He obtained inverted tuning curves in a
bottlenose dolphin *Tursiops truncatus* around a masker frequency of 70 kHz
at two masker levels, in conditions of simultaneous masking. As predicted by
the tuning-curve paradigm, most of masking (the highest probe thresholds)
occurred when the probe and masker frequencies were little different. With
increasing the difference, masked thresholds fell steeply, about 30 dB at 10-
kHz difference (Fig. 2.54).

However, when the probe and masker frequencies were almost equal,
there was a dip in the tuning curve; it is clearly visible in Fig. 2.54. It is a
commonly known effect in psychoacoustics, and its nature is also well

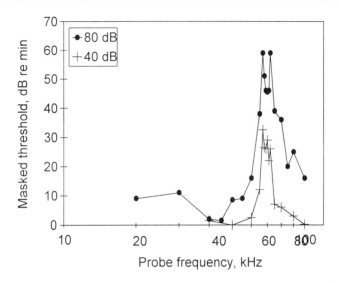

Figure 2.54. Psychophysical tuning curves in a bottlenose dolphin, *Tursiops truncatus.* The curves are obtained as probe level dependence on frequency, keeping masker frequency and level constant (40 and 80 dB above threshold, as indicated). Plotted by table data in Johnson (1971).

known. This effect is assigned to beats arising when the probe and masker tones are superimposed. The rate of these beats is equal to the frequency difference between the probe and masker. When this difference is large enough, the rate of beats is too high to be detected; therefore, the beats do not influence the probe detection. When the probe frequency becomes close to the masker frequency, beats become detectable and serve as an additional cue to detect the probe. This additional cue decreases the masked threshold. The result is a distortion of the tuning curve which looks like a dip just in the region of the tuning-curve tip, the region of the most interest.

To avoid or diminish the beat cue, narrow-band noise is used instead of tone as a probe in human psychoacoustics; envelope fluctuations intrinsic in narrow-band noise mask the beats. Note that the human upper limit of fluctuation detection is not higher than 50–70 Hz; therefore, beats arising at difference between the probe and masker frequencies more than 50–70 Hz are not heard. Dolphins are able to detect sound modulations with a rate near 2000 Hz (see Sections 2.5.6.–2.5.7). Indeed, the dips at Johnson's tuning curves are bounded at frequencies of 67 and 72 kHz – that is, ±2.5 kHz from the probe. So psychophysical tuning-curve measurements did not succeed in evaluating frequency tuning in dolphins.

2.6.2.2. ABR Tuning Curves

Supin et al. (1993) studied frequency tuning in bottlenose dolphins *Tursiops truncatus* using the tuning-curve paradigm in conjunction with ABR recording. This approach has already demonstrated its productivity in many studies in other animals and humans. Tuning curves were obtained in humans and some experimental animals using evoked potentials such as cochlear action potentials (AP) (Dallos and Cheatham, 1976; Eggermont, 1977; Harris, 1978; Abbas and Gorga, 1981; Gorga and Abbas, 1981; Harrison et al. 1981) and ABR (Mitchell and Fowler, 1980; Gorga et al., 1983; Salvi et al., 1982; Brown and Abbas, 1987). So data on frequency tuning in dolphins obtained using this method can be compared directly with data obtained by the same method in other subjects.

In evoked-potential studies, the tone-tone masking paradigm was used in various versions that differed in masked threshold criteria and succession of the masker-probe presentation. A difference in masked threshold criteria means that tuning curves can be defined as the masker levels required to render a probe response either just-detectable (near-complete masking criterion) or partially suppressed (partial masking criteria). Difference in succession of the masker-probe presentation means that the masker and probe either coincide in time (simultaneous masking) or the probe is presented immediately after the masker (forward masking). In making decisions about a masking paradigm for studies in dolphins, the following reasoning was taken into account:

(i) There is some problem in judging the masker level required for near-complete masking because it is difficult to measure a just-detectable response that becomes comparable with recording noise. Therefore, partial masking criteria were used in many studies (response amplitude reduction of 25 to 75%, in many cases 50%). Moreover, in some studies, a just-detectable masking criterion was used – that is, the masker level that resulted in just-detectable reduction of the probe response (Pantev and Pantev, 1982; Pantev et al., 1985; Salt and Garcia, 1990). However, using the near-complete masking criterion results in sharper tuning curves than partial masking (Abbas and Gorga, 1981; Gorga and Abbas, 1981) and just-detectable masking (Salt and Garcia, 1990). This indicates a better measure of hearing tuning abilities with the use of the near-complete masking criterion. This criterion is also preferable for comparison with psychophysical data. All these reasons inclined us to choose the near-complete masking criterion in frequency-tuning studies. As to the problem of detection and measurement of low-amplitude responses, it may be overcome by using more sophisticated methods of response-amplitude measurements, such as match-filtering of ABR or Fourier analysis of EFR (see Sections 2.3.1, 2.3.2).

(ii) It is generally accepted that the forward masking paradigm has some advantages as compared to simultaneous masking. Indeed, simultaneous presentation of both probe and masker signals may result in some undesirable effects, such as beats between two tones of close frequencies. However, it was difficult to use forward masking in studies of the dolphin's hearing because of the very short recovery time of the dolphin's ABR; as shown above, a probe stimulus releases from the influence of a preceding masking stimulus after a time as short as a few milliseconds. Presenting a probe so shortly after the masker requires an abrupt fall of the masker and rise of the probe. Such quick changes of the masker and probe levels result in a significant spectrum splatter, which contradicts the task of the study: in order to make the measurements precise enough, the spectra of both masker and probe must be as narrow as possible. Therefore, in studies of frequency tuning in dolphins, the simultaneous masking paradigm was used.

In the study performed using the principles described above (Supin et al., 1993), tuning curves were obtained in bottlenose dolphins *Tursiops truncatus* at probe frequencies from 16 to 128 kHz. Probe stimuli were tone bursts with a cosine envelope – that is, the envelope was one cycle of a function 1– cos(*t*). Masking signals were continuous tones of various frequency and level. Probe and masker were not coherent.

Figure 2.55 shows ABR evoked by a probe without a masker and in the

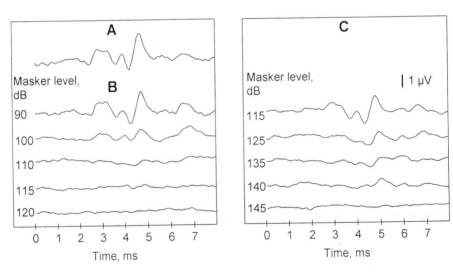

Figure 2.55. Frequency-selective ABR suppression by tone maskers in a bottlenose dolphin *Tursiops truncatus.* **A.** Unmasked ABR evoked by a probe stimulus of 90 kHz frequency, 0.5 ms duration, 110 dB re 1 μPa level. **B.** Responses to the same probe in the background of the 90-kHz masker. **C.** The same with the 80-kHz masker. Masker levels (dB re 1 μPa) are shown near the records.

presence of various maskers. ABR waveform looks just as described above, and its amplitude is not very high at the used probe parameters. A probe of a moderate intensity (100 to 110 dB re 1 μPa) evoked a response of about 3 μV amplitude when measured between the largest positive and negative peaks (A). The same stimulus in a masker background evoked a smaller response. As the masker level increased, ABR diminished until it disappeared.

The figure shows manifestation of frequency specificity of the masking effect. The masking was the most effective when the probe and masker frequencies were equal (90 kHz in Fig. 2.55B). In this case, the complete masking required the masker level of about 10 dB above the probe level (120 dB re 1 μPa). When the masker and probe frequencies were different, the masking was less effective: at a 10-kHz difference, the masking required masker levels 25 to 30 dB higher than in the preceding case; the near-complete masking required the masker level of 145 dB (Fig. 2.55C).

For obtaining a tuning curve, the probe frequency and intensity were fixed, and masker frequency was varied. At each masker frequency, its level was varied, and the level required for near-complete masking of the probe response was determined. This procedure was repeated at each frequency of the masker to yield a complete tuning curve.

A family of tuning curves obtained in the way is shown in Fig. 2.56A. These tuning curves demonstrate many of features known in other mammals. In particular, there is a sharp tip segment, an elongated low-frequency tail, and steep rise at high frequencies. All the curves peak at or near the probe frequency. At this frequency, the near-complete masking required a masker level of 2.5 to 7.5 dB above the probe level.

The tuning curves obtained by the evoked-potential method never displayed a dip at their center, which is characteristic of psychophysical tuning curves because of beats between the probe and masker. Of course, overlapping of masking and probe tones produced beats in evoked-potential experiments also. However, ABR were collected by averaging of a few hundreds samples, and it is noteworthy that the masker tone was not coherent with the averaged sweeps. Therefore, beats were not coherent with the sweeps either, so their effects on averaged evoked potentials were negligible. The result is that tuning curves obtained by this method were not distorted by the beats effect, and the curves could be used to estimate the frequency tuning.

The curves demonstrated graduation of tuning depending on frequency. Tuning was sharper (the curves were narrower) at higher frequencies and less sharp at lower frequencies. The Q_{10} values (the center frequency divided by the bandwidth at the level 10 dB above the tip of the curve) are a convenient measure of the acuteness of tuning. Within the tested frequency range, Q_{10} values varied with frequency with a certain trend: from 4.7 at 16 kHz to

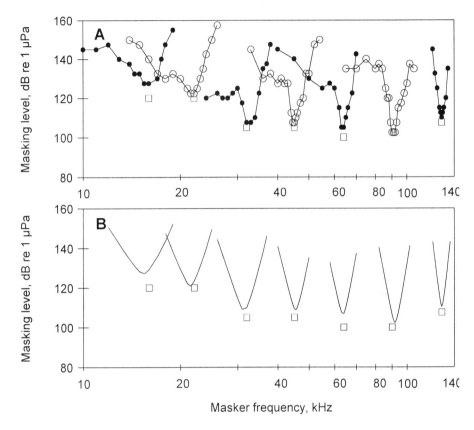

Figure 2.56. Tuning curves obtained at various probe frequencies in a bottlenose dolphin *Tursiops truncatus.*. Probe frequencies (for curves from left to right) are 16, 22, 32, 45, 64, 90, and 128 kHz. Probe duration is 0.5 ms. The square symbols represent probe frequency and level (40 dB above the ABR thresholds). **A**. Original experimental data. **B**. Approximation of the data by *roex* functions.

18.2 at 128 kHz. This trend is significant but less steep than frequency-proportional (Fig. 2.57).

Another generally adopted metric of the filter quality is its equivalent rectangular bandwidth (ERB) and corresponding equivalent rectangular quality $Q_{ER} = f_0/ERB$. It should be noted that most of psychophysical studies of frequency tuning in humans and animals present the results not in the Q_{10} but in ERB and Q_{ER} measure; so the latter metric is more universal for comparison. ERB and Q_{ER} are defined as the bandwidth and quality of an idealized rectangular filter centered at the same frequency and transferring the same power of a wide-band signal as the real filter.

ERB can be computed as

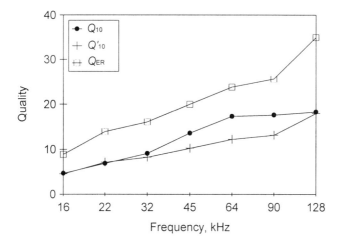

Figure 2.57. Q_{10} and Q_{ER} dependence on frequency in the bottlenose dolphin *Tursiops trunca-tus* (averaged data from two animals). Q_{10} values were obtained directly from experimental tuning curves, Q'_{10} and Q_{ER} values were obtained from approximating *roex* functions.

$$B_{ER} = \frac{1}{W_0} \int_{-\infty}^{x} W(f)df ,$$

(2.24)

where B_{ER} is ERB, $W(f)$ is the power transferred by filter as a function of frequency, f is frequency, and W_0 is the value of $W(f)$ at the filter center frequency. Basing on ERB, the filter tuning can be specified as the equivalent rectangular quality Q_{ER}:

$$Q_{ER} = f_0 / B_{ER} ,$$

(2.25)

where f_0 is the filter center frequency. If the filter form $W(f)$ is known, the filter bandwidth B_{10} at the -10 dB level can also be found by finding values of frequency f, which give

$$W(f) = 0.1W_0 .$$

(2.26)

Thus, the ratio of Q_{10} to Q_{ER} depends on the filter form $W(f)$, which should be known. There were many attempts to find a simple analytical expression that satisfactorily describes the form of auditory frequency-tuned filters. The use of the resonance equation

$$W(f) = \cfrac{W_0}{1 + \left[Q\left(\cfrac{f}{f_0} - \cfrac{f_0}{f}\right)\right]^2} , \qquad (2.27)$$

where the only parameter Q (quality) determines the filter quality, seemed to be reasonable since it describes behavior of a wide variety of frequency-tuned devices. However, numerous psychophysical data have shown that this equation is not the best approximation of experimental data. As a rule, at a certain slope of side branches of a real tuning curve, its tip is more rounded than predicted by the resonance equation. Such shape of a tuning curve may be described by superposition of a few resonance curves slightly shifted relative to one another. Very probably, such description simulates real events in the auditory system. However, it is not favorable for practical use.

Therefore, a few empirical formulae were suggested to describe the form of auditory filters, such as cubic rounded exponential function (Patterson and Nimmo-Smith, 1980) and rounded exponential function (*roex*, Patterson et al., 1982). The latter gives rather simple expression presenting the filter form $W(g)$ as

$$W(g) = (1 - r)(1 + pg)e^{-pg} + r , \qquad (2.28)$$

where g is the normalized frequency deviation from the center of the filter. The term p determines the filter tuning in the passband, and the term r approximates the shallow tail section of the filter shape. Neglecting the tail section, the expression can be reduced to a simple form describing the main central section of the filter form:

$$W(g) = (1 + pg)e^{-pg} . \qquad (2.29)$$

This simple function has a rounded tip at small g ($pg \ll 1$) and exponential side branches at large g ($pg \gg 1$). It approximates satisfactorily the auditory filter form in humans. It should be noted, however, that the *roex* function does not describe any real events in the auditory system that provide the frequency tuning; it is a truly empirical expression designed for easy approximation and processing of frequency-tuning data.

It is easy to compute that for the *roex* function, $B_{ER} = 4f_0/p$ (where f_0 is the center frequency), and $B_{10} \approx 1.96 B_{ER}$. Thus, $Q_{ER} = p/4$ and $Q_{10} \approx Q_{ER}/1.96$.

Since the *roex* function is widely adopted in psychoacoustics for approximation of the auditory filter forms, we used it also to approximate the tip segment of tuning curves in dolphins. For approximation, the parameter p of each *roex* curve and its shift along the ordinate axis were adjusted in such a way as to fit experimental points according to the least-square criterion.

Figure 56B shows *roex* approximations of the tuning curves presented in Fig. 56A. Q_{ER} of these curves are presented in Fig. 2.57 along with Q_{10} as a function of frequency. Q_{ER} features the same trend as Q_{10}: from 8.9 at 16 kHz to 35 at 128 kHz (about 4 times). Approximation of tuning curves by analytical functions was used also to recalculate Q_{10} values basing on a number of experimental points of each curve, not only those around the 10-dB level. These estimates of Q_{10} are also presented in Fig. 2.57; they differ very little from those obtained directly from experimental tuning curves.

The degree of frequency tuning at high frequencies (Q_{10} more than 18 and Q_{ER} as high as 35) is a remarkable feature of the dolphin's hearing. These values are not characteristic for most other mammals including humans. In the majority of studies in terrestrial mammals and humans, the maximal obtained Q_{10} values were within a range from 4 to 6 for both cochlear AP and ABR tuning curves (Dallos and Cheatham, 1976; Mitchell and Flower, 1980; Harrison et al., 1981), and in rare cases the values more than 10 were reported (Brown and Abbas, 1987). Respectively, Q_{ER} values in humans in the upper part of the frequency region of hearing are about 9 (Glasberg and Moore, 1990).

Since the difference between frequency tuning in dolphins and terrestrial mammals is very significant, a question arises whether this difference reflects peculiarities of the dolphin's auditory system or a difference in methods. In particular, the choice of the masking criterion is of importance. In the study described above, the near-complete masking criterion was used, whereas in many other studies, the 50% masking criterion was applied. Masking criterion can influence the tuning curve sharpness to some extent (Abbas and Gorga, 1981; Gorga and Abbas, 1981). Is the near-complete masking really the best one to assess the true frequency tuning?

To answer this question, the dependence of tuning curves on masking criterion was tested (Supin et al., 1993). Tuning curves were obtained for various response amplitudes: from 95% response decrement (that is, criterion response amplitude was 5% of the control value) to 50% response decrement. Figure 2.58 shows that the tuning curves were dependent on the masking criterion. Apart from the self-evident shift of the curves along the ordinate axis (the deeper masking, the higher masker level was required), their sharpness depended on the criterion as well. This dependence was significant when a high-level probe was used, which evoked a large response (Fig. 2.58A). The tuning was the sharpest for the near-complete masking (Q_{10} = 14.3 for 95% response decrement) and became broader with the masking criterion decrease (Q_{10} = 5.5 for 50% response decrement). However, quite

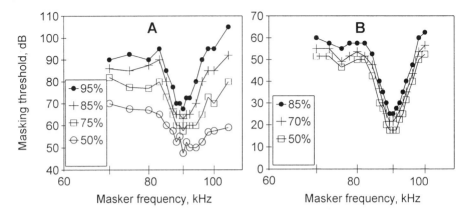

Figure 2.58. Tuning curves obtained using various response-decrement criteria in a bottlenose dolphin *Tursiops truncatus*. The masking criterion is indicated in the legend as the response decrement. **A**. Probe level 120 dB re 1 μPa (60 dB above response threshold), the unmasked response amplitude 5 μV. **B**. Probe level 80 dB re 1 μPa (20 dB above response threshold), the unmasked response amplitude 1.6 μV.

another picture was observed when a low-level probe was used, which evoked a low-amplitude response. In this case, the curve characteristics were little dependent on the masking criterion (Fig. 2.58B). All the curves demonstrated sharp tuning without systematic dependence of Q_{10} on the response decrement criterion.

The difference between results obtained with high- and low-level probes disappears if the masking criterion is expressed as a remainder absolute response amplitude (μV) instead of an amplitude decrement percentage. Independently of the decrement percentage, the curves demonstrated sharp tuning when the criterion response amplitude was low (less than 1 μV). It was observed both in conditions of deep masking of a high-level probe and in conditions of deep or light masking of a low-level probe. The tuning became broader when the criterion response amplitude became higher – that is, in conditions of light masking of a high-level probe.

It is noteworthy that the influence of masking criterion is insignificant when a low-level probe is used. It is this range of the probe levels (10 to 20 dB above the response threshold) that was used in the majority of studies in terrestrial mammals and humans (Dallos and Cheatham, 1976; Eggermont, 1977; Abbas and Gorga, 1981; Gorga and Abbas, 1981; Harrison et al., 1981). Using low-level probes, Dallos and Cheatham (1976) observed little influence of masking criterion on the curve sharpness. Thus, it seems unlikely that tuning sharpness in terrestrial mammals was underestimated due to inadequate masking criteria. It suggests that much more sharp frequency

quency tuning in dolphins, as compared with terrestrial mammals, reflects real peculiarities of their auditory system.

Moreover, we cannot exclude some underestimation of frequency tuning in dolphins in the data presented above. In these experiments, the simultaneous masking paradigm was used. Simultaneous masking may result in less acute frequency tuning curves than forward masking because of complex interactions between simultaneously presented probe and masker, particularly, two-tone inhibition. Thus, frequency tuning in dolphins may be even sharper than the data presented above show.

A similar study was carried out in a common dolphin *Delphinus delphis* (Popov and Klishin, 1998); however, only limited data were obtained in one animal. Measurements with probes of 64 and 90 kHz frequency resulted in tuning curves of Q_{10} of 10.7 and 9.5 – that is, almost twice less acute than in the bottlenose dolphin at the same frequencies. Taking into account that the subject was not normal (it was a sick animal that was found ashore and died a few days later in spite of intensive treatment), these data can hardly be adopted as representative for the species.

Of course, it is of great interest to compare the evoked-potential data presented above with similar data obtained by psychophysical method (Johnson, 1971). However, it is very difficult to do because of distortion of psychophysical tuning curves by a beat-caused dip on their tip, the effect described and discussed above.

An important parameter of a probe used in the tuning-curve paradigm is the probe duration. The main problem associated with the use of ABR in frequency-tuning measurements arises from the fact that these responses are evoked by transient acoustic stimuli, such as a short sound burst, a sound onset or offset, or a quick change of level. Transient stimuli feature broader frequency spectra than long-duration, slowly rising and falling stimuli. This limits their use as probes for measuring the frequency selectivity. The problem is that we do not know exactly which part of a stimulus is actually effective to elicit the evoked response. Hence we do not know how broad the effective spectrum of the stimulus actually is. Even with the use of a slowly rising and falling stimulus, only its initial part may be effective enough to evoke the response, and the spectrum of this initial part may be wider than the spectrum of the whole burst, as Fig. 2.59 illustrates.

This effect may be the most affecting results at low probe frequencies. If the auditory filter bandwidth is proportional to the frequency, it becomes narrower with frequency decrease. Thus at low frequencies, the filter bandwidth may be narrower than the probe spectrum. Under these conditions, tuning curves may reflect probe spectrum rather than real filter form, and the filter quality may be markedly underestimated. One can suppose, in particular, that the filter quality decrease at low frequencies (see Fig. 2.57) is a re-

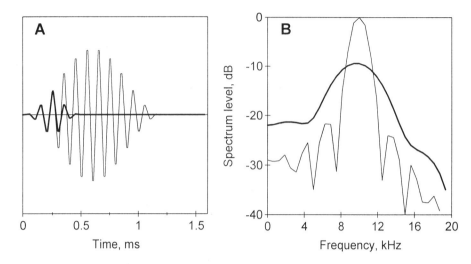

Figure 2.59. Schematic presentation of difference between spectra of a total sound burst and its effective part. **A**. Waveforms of a cosine-enveloped burst (thin line) and its arbitrarily isolated initial part (solid line). **B**. Frequency spectra of the burst and its initial part (thin and solid lines, respectively).

sult of this effect rather than a real filter-quality trend. Therefore, other methods were searched to obtain tuning curves in dolphins. The searched-for method should be as effective as ABR recording but make it possible to use narrow-band probes. There appeared an idea to use the envelope-following response (EFR) evoked by sinusoidally amplitude-modulated (SAM) tone as a probe response for tuning-curve measurements.

2.6.2.3. EFR Tuning Curves

The expected advantage of the use of SAM stimuli is that, contrary to the short tone pips or clicks that are used to evoke ABR, SAM tones have a strictly limited spectrum bandwidth that is either $2f$ or f depending on the modulation type, where f is the modulation rate.

Of course, the transfer of amplitude-modulated signal to rhythmic evoked response is not a precisely linear process; therefore, this transfer may result in spectrum splatter and appearance of higher harmonics. It may be the same process that makes the spectrum of an effective part of a tone pip wider than the spectrum of the whole pip. However, an advantage of a rhythmic response like EFR is that a simple Fourier analysis allows us to distinguish the response fundamental from higher harmonics. One can be sure that the response fundamental is evoked by the fundamental of the stimulus envelope – that is, by a signal that has a spectrum band not wider than f or $2f$.

This spectrum width can be narrow enough allowing to use rhythmic stimuli for frequency-tuning measurements in dolphins. It was shown above that EFR in dolphins can be evoked by modulation rates of an order of hundreds of Hz. Thus, EFR is evoked by stimuli of much narrower bandwidth (hundreds of Hz) than those evoking ABR. Such narrow-band stimuli are convenient probes to obtain tuning curves.

In a tuning-curve study based on the use of SAM probes and EFR recording (Popov et al., 1995), just as in the ABR-based studies described above, the simultaneous masking paradigm was used to obtain tuning curves. Forward masking could not be used at all in conjunction with rhythmic amplitude-modulated stimuli since amplitude-modulated sound bursts inevitably must be long enough, at least around 10 ms or longer, whereas recovery time of ABR in the dolphins is as short as a few milliseconds or less. So no significant forward masking of such long probes might be expected. As to the masking criterion, the near-complete masking was used in this study as well as in the preceding one.

In that study, probes were SAM tone bursts lasting more than 10 ms. Tone carrier was not coherent with sweeps of ABR collection. Modulation signals always started from a zero phase in coherence with evoked-response collection. To avoid adaptation to probes, they were presented at a rate of 10 s^{-1} – that is, interburst intervals were of an order of magnitude longer than probes. A modulation rate of 600 Hz was used since this rate was among the most effective ones and provided a narrower probe bandwidth then other equally effective rates (1000–1200 Hz). Modulation depth was always maximal – 100%. As to masking signals, they were continuous tones of various frequency and level. Probe and masker were produced by different generators and were not coherent. Probe and masker were mixed and emitted through one and the same transducer.

It should be noted that two types of SAM stimuli are widely used: signals containing three harmonic components in the spectrum (the central carrier frequency F and two side component $F–f$ and $F+f$) (Fig. 2.60A and B) and signals containing two components ($F–f/2$ and $F+f/2$) (Fig. 2.60C and D), where F and f are the carrier frequency and the modulation rate, respectively. Both stimulus types were found to be equally effective to evoke EFR (Fig. 2.61). Amplitude-modulated tone bursts of the both modulation types evoked robust EFR, and responses were of almost the same amplitude at both stimulus types when the modulation depth was equal (100%). It is noteworthy that the two-component stimulus with separation f between the components evoked EFR with a wave rate equal to f (that is, with the rate of the stimulus

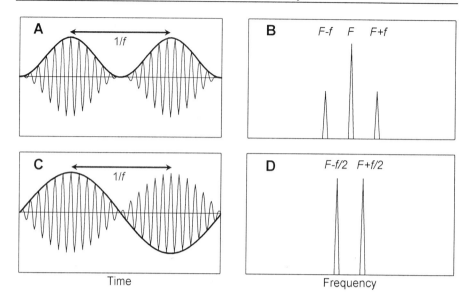

Figure 2.60. Probe stimuli used to evoke EFR in tuning-curve experiments. **A** and **B**. A three-component signal. **C** and **D**. A two-component one. **A** and **C**. Stimulus waveforms (solid lines show the modulation waveform). **B** and **D**. Stimulus spectra (F - carrier frequency, f - modulation rate).

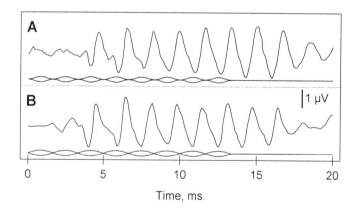

Figure 2.61. EFR evoked by different types of SAM stimulus in a bottlenose dolphin *Tursiops truncatus.* **A**. Three-component stimulus. **B**. Two-component stimulus. Carrier tone 90 kHz, 120 dB re 1 μPa. Stimulus envelopes are shown under the records.

envelope), although the actual modulation frequency of this signal is half as much (that is, $f/2$, as Fig. 2.60C shows).

Since both the two- and three-component SAM signals were equally effective in provoking EFR, the two-component one had some advantage as a probe for tuning measurements since its bandwidth is twice as narrow as the three-component bandwidth. The narrower probe was preferable for frequency-tuning measurements. Therefore, two-component SAM probes were used in these measurements.

Using these stimulus parameters, masking experiments were performed as shown in Fig. 2.62. The figure exemplifies EFR evoked by a SAM probe without a masker and in the presence of various maskers. The records were made with a probe stimulus of a moderate intensity of 100 dB re 1 μPa RMS; it was about 40 dB above the response threshold. Such a stimulus evoked a response of less than 1 μV peak-to-peak amplitude (Fig. 2.62A). The same stimulus in a masker background evoked a smaller response. As the masker level increased, EFR diminished until it disappeared in noise. This masking was the most effective when the probe carrier frequency and masker frequency were equal (90 kHz in Fig. 2.62B). In this case, the complete masking required the masker level of about 15 dB above the probe level (115 dB re 1 μPa). When the masker and probe frequencies were different, the masking was less effective: in Fig. 2.62C, the masking frequency was 85 kHz while the probe carrier frequency was 90 kHz, and the masking

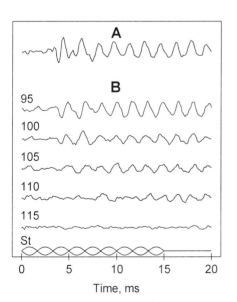
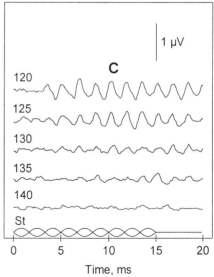

Figure 2.62. EFR suppression by maskers in a bottlenose dolphin *Tursiops truncatus*. **A**. Unmasked EFR evoked by a probe stimulus of 90 kHz carrier frequency, 600 Hz modulation rate, 100 dB re 1 μPa level. **B**. Responses to the same probe in the background of 90-kHz masker. **C**. The same as (**B**) but with 85-kHz masker. Masker levels (dB re 1 μPa) are shown near the curves. The stimulus envelope (St) is shown at the bottom of the panels.

required the masker levels 25 dB higher than in the preceding case. In particular, the near-complete masking occurred at a masker level of 80 dB above the probe (140 dB re 1 μPa).

To evaluate EFR magnitude, a whole number of EFR cycles in the last 12-ms part of the response were Fourier transformed in order to find the weight of the fundamental component at the modulation frequency. The peak-to-peak amplitude of this component was taken as the EFR magnitude measure.

To obtain tuning curves, the probe-carrier frequency and intensity were fixed and masker frequencies and intensities were varied. For each masker frequency, the level required for a near-complete masking of the probe response was determined. A response magnitude of 0.05 to 0.07 μV (about 10% of the unmasked response and obviously above the background noise) was adopted in this study as a criterion of near-complete masking. The masker level required for the near-complete masking could be identified with an accuracy of 2 to 3 dB. This procedure was repeated at each frequency of the masker to yield a complete tuning curve.

A family of tuning curves obtained in such a way is shown in Fig. 2.63A. These tuning curves possessed many typical features common to tuning curves obtained by ABR recording. They had a relatively sharp tip segment, steep rise at the high-frequency branch, and similar or less steep rise at the low-frequency branch. All the curves peaked at or near the probe carrier frequency. At this frequency, the near-complete masking required a masker level of 5 to 15 dB above the probe level.

Similarly to ABR-data, the tuning curves obtained using EFR do not feature a dip when the masker and the probe were of a similar frequencies. Of course, beats between the probe and masker arose in SAM-stimulation experiments as well, these beats and the probe amplitude modulation superimposed. However, the averaging procedure during EFR recording revealed only responses to sound modulations coherent with the evoked-potential collection. Noncoherent modulations of other origin did not manifest themselves in the averaged responses. Therefore, the records never demonstrated any rhythmic response with a rate of beats between the probe and masker.

For the curves presented in Fig. 2.63A, the Q_{10} values (as estimated directly from the experimental curves) varied with a slight trend within the frequency range from 11.2 to 90 kHz, from 12.2 at the 11 kHz probe frequency to 19.6 at the 90 kHz frequency; at 110 kHz, Q_{10} was a little lower again. At probe frequencies of 8 kHz and lower, masking was little dependent on the masker frequency (8 kHz probe in Fig. 2.63A), so it was impossible to obtain real tuning curves and evaluate their Q_{10} at such low probe frequencies. Similar results gave approximation of the tuning curves by *roex* functions (Fig. 2.63B): Q_{ER} values derived in such a way drifted from 19.5 at

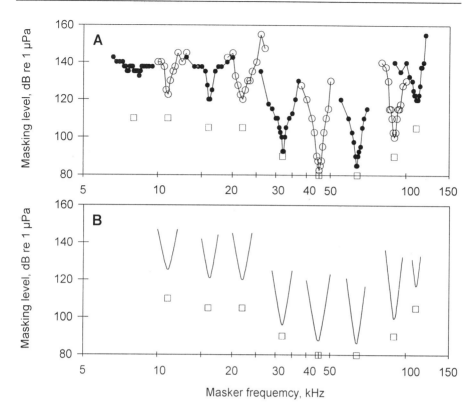

Figure 2.63. Tuning curves obtained in a bottlenose dolphin *Tursiops truncatus* at various probe frequencies using SAM probes. Probe frequencies (for curves from left to right) are 8, 11.2, 16, 22.5, 32, 45, 64, 90, and 110 kHz. The square symbols represent probe frequencies and levels (30 dB above the EFR thresholds). **A**. Original data. **B**. Approximation by *roex* functions.

11 kHz to 34.3 at 90 kHz, and Q_{10} from 10 to 17.5 within the same frequency range. Figure 2.64 presents Q_{10} and Q_{ER} as functions of center frequency.

Thus, within a high-frequency range (64–90 kHz), the tuning curves obtained using SAM probes demonstrate almost the same tuning quality as those obtained with tone-pip probes: Q_{10} of 17 to 19 and Q_{ER} more than 30. At the same time, a distinctive feature of the SAM-probe tuning curves is their sharp tuning within a wide frequency range, including rather low frequencies. Although at lower frequencies the tuning was somewhat less acute than at higher frequencies, it remained still high: Q_{10} of 10 to 12 and Q_{ER} near 20 at the probe frequency as low as 11.2 kHz. This is much higher than those obtained with low-frequency short pip probes (Q_{10} less than 5 and Q_{ER} less than 10 at the 16 kHz probe frequency). These results agree with the suggestion that the rather high bandwidths of the tuning curves obtained us-

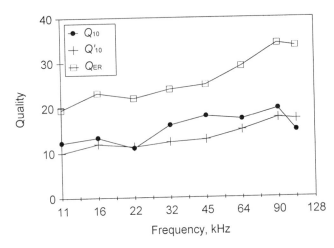

Figure 2.64. Dependence of Q_{10} and Q_{ER} on probe frequency in the bottlenose dolphin *Tursiops truncatus* obtained using SAM probes (average data from two animals). Q_{10} values were obtained directly from experimental tuning curves, Q'_{10} and Q_{ER} values were obtained from approximating *roex* functions.

ing pip probes reflected the bandwidth of a short ABR-eliciting part of the probe. The results confirm that SAM tones can serve as convenient narrowband probes to obtain tuning curves by EFR recording.

Even SAM probes can be used only when the tested filter bandwidth is wider than the probe spectrum. This limitation explains why tuning curves could not be obtained at frequencies of 8 kHz and lower. Indeed, at the probe frequency of 11.2 kHz, the Q_{10} value of 10 to 12 corresponds to the filter bandwidth less than 1 kHz at the 10-dB level; it is still wider than the probe spectrum width (600 Hz). But at the 8-kHz frequency, a filter bandwidth anticipated basing on the Q_{10} trend may be as narrow as 550 to 650 Hz at the 10-dB level. This bandwidth is comparable to the width of the SAM probe spectrum. Filters of narrower bandwidths cannot transfer amplitude modulations as frequent as around 600 Hz, thus the SAM signal cannot serve as a probe for measurements of transfer properties of these filters. However, at a frequency region higher than 8–11 kHz, SAM probes seems to be more appropriate for measurements of frequency tuning than short pip probes.

Using ERB and Q_{ER} values of *roex*-approximated tuning curves, frequency tuning of dolphins can be compared with a huge body of psychophysical data on frequency tuning in humans. Not to mention numerous particular studies of frequency tuning in humans, several analytical expressions can be used that have been proposed to describe ERB of the auditory filters as a function of frequency. For example, a simple equation given by Glasberg and Moore (1990) may be used:

$$B_{ER} = 24.7(4.37f + 1),\qquad\qquad(2.30)$$

where B_{ER} is ERB in Hz, and f is the center frequency in kHz. At high frequencies ($f \gg 24.7$ Hz) this equation may be reduced to $B_{ER} = 0.108f$, thus $Q_{ER} = f/B_{ER} = 9.3$. This is almost four times less than Q_{ER} values in the high-frequency range in dolphins.

Moreover, it should be noted again that both ABR and EFR tuning curves in dolphins might somewhat underestimate the frequency tuning in dolphins due to the use of the simultaneous masking paradigm. This paradigm may result in broader tuning curves than forward masking because of the two-tone inhibition. Thus, frequency tuning in dolphins may be even sharper than the present data show.

The data presented above were obtained in the bottlenose dolphin, which was a subject of the most detailed studies. Similar tuning curves (based on SAM probes and EFR recordings) were obtained in a beluga whale *Delphinapterus leucas* (Klishin et al., 2000). Masked thresholds were estimated in the same manner as described above – that is, EFR records were Fourier transformed, the peak magnitude at the SAM frequency (1000 Hz) was measured, and the masker level was found that suppressed the probe response to a 0.01 μV criterion level. This procedure was repeated at various masker frequencies to yield a complete tuning curve. To draw the curve, the experimental points were approximated by the *roex* function. The *roex* curve was adjusted to experimental data, for which purpose the frequency and level of the curve peak and the parameter p were adjusted to minimize the mean-square deviation from the data.

A family of tuning curves obtained in such a way is shown in Fig. 2.65. The curves were obtained at probe frequencies varied by quarter-octave steps within a range of 32 to 108 kHz. All the curves peaked at or near the probe frequency. The parameter p varied from 88.2 (at 45 kHz probe frequency) to 198.7 (at 108 kHz). The found values of the parameter p correspond to Q_{ER} of 22.1 to 49.7 and Q_{10} of 11.3 to 25.5 (Fig. 2.66). The tuning dependence on probe frequency exhibited an obvious trend (except a small dip at 45 kHz): from $Q_{ER} = 27.2$ ($Q_{10} = 14.0$) at 32 kHz to $Q_{ER} = 49.7$ ($Q_{10} = 25.5$) at 108 kHz.

Thus, according to these data, frequency tuning of the beluga whale was extremely acute: at the highest of the studied frequencies (108 kHz), Q_{ER} almost reached 50, which was markedly higher than in the bottlenose dolphin and about five times higher than in any part of the hearing frequency range in humans. It is possible that this acute frequency tuning explains bet-

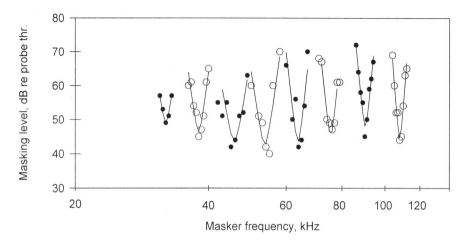

Figure 2.65. Tuning curves obtained in a beluga whale *Delphinapterus leucas* at various probe frequencies using SAM probes. Probe frequencies (for curves from left to right) are 32, 38, 45, 54, 64, 76, 90, and 108 kHz. Experimental points are approximation by *roex* functions.

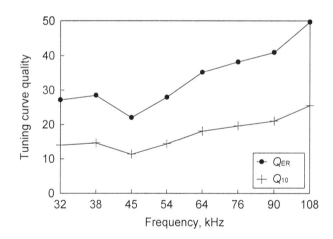

Figure 2.66. Tuning curve quality (Q_{ER} and Q_{10} measures) as a function of probe frequency in the beluga whale *Delphinapterus leucas.*

ter detection of echo-signals in noise (Turl et al., 1987) and reverberation (Turl et al., 1991) by belugas rather than by bottlenose dolphins.

All the tuning-curve measurements described above were performed using ABR or rhythmic ABR sequence, EFR. There was also an attempt to use for this purpose another EP type – namely, the auditory cortical response (ACR). Contrary to ABR, ACR are effectively provoked by stimuli with rise-fall times of an order of a few milliseconds – that is, narrow-band stim-

uli. This peculiarity of ACR is favorable for use as a probe response in fre-
quency-tuning-measurements. In a study of Popov and Supin (1986) and Supin and Popov (1990), ACR was used to obtain inverted tuning curves: the
tonal masker level was kept constant (60 dB re 1 μPa) and the probe (a pip
with 1-ms linear rise, 1-ms plateau, and 1-ms fall) varied in level to find the
masked threshold. A family of tuning curves was obtained with center frequencies of 25, 35, 50, 70, and 100 kHz. They looked like typical tuning
curves, though inverted. Their acuteness, however, was noticeably less than
those described above. The bandwidth at the −3 dB level (which is close to
ERB) was specified as 0.06 to 0.07 of the center frequency – that is, Q_{-3dB}
was 14 to 16. Q_{10} was not specified in that paper, but it can be estimated as 7
to 8. Thus, both the quality estimates of ACR tuning curves were approximately twice less than corresponding estimates of ABR tuning curves.

   A possible explanation of this data divergence is that the auditory cortex
realize more complex functions than simple frequency analysis; therefore,
ACR do not manifest the frequency tuning in full measure.

2.6.3. Notch-Noise Masking

Tone masking is a convenient and demonstrative method for measuring frequency tuning since tuning curves obtained in such a way directly show the
form of the auditory frequency-tuned filters. However, this method has an
important disadvantage: tuning curves obtained with the use of tone-tone
masking may overestimate the frequency tuning because of the so-called off-
frequency listening effect.

   Indeed, the masking paradigm is based on an assumption that the response to a narrow-band probe (tone or narrow-band noise) is evoked by a
signal from the peripheral auditory filter, which is centered just at the probe
frequency. This assumption seems almost self-evident since the probe-
centered filter provides the highest output signal. This is really true in no-
masking conditions; however, this may not be true in the presence of a
masker.

   This is illustrated in Fig. 2.67A. The output signal P_a of a filter a centered
at the probe frequency f_0 is really much higher than the signal P_b of a filter b
centered on another frequency. However, in the presence of a masker, the
filter b has some advantage in transferring the probe signal if the masker frequency is slightly different from the probe frequency f_0. Due to the rounded
tip of the filter transfer function, both the probe and masker are almost
equally transferred through the filter a (values P_a and M_a, respectively), thus
resulting in significant masking of the signal P_a. On the contrary, due to
rather steep slope of the filter transfer function, the probe signal is trans-

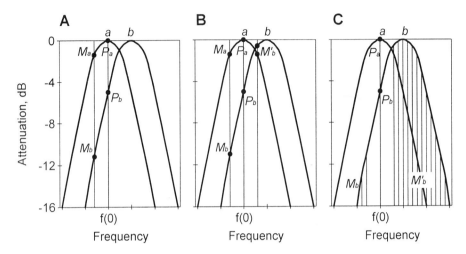

Figure 2.67. **A**. The nature of the off-frequency listening effect with the use of a tone masker. *a* and *b* – filters tuned to adjacent frequencies. P_a and P_b – responses of filters *a* and *b* to a probe of the frequency f_0. M_a and M_b – responses of these to filters to a masker below the probe. **B**. Absence of the off-frequency listening with the use of a two-tone masker; M'_b – response of the filter *b* to the masker above the probe; other designations are the same as in (**A**). **C**. Absence of the off-frequency listening with the use of a notch-noise masker; hatched areas M_b and M'_b – responses of the filter *b* to noise below and above the notch, respectively.

ferred through the filter *b* (value P_b) markedly better then the masker (value M_b), thus resulting in weaker masking of the signal P_b. As a result, the probe signal can be detected due to output signal from the filter *b*, though it is not detectable at the filter *a*. This is the off-frequency listening effect.

Due to rather steep side branches of the filter transfer function, the ratio of the probe to masker output signals of side filters steeply increases with increasing the difference between the probe and masker frequencies. Because of this effect, the tuning curve becomes sharpened around its tip if the detection of the probe is determined by a side filter; the resulting tuning curve does not reproduce precisely the shape of the filter centered at the probe frequency.

To avoid the off-frequency listening effect and make the masking curve closer to the true filter form, the masker spectrum must be symmetrical relative to the probe frequency. The simplest case is a masker consisting of two tones with frequencies equally spaced on either side of the probe frequency (Fig. 2.67, B). When a masker of such type is used, the off-frequency listening effect disappears because the response of the side filter is effectively masked by one of the two masking tones. For example, in Fig. 2.67B, a weak masking of the filter *b* by the tone below the probe (value M_b) is compensated by strong masking by the tone above the probe (value M'_b). Therefore, the tuning curve reflects the response of the probe-centered filter and does not overestimate its tuning.

The two-tone masker was used for frequency-tuning measurements in a number of psychophysical studies (Green, 1965; Patterson and Henning, 1977; Nelson, 1979; Johnson-Davis and Patterson, 1979; O'Loughlin and Moore, 1981; Glasberg et al., 1984). This method was also used to study frequency tuning in dolphins (Klishin and Popov, 1996). However, it was found that masking efficiency was influenced markedly by masker envelope fluctuation arising because of beats between two tones. This effect caused no serious complications for studies in humans since the beats were not perceived at two-tone separations of more than 50–100 Hz. However, dolphins are able to resolve sound fluctuation rates up to 2 kHz. These responses to two-tone masker beats interfered with the probe response and influenced the probe-detection efficiency.

A much more popular masker type for studying frequency tuning is the notch-noise (Patterson, 1976; Patterson and Nimmo-Smith, 1980; Patterson and Moore, 1986; Glasberg and Moore, 1990). This method is based on the use of a masker, which is a noise with a spectral notch (narrow stop band) symmetrical around the probe frequency. The variation of masking threshold as a function of notch width can be used to estimate the frequency tuning. Since the frequency spectrum of this masker is symmetrical relative to the probe frequency, it effectively prevents the off-frequency listening. For example, in Fig. 2.67C, weak masking of the filter *b* by noise below the probe (hatched area M_b) is compensated by effective masking by the noise above the probe (hatched area M'_b). An additional advantage of this masker type is that a broad-band masker have less prominent envelope fluctuations than a two-tone masker.

In humans, the notch-noise-masking was used not only in psychophysic studies but also in conjunction with ABR recording (Abdala and Folsom, 1995), although very low amplitude of masked ABR in humans was not favorable for precise measurements. The use of the notch-masking technique in conjunction with ABR recording seemed reasonable for measurements of frequency tuning in dolphins.

It was made in a study (Popov et al., 1997a) that provided estimates of the frequency tuning in dolphins using the evoked-potentials method and notch-noise-masking paradigm. Similarly to preceding studies, only simultaneous masking was used since very short suppression of probe ABR after the preceding stimulus made it difficult to obtain deep forward masking. As to the probe stimuli, rhythmic amplitude-modulated sounds could not be used in that study because of their complicated interactions with wide-band maskers, which are described below (see Section 2.9.2). Therefore, probe stimuli were tone pips with a cosine-wave envelope – that is, the envelope was one cycle of a function $1-\cos(2\pi t/\tau)$. Duration of the pip (cosine cycle τ) was 1 ms. As shown above, probes of this type are adequate for frequency-tuning measurements in dolphins only in a high-frequency range (above 64 kHz). Therefore, tested probe frequencies were 64, 76, 90, and 108 kHz (quarter-octave steps).

A feature of that study was generation of masking noise with a precisely defined notch in its spectrum. The noise was digitally generated at a 500-kHz sampling rate. The generation procedure involved a digital filtering of a wide-band signal (random digital sequence). The filter passband had cutoffs at 12.5 kHz below and 12.5 kHz above the probe frequency, with 1-kHz ramps (Fig. 2.68A). The notch of 1 to 12 kHz width was centered at the probe frequency and had 1-kHz ramps also. The notch width was specified at the –6-dB level. Filters with no notch (zero notch width) were used as well.

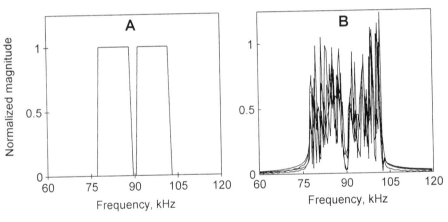

Figure 2.68. Notch-noise spectra. **A**. An example of digital notch-filter shape centered at 90 kHz, notch width is 2 kHz. **B**. Fourier transforms (superimposed) of four randomly chosen digital samples, each 2 ms long, of the notch noise obtained using the filter shown in (**A**).

Of course, only long-term spectra of generated noise reproduced precisely the given filter shape. The spectra of short noise samples deviated from the filter form because of random fluctuations inherent in noise (Fig. 2.68B). Nevertheless, even in short samples, narrow notches were noticeable. Pregenerated digital noise sequences were stored in computer memory and played through a transducer. Frequency response of the transducer is of importance for noise-masking experiments to keep the masker spectrum flat enough in the passband. So it should be noted that although the transducer frequency response varied by 16 dB within the frequency range of 50 to 120 kHz, this variation did not exceed ± 3 dB within any of the used ± 12.5-kHz noise bands.

Another noteworthy problem arises when the noise is digitally pre-generated off-line, as it was done in that study. Being stored in computer memory, one and the same noise sample was played repeatedly during the ABR collection. Therefore, special precautions were necessary to make the masking effect independent of the particular noise fragment coinciding with the probe. To do this, the noise burst was presented noncoherently with the probe and ABR collection sweeps, so the probe coincided randomly with various parts of the noise burst.

Successive stages of data collection and processing in notch-noise-masking experiments are shown in Figs. 2.69 to 2.73. Figure 2.69 exemplifies ABR evoked by a probe without masker (A) and in the presence of various maskers (B–D). In the exemplified experiment, the probe frequency was 90 kHz, and the noise notch was centered at the same frequency. As mentioned above, ABR in dolphins may be of high amplitude at appropriate stimulus parameters; however, probe stimuli used for frequency selectivity

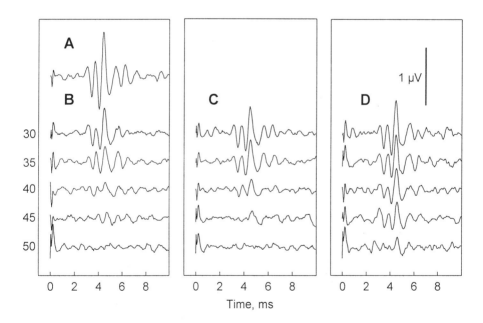

Figure 2.69. ABR suppression by notch-noise masker in a bottlenose dolphin *Tursiops truncatus*. **A**. Unmasked ABR evoked by a probe of 90 kHz frequency, 80 dB re 1 μPa level. **B**. Responses to the same probe in the background of the no-notch masker. **C**. The same with 1-kHz notch in the masker. **D**. The same with 5-kHz notch. Masker levels (dB re 1 μPa^2/Hz) are shown on the left.

measurements were of rather long rise time (1 ms duration of the cosine envelope) and low level. In the presented case, the probe level was 80 dB re 1
µPa which was 20 dB above the threshold. Therefore, ABR amplitude was
rather low even without masking, about 1 µV peak-to-peak (Fig. 2.69A).

The same stimulus in a masker background evoked smaller responses
(Fig. 69.2B–D). As the masker level increased, ABR amplitude diminished
until the response disappeared in the background noise. The masking effect
depended on the notch width. Even at a notch as narrow as 1 kHz (C), ABR
amplitudes were slightly but noticeably higher than with no-notch masker
(B) – that is, masking effect decreased. Further notch widening resulted in a
further increase of the ABR amplitude – that is, decrease of the masking effect (D).

Near the masking threshold, ABR amplitude was comparable with the
background noise. Therefore, the match-filtering procedure (see above) was
used to evaluate the weight of the ABR waveform in the records and thus to
estimate the near-threshold ABR amplitude.

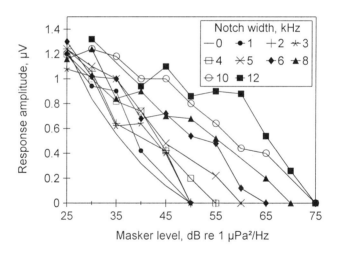

Figure 2.70. Response amplitude as a function of notch-noise masker level, with notch width
as a parameter.

Figure 2.70 presents ABR amplitude as a function of masker level taking
notch width as a parameter. In spite of some data scatter, the general trend is
obvious: the wider notch, the higher masker levels required to reduce ABR
amplitude to a certain level. Therefore, the next step of the data processing
was to present the masker level required for a certain degree of ABR reduction as a function of notch width. These functions are presented in Fig. 2.71

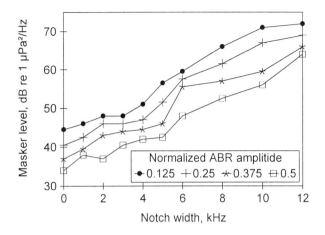

Figure 2.71. Masker level required to a certain reduction of ABR amplitude as a function of notch width, taking response amplitude as a parameter. ABR amplitude is shown as a fraction of the unmasked response amplitude.

for several degrees of ABR reduction – namely, for response amplitudes of 12.5, 25, 37.5, and 50% of the unmasked response. Since all the plots run mostly in parallel, it was reasonable to average them in order to reduce data scatter and express the masked threshold in dB relative to that at the zero notch width.

The resulting masked thresholds are presented in Fig. 2.72A as a function of the notch width for probe frequencies of 64, 76, 90, and 108 kHz. The

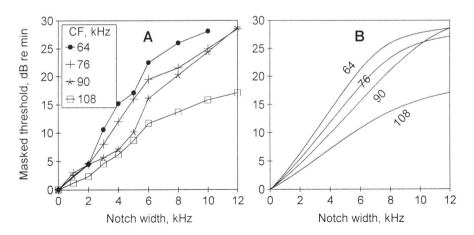

Figure 2.72. **A.** Masked threshold as a function of notch width for four center frequencies, 64 to 108 kHz. **B.** Approximation of data by integrated *roex* functions.

overall trend was clear: the noise level required for a certain degree of masking rose with widening the notch. All the plots were of similar shape: within
a range of notch widths up to 6 kHz, the data form essentially a straight line
when plotted on dB-level versus linear-frequency coordinates; at notch
widths above 6 kHz, all the plots tend to flatten.

We adopted the auditory filter form described by the rounded-exponential
(*roex*) function (see above)

$$W(g) = (1-r)(1+pg)e^{-pg} + r,$$ (2.31)

where g is the normalized frequency deviation from the center of the filter,
the term p determines the filter tuning in the passband, and the term r approximates the shallow tail section of the filter shape. Thus the power of
notched masker transmitted by each side of the filter when the notch is centered on the filter – that is, the power transmitted beyond a given $g = N/2f$ (N
is the notch width and f is the center frequency) is

$$W_m(g) = \int_g^L W(g,r) = (1-r)\left[\left(\frac{2}{p}+g\right)e^{-pg} - \left(\frac{2}{p}+L\right)e^{-pL}\right] + r(L-g),$$ (2.32)

where W_m is the transmitted masker power and L is the integration limit that
is necessary to keep the integral bounded at $r \neq 0$. When $L >> 1/p$ ($e^{-pL} \to 0$)
and $r << 1$ ($1 - r \to 1$), the latter expression may be reduced to

$$W_m = \left(\frac{2}{p}+g\right)e^{-pg} + r(L-g).$$ (2.32)

With the use of this equation, the dB-ratio R between no-notch noise
power ($g = 0$) and notch-noise power transmitted by the filter is

$$R = 10\log\left(\frac{2}{p}+rL\right) - 10\log\left[\left(\frac{2}{p}+g\right)e^{-pg} + r(L-g)\right].$$ (2.34)

The last expression predicts how masker level required for a certain
masking degree depends on notch width. To estimate frequency tuning on
the basis of this model, the lines drawn according to the last equation were
fitted to the data presented above. For each curve, the values of the terms p
and r were found that minimized the mean-square deviation of the predicted
values from the data; the integration limit L was chosen to keep $fL = 12.5$
kHz (this was the half-width of the noise band). The obtained lines are

shown in Fig. 2.72B as functions of notch width. For all the filters, p values were from 140 to 146.

The underlying auditory filter forms were derived by substitution the found p and r values into the *roex* equation. The results are shown in Fig. 2.73, which presents filter attenuation in dB as a function of frequency. Being presented at the logarithmic frequency scale, all the filters seems similar, except a shorter dynamic range of the filter centered at 108 kHz (a higher r value). The filter forms are symmetric by definition because measurements were restricted to conditions in which the masker spectrum was symmetric about the center frequency.

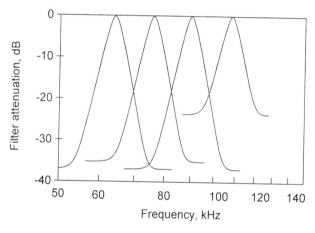

Figure 2.73. Auditory filter forms (*roex* functions) derived for four center frequencies, from left to right: 64, 76, 90, and 108 kHz, in the bottlenose dolphin *Tursiops truncatus*.

As shown above, for the *roex* function, $Q_{ER} = p/4$ and $Q_{10} \approx p/7.78$. Thus, the found values of the parameter p of 140 to 146 correspond to Q_{ER} of 35 to 36.5 and Q_{10} of 18 to 19. These estimates of frequency tuning are in very close agreement with those obtained in tone-masking experiments. At first sight, this coincidence seemed surprising since tone-masking data could be affected by off-frequency listening, whereas notch-noise-masking data must not be noticeably affected. A probable explanation of this unexpected coincidence is that the off-frequency listening did not manifest itself in the tone-masking experiments. It does not seem impossible. Contrary to psychophysical threshold measurements, EP were recorded at a certain level above their threshold. At a suprathreshold level, the filter tuned to the probe frequency gives higher response than side filters; this may make EP less sensitive to the off-frequency listening effect.

Thus, the notch-noise-masking data confirmed the high acuteness of the frequency tuning in dolphins: it is at least several times better than in a number of other mammals including humans. The notch-noise-masking data allow also a direct comparison with psychophysical data obtained with the use of the same masking paradigm. There is a large body of data obtained in humans using the notch-noise-masking method. Normally, the parameter p of *roex* filters in humans was 20 to 50 depending on the listener age, probe values, and so on (Patterson et al., 1982; Glasberg and Moore, 1984, 1990; Moore and Glasberg, 1995), contrary to about 140 in the dolphin. Correspondingly, ERB calculated from these data in humans was several times wider (that is, Q_{ER} was several times less) than in dolphins.

2.6.4. Frequency-Discrimination Limens

An additional estimate of frequency tuning is the frequency-modulation sensitivity, which is measured by the lowest detectable shift of pure-tone frequency (frequency-discrimination limen). This measure does not coincide with the critical band or the peripheral auditory filter bandwidth. For example, the frequency-discrimination thresholds in humans are mainly below 1%, depending on frequency, sound level, and so on (Weir et al., 1976), whereas critical bands in a major part of the hearing frequency region are more than 10% (Moore and Glasberg, 1983; Glasberg and Moore, 1990).

The nature of this difference can be explained by the effect of the side branches of the auditory filters (Fig. 2.74). Suppose a tone of a frequency f_1 changes its frequency to f_2. Even if this change is much less than the filter bandwidth, the signal transferred by a filter a (centered below the tone) markedly decreases (from a level a_1 to a level a_2), whereas the signal transferred by a filter b (above the tone) increases (from b_1 to b_2). This difference in the levels of signals transferred by filters can be detected at a small shift of the tone frequency.

This high sensitivity to frequency shifts can appear only in idealized conditions of pure-tone stimulation. Real discrimination of components of a complex sound is possible only if they differ by more than the filter bandwidth (critical band). Nevertheless, the frequency-discrimination limen is a valuable indicator of frequency-discrimination abilities. It can be expected that higher frequency tuning of hearing manifests itself, in particular, in better frequency-discrimination limens.

Several attempts have been made to measure frequency-discrimination thresholds in dolphins (Jacobs, 1972; Herman and Arbeit, 1972; Thompson and Herman, 1975). In those studies, the animal had to discriminate between a low-rate frequency-modulated tone and unmodulated tone of the same

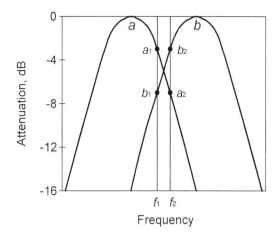

Figure 2.74. A mechanism of frequency-discrimination ability. *a* and *b* – two auditory filters, f_1 and f_2 – tone frequencies, a_1 and a_2 – signal levels transferred by the filter *a* at frequencies f_1 and f_2 respectively, b_1 and b_2 – the same for the filter *b*.

center frequency. It was assumed that the least frequency deviation that allows discrimination of modulated and unmodulated tones represents the frequency-discrimination limen. The results of these studies are summarized in Fig. 2.75, which presents the frequency-discrimination limens as a function of center frequency. Limens are expressed as a Weber's fraction $\Delta f / f_0$, where f_0 is the center frequency and Δf is the detectable frequency shift. Threshold estimates obtained in those studies varied to a significant extent; this is understandable given that each of those studies was carried out on only one subject. In general, the studies yielded moderate values of frequency-discrimination limens: they were not less than 0.2% (except one point in the study of Herman and Arbeit, 1972); at higher and lower frequencies, the limens were as high as 0.5 to 1.5%. These values are of the same order as those in humans (Weir et al., 1976). Thus, there is some disagreement with the data on the frequency tuning in dolphins (see above) which indicated markedly better tuning in dolphins than in humans.

In a study of Supin and Sukhoruchenko (1974), a single shift of the tone frequency was used as a stimulus (the shift time was around 0.1 s), and electrodermal, breathing, and heart-beat responses were evoked by this stimulus due to the Pavlovian conditioning in harbor porpoises *Phocoena phocoena*. Rather low frequency-shift thresholds were found in that study: around 0.1% within a frequency range of 11 to 90 kHz and up to 0.2–0.4% at frequencies from 5.6 to 150 kHz. This is markedly better than was obtained in operant-conditioning experiments with bottlenose dolphins. It remained uncertain,

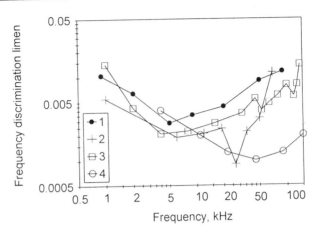

Figure 2.75. Frequency-discrimination limens in bottlenose dolphins *Tursiops truncatus* (1–3) and harbor porpoises (4) obtained in psychophysical studies: 1 – Jacobs (1972); 2 – Herman and Arbeit (1972); 3 – Thompson and Herman (1975); 4 – Supin and Sukhoruchenko (1974). Data are presented as Weber's fraction: $\Delta f/f_0$. Plotted by table data presented in the papers referred to.

whether this difference reflected a real interspecies difference of frequency selectivity or arose due to difference in methods and experimental conditions.

An evoked-potential study of frequency-discrimination thresholds was undertaken by Supin and Popov (1976a). It was found in bottlenose dolphins that a steep (during a few ms) shift of tone frequency provokes intracortical EP recorded through implanted electrodes (Fig. 2.76A). Their amplitude depended on the frequency shift, so the threshold frequency shift corresponding to zero EP amplitude could be found in a usual manner – for example, as a point of interception of the regression line with zero level (Fig. 2.76B). Frequency-discrimination threshold found in such a way were below 0.25%.

Further study of this problem was performed using non-invasive evoked-potential technique (Supin and Popov, 2000). In that study, the fact was exploited that EFR can be evoked not only by amplitude-modulated (AM) but also by frequency-modulated (FM) stimuli. It was found that at moderate modulation depths (an order of a few percent), FM stimuli evoked robust EFR – that is, a rhythmic sequence of potential waves that reproduced the modulation rate.

Figure 2.77 exemplifies EFR at a few modulation rates and shows that the most prominent EFR arises at the same modulation rates as AM-evoked EFR, namely, at rates of 600 to 650 Hz (625 Hz in Fig. 2.77) and 1000 to 1200 Hz. The rates of 600 to 650 Hz and 1000 to 1200 Hz are convenient for threshold measurements as evoking the most prominent EFR. Among these

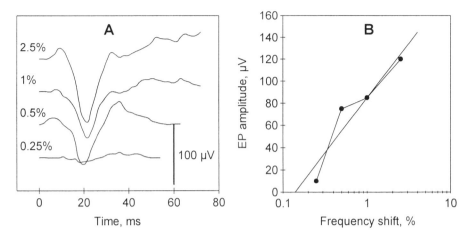

Figure 2.76. EP responses in the cerebral cortex of a bottlenose dolphin *Tursiops truncatus* to tone frequency shift. Initial tone frequency 20 kHz. **A**. EP waveforms at frequency shifts from 2.5 to 0.25%, as indicated. **B**. EP amplitude as a function of frequency shift. The straight line is a regression line approximating the plot; it intercepts with the abscissa axis at a point of 0.14%.

two regions, the lower is preferable to minimize the stimulus bandwidth. Therefore, the modulation rate of 625 Hz was used for the measurements.

EFR amplitude was dependent on modulation depth; the lower the depth, the less amplitude of the response (Fig. 2.78). Modulation depth here is defined as maximal deviation of frequency from its mean level:

$$m = (f_u - f_l) / (f_u + f_l), \qquad\qquad (42)$$

where m is the modulation depth, f_u and f_l – the upper and lower frequency limits.

Noticeable responses were observed at modulation depths far below 1%. To quantify EFR amplitude and find its threshold, the record was Fourier transformed. The temporal window taken for the transform covered a major part of the EFR and was equal to a whole number of stimulus cycles.

The result of Fourier analysis of records is presented in Fig. 2.79. All the spectra had prominent peaks at the modulation-rate fundamental (625 Hz) and harmonics. At large FM depths, harmonics were significant; at small depths, they became negligible. A fundamental peak was well detectable at a modulation depth as low as 0.1%.

Since spectrum level between the fundamental and harmonics was low, a spectral peak magnitude as low as of 0.05 µV could be identified confi-

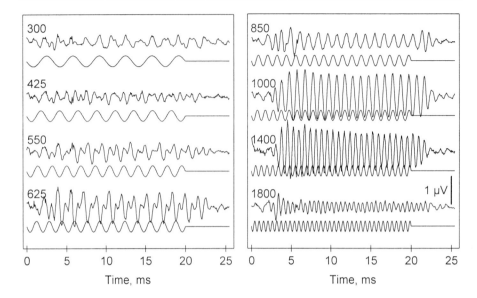

Figure 2.77. EFR in a bottlenose dolphin *Tursiops truncatus* to FM stimuli at various modulation rates. The stimulus carrier is 90 kHz, 120 dB *re* 1 µPa, modulation depth 5.6%. In each pair of records, the upper one is the EFR record, the lower one is the modulating function (frequency deviation) record. Modulation rates (Hz) are indicated near the curves.

dently. The 0.07 µV level was chosen as an arbitrary criterion for threshold estimation. To find a threshold, EFR was recorded when the modulation depth varied in 5-dB steps – 10%, 5.6%, 3.2%, 1.8%, 1%, and so on. Thresholds were calculated by linear interpolation between modulation depths evoking EFR fundamental magnitudes just above and just below the criterion level.

Using this technique of threshold estimation, FM thresholds were measured as a function of sound level at a variety of carrier frequencies, from 32 kHz to 128 kHz. Results obtained in a bottlenose dolphin are presented in Fig. 2.80A. At high sound intensities (130–150 dB re 1 µPa), FM thresholds were less than 0.1% at carrier frequencies between 54 and 108 kHz .

The data are summarized in Fig. 2.80B as the lowest FM thresholds found at each of the tested frequencies. Within a rather wide frequency range (from 45 to 108 kHz), FM thresholds were very low: modulation depth about 0.05%. Outside this range, FM thresholds increased although remained rather low within all the frequency range under study (0.6% at 32 kHz).

An important question is: were the responses to FM stimuli really evoked by frequency modulation? Indeed, any frequency change could produce an accompanying amplitude modulation because of irregularities of frequency responses of both the transferring and receiving systems. As to the sound-

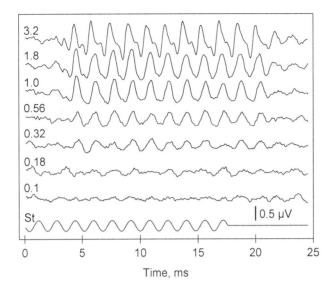

Figure 2.78. EFR of a bottlenose dolphin *Tursiops truncatus* to FM stimuli at various modulation depths. The stimulus carrier was 76 kHz, 150 dB *re* 1 μPa, modulation rate 625 Hz. Modulation depth was varied by 5-dB steps and is indicated near the curves (%); St (stimulus) is the modulating function (frequency deviation) record.

emitting system, its frequency response was known and was compensated by corresponding amplitude modulation applied to the FM signal. However, it is difficult to quantify and compensate the frequency-response irregularities of the sound-conducting pathways and the receiving structures of the subject. To some extent, it can be done basing on the animal's audiogram; however, there is no assurance that at the suprathreshold sound levels these irregularities reproduce the audiogram.

This problem can be solved by comparing FM and AM thresholds. For comparison, EFR to AM stimuli were recorded under the same conditions as EFR to FM stimuli; they were analyzed in the same manner, using the same threshold criterion. Contrary to FM stimuli, AM stimuli required modulation depths of about an order of magnitude higher; robust responses appeared at modulation depths of tens of percents while threshold responses appeared at depths of about 1%. The lowest AM thresholds were about 1% (Fig. 2.80B). Almost the same relation between the FM and AM thresholds was reported for humans (rev. Kay, 1982). If responses to FM stimuli were evoked not by the FM proper but by accompanying AM, this would mean that small frequency shifts resulted in 10 times larger amplitude shifts (expressing both in terms of modulation depth):

$$\delta A \, / \, A_0 \approx 10 \delta f \, / \, f_0, \tag{2.36}$$

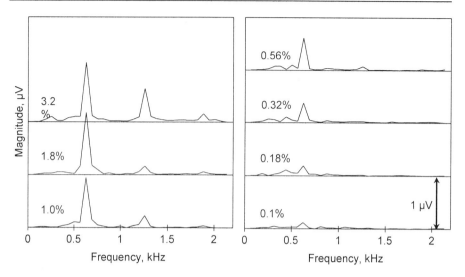

Figure 2.79. Fourier transforms of records presented in Fig. ##. Modulation depth (%) is indicated near the spectra.

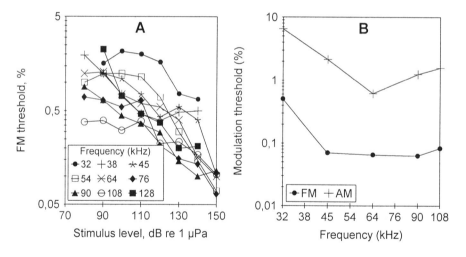

Figure 2.80. **A**. FM thresholds of a bottlenose dolphin as a function of stimulus level with carrier frequency as a parameter. **B**. Lowest FM and AM threshold dependence on carrier frequency. Averaged data from two bottlenose dolphins *Tursiops truncatus*.

where, A_0 and f_0 are the mean amplitude and frequency and δA and δf are the amplitude and frequency shifts, respectively. It can be shown that if $\delta \ll 1$, then $(1 + \delta)^n \approx 1 + n\delta$, thus

$$1 + \delta A / A_0 \approx (1 + \delta f / f_0)^{10} \tag{2.37}$$

– that is, to explain FM thresholds by accompanying amplitude modulations, one must assume that signal amplitude is approximately the 10th power of frequency (60 dB/octave). Such steep dependence seems unrealistic; at least there was no indication that such a degree of dependence can be expected. Thus, there is a good reason to believe that responses to FM stimuli were evoked by the frequency shifts proper.

Based on this conclusion, the data presented herein can be used for estimation of frequency-discrimination thresholds. At high stimulus levels, the threshold modulation depth was as low as about 0.05% at frequencies between 45 kHz and 108 kHz. However, the modulation depth is the frequency deviation from the middle frequency value; the peak-to-peak frequency deviation is twice as large: from $f_0 - \delta f$ to $f_0 + \delta f$. Thus, the lowest frequency-discrimination thresholds can be estimated as 0.1%. These values are close to the lowest estimates obtained in psychophysical experiments (Herman and Arbeit, 1972; Supin and Sukhoruchenko, 1974) and markedly lower than the standard frequency-discrimination limen in humans (about 0.3%).

Moreover, it should be pointed out that the frequency-discrimination threshold estimates in dolphins were obtained using an arbitrary threshold criterion that was limited by response detection in noise. Thus, a possibility cannot be excluded that at lower noise levels, threshold estimates would be even lower. Additionally, in experiments on dolphins, a modulation rate as high as 625 Hz was used since lower rates evoked poor EFR. It is known, however, that in humans, frequency discrimination deteriorates at modulation rates higher than 2–3 Hz (rev. Kay, 1982). Although the dolphin's auditory system is capable of reproducing high-rate modulations (see Section 2.5.6), it cannot be ignored that at lower modulation rates the dolphin's frequency-discrimination capability may be better. Therefore, the obtained estimates should be considered as rather conservative: frequency-discrimination thresholds in dolphins are at least as low as those reported above. Even such conservative estimates show a very acute frequency selectivity in dolphins.

Thus, the frequency-discrimination data can be taken as an additional indirect indication of acute frequency tuning in dolphins. It is noteworthy that these lowest thresholds were found in a frequency region of the best hearing sensitivity.

2.6.5. Frequency Resolving Power

All the tests described above used indirect approaches to measure the acuteness of the frequency tuning of the dolphin's hearing. These measures are based either on the frequency specificity of masking effects or on the ability to discriminate one pure tone from another in idealized stimulation conditions. Do these data indicate the real ability of the auditory system to analyze complex sounds and discriminate their frequency compositions? The answer to this question can be provided by the use of tests that directly show abilities to discriminate complex frequency spectra of sounds. The ability to discriminate (resolve) fine spectrum patterns is referred to as the frequency resolving power (FRP).

In studies of Popov and Supin (1984) and Supin et al. (1994, 1997, 1998), a FRP test was suggested based on the use of so-called rippled noise. Rippled noise is a signal with the frequency spectrum containing periodically alternating maxima and minima (peaks and valleys) – that is, the spectrum features a rippled pattern (see Fig. 2.81). Such a "grid" spectrum is a quantitatively defined version of complex-form sound spectra like many natural ones. It may be a reliable probe to measure frequency resolution: the finer the spectrum pattern (higher ripple density) is resolvable, the better the frequency resolution. The highest ripple density at which the fine spectrum structure is resolvable is a good measure of FRP.

Rippled noise was widely used in neurophysiological single-unit studies (Evans and Wilson, 1973; Bilsen et al., 1975; Evans, 1977), in psychophysical studies as a masker (Houtgast, 1974, 1977; Pick et al., 1977; Pick, 1980), and in studies of pitch perception of complex sounds (Yost et al., 1977; Yost and Hill, 1979; Bilsen and Wieman, 1980; Yost, 1982). The latter studies focused on the so-called time separation pitch (TSP). TSP is a pitch sensation arising at the action of rippled-spectrum stimuli – rippled noise or paired pulses with a constant frequency spacing between ripples throughout the spectrum. The perceived pitch corresponds to a frequency equal to δf, the frequency spacing between adjacent ripples.

Au and Pawloski (1989) studied the ability of a bottlenose dolphin *Tursiops truncatus* to discriminate between rippled and nonrippled noise. The experiments were performed in a go/no-go paradigm: the animal was required to respond yes to a burst of rippled noise and no to a burst of nonrippled noise. The noise was wide-band (from almost zero to 200 kHz) with a constant ripple spacing throughout the band. This version of rippled noise can be generated in a simple way by adding wide-band noise to a delayed version of itself; in this case, the ripple spacing $\delta f = 1/\tau$, where τ is the delay. It was found that the dolphin performed above 90% correct for delays between 50 and 400 µs, which corresponds to ripple spacing from 20 to 2.5

kHz, respectively. The upper 75% correct-response threshold was 500 μs (ripple spacing of 2 kHz) for the so-called cos– spectra with ripples starting from a dip at zero frequency. For cos+ spectra (ripples start from a peak at zero frequency), the upper delay threshold was 190 μs (around 5 kHz ripple spacing).

Authors discussed their results in terms of the involvement of TSP to discriminate the rippled spectra. However, stimuli of the kind used in this study have little to offer as a probe to measure frequency tuning since constant ripple spacing throughout a wide spectrum band results in large variety of relative ripple density across the band. In the low-frequency part of the band, the ripple density is low (the ripple spacing is large as compared with the center frequency of a ripple), whereas at the upper boundary of the band, the ripple density is high (the same ripple spacing constitute only a small fraction of the ripple center frequency).

As shown in bottlenose dolphins by Dubrovskiy et al. (1992), the rippled pattern (microstructure) of spectra of double-pulse sounds is an important cue to discriminate acoustic signals, though of lower priority than the macrostructure.

For testing frequency resolution (FRP), narrow-band rippled spectra are more appropriate since they allow the result of measurement to be attributed to a certain narrow frequency band (Fig. 2.81). Apart of that, in studies of Popov and Supin (1984), Supin et al. (1994, 1997, 1998), FRP was measured with the use of rippled noise in conjunction with a ripple phase-reversal test to find the highest resolvable ripple density and to estimate FRP. The test principle is as follows. Continuous rippled noise of a certain ripple density is presented to a subject. At some instant, this noise is replaced by another one of the same intensity, frequency band, and ripple density but of the opposite position of spectral peaks and valleys at the frequency scale (compare Fig. 2.81A and B). This is the phase-reversal stimulus. During the phase reversal, the subject detects some change in the noise timbre. This change can be detected only if the fine spectrum structure is resolvable. If the ripples are spaced too densely to be discriminated (Fig. 2.81C and D), the change cannot be detected because the noise before and after the switch is quite the same in all respects except the peak and valley positions. Thus the highest ripple density at which the phase reversal is detectable can be taken as a measure of FRP.

More detailed study is possible by varying both the ripple density and ripple modulation depth (spectral "contrast"). If the modulation depth is too low, the spectra before and after the phase reversal cannot be discriminated either (Fig. 2.81E and F). In such a way, "contrast" thresholds at various ripple densities can be found using the phase-reversal test.

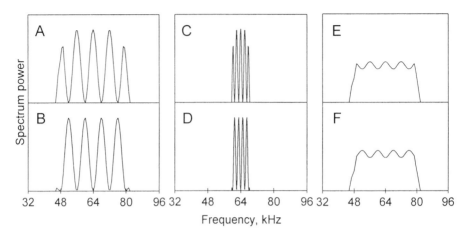

Figure 2.81. Rippled spectra and the phase-reversal test used in FRP measurements. **A** and **B**. Spectra with low ripple density (8 relative units), large ripple depth (100%), with opposite positions of peaks and valleys at the frequency scale. **C** and **D**. The same with high ripple density (32 units). **E** and **F**. The same with low ripple density (8 units) and small ripple depth (10%). All spectra are centered at 64 kHz and have a constant number of ripples.

It is noteworthy that this principle is similar to that widely used to measure the visual acuity and contrast sensitivity. In those studies, grids of varying spatial frequency are used as test stimuli, and the phase reversal of the grids is used to find the highest resolvable spatial frequency and (or) the lowest resolvable contrast.

Using this test, psychophysical measurements of FRP in bottlenose dolphins *Tursiops truncatus* were carried out (Supin et al., 1992a,b; Tarakanov et al., 1996). In those studies, the rippled noise was narrow-band with 4 ripples within the band (Fig. 2.81). The noise was produced by mixing a quasi-random binary sequence with a delayed version of itself (Narins et al., 1979). This results in spectral ripples with frequency spacing of $\delta = 1/\tau$, where δ is the ripple spacing and τ is the delay. Ripple density of narrow-band noise is characterized by its relative measure:

$$D = F / \delta = F\tau,\tag{2.38}$$

where D (relative units) is the relative ripple density, δ (kHz) is the ripple spacing, and τ (ms) is the delay. Narrow-band noises were used with center frequencies varying in half-octave steps from 2 to 128 kHz, and ripple densities were varied using the steps as follows: 4, 5, 6, 8, 10, 12, 16, 20, 24, 32, 40, 48, and 64 relative units – that is, three steps per ripple-density doubling. As to the noise bandwidth, it covaried with the ripple density in such a way as to hold the number of ripples constant – that is, two ripples above and two below the center frequency (compare Fig. 2.81A–D). The variation of noise bandwidth together with ripple density was used to make the bandwidth as narrow as possible, which is necessary to attribute measurement results to a certain center frequency but provide at least several spectrum ripples within the passband.

In order to measure the resolvable ripple density, one of two signal types was presented to the animal. One signal type called *constant* was a rippled noise of a certain ripple density without any changes during the presentation to the animal. The other type called *alternating* was a rippled noise containing ripple-phase reversals – that is, ripple peaks and valleys periodically (every 0.5 s) interchanged their position at the frequency scale. The task of the animal was to discriminate the alternating and constant stimuli. It was assumed that the discrimination is possible only if the ripple pattern of the noise spectrum was resolvable since the alternating and constant stimuli were identical in all respect except the constant or alternating positions of ripple peaks and valleys.

Measurements were performed in the go/no-go paradigm. Staying at a start position, the animal had to listen to the stimulus and remained at the position if the stimulus was constant or go to a paddle and touch it if the signal was alternating. The noise-ripple density was varied according to the adaptive "one-down, two-up" procedure: when the animal detected the alternating signal and responded correctly twice, the ripple density of the next stimulus (either alternating or constant) was increased by one step; when the animal missed the alternating signal, the ripple density of the next signal was decreased by one step. Responses to constant stimuli did not influenced the ripple density of the next stimulus.

The performance data were processed using the ROC-format (see Section 1.4.3); correct-detection probability was plotted against false-alarm probability. The ROC-line meeting the point of 0.75 correct-detection and 0.25 false-alarm probability was adopted as a resolution-limit criterion. The intersection of the experimental plot with this line was adopted as the resolution limit.

Measurement revealed the typical psychometric dependence of the animal's performance on ripple density: increasing ripple density diminished the probability of correct detections and enlarged the probability of false alarms. A few examples of results obtained at various center frequencies are presented in Fig. 2.82. It shows also the procedure of threshold evaluation. For example, at a frequency of 2 kHz, the experimental plot crossed the criterion line between the points of ripple densities of 8 and 10; interpolation resulted in a threshold estimate of 9.0. Similarly, at frequencies of 16 and 64 kHz, threshold estimates were 10.2 and 23.3, respectively.

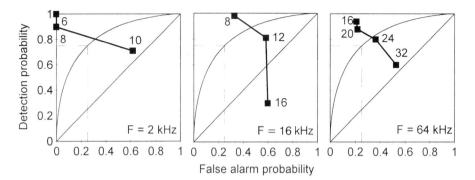

Figure 2.82. FRP measurements in a bottlenose dolphin *Tursiops truncatus* at three center frequencies of narrow-band rippled noise. Data are presented in ROC-format. Panels from left to right – center frequencies of 2, 16, and 64 kHz, as indicated. Dashed straight lines show the point of 0.75 correct-detection probability and 0.25 false-alarm probability; the curve meeting this point is the criterion ROC-curve. Polygonal plots show experimental data; ripple densities are indicated near experimental points.

Overall results of ripple-density resolution limits found as described above are presented in Fig. 2.83. The mean ripple-density resolution limit varied from 7.0 at 2-kHz center frequency to 48.3 at 128 kHz. Being presented in the log-log scale, resolution dependence on frequency is close to a straight line and can be approximated by a regression line from 7.1 at the 2 kHz frequency to 45.8 at 128 kHz.

A question now arises of how the FRP data relate to frequency-tuning estimates obtained by other methods. Supposing that FRP is dictated by frequency-tuned auditory filters, it is possible to compute the filter passbands and thus compare FRP data with other estimates of the tuning. The principle of the computation is presented in Fig. 2.84.

Suppose the filter form is as shown by a plot (1) in Fig. 2.84. When an input signal has a rippled spectrum with a peak centered on the filter (2), the output spectrum (4) is a product of the functions (1) and (2); the overall transferred power is proportional to the area under the curve (4). Respectively, when a valley of an input signal is centered on the filter (3), the output spectrum is (5), and the transferred power is proportional to the area under this curve. When the ripple density is low as compared to the filter passband (A), the two output signals markedly differ in power (areas under the curves 4 and 5); thus the change from one input signal to the other (the phase reversal) is easily detectable. When the ripple density is high (B), the areas under the curves (4) and (5) becomes almost equal, so the phase reversal becomes undetectable.

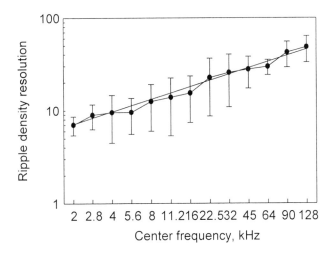

Figure 2.83. Ripple density resolution limit (relative units) as a function of center frequency of narrow-band rippled noise in the bottlenose dolphin *Tursiops truncatus.* Averaged data of four animals, means ± SD. The straight line shows linear approximation of the data.

To describe this model quantitatively, let us approximate the filter transfer function by a certain analytical function – for example, the *roex* function:

$$W(g) = (1 + pg)\exp(-pg),\qquad(2.39)$$

where W is the transmitted power, g is the relative deviation from the center frequency, and p is a parameter determining the filter tuning. The power spectrum of rippled noise at the filter input $P_{in}(g)$ may be defined as

$$P_{in}(g) = 1 \pm \cos 2\pi Dg,\qquad(2.40)$$

where D is the ripple density and the $\pm$ sign determines two version of the spectrum: with peak or valley centered at the filter (Fig. 2.84).

The power $P_{out}(g)$ transmitted by the filter is

$$P_{out}(g) = 2\int_{0}^{\infty} P_{in}(g)W(g)dg = 4\left[\frac{1}{p} \pm \frac{1}{p(1 + 4\pi^2 D^2 / p^2)^2}\right].\qquad(2.41)$$

At ripple phase reversals, the dB difference δI at the filter output is

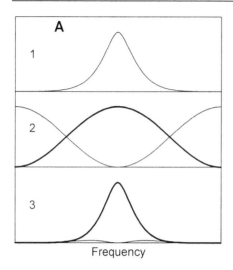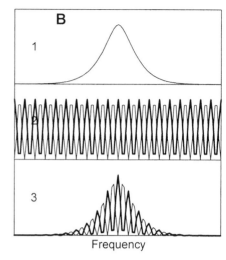

Figure 2.84. Transfer of rippled spectra by a frequency-tuned filter. 1 – filter form, 2 (solid line) – input spectrum with a peak centered on the filter, 3 (thin line) – input spectrum with a valley centered on the filter, 4 and 5 – output spectra for input signals 2 and 3, respectively. **A.** Low ripple density. **B.** High ripple density.

$$\delta I = 10 \log \frac{(1 + 4\pi^2 D^2 / p^2)^2 + 1}{(1 + 4\pi^2 D^2 / p^2)^2 - 1}. \tag{2.42}$$

Suppose that δI becomes undetectable below a certain threshold. This threshold may be set as the intensity-discrimination threshold which, is around 1 dB in humans (Jesteadt et al., 1977) and many mammals (Fay, 1988); as shown below, intensity-discrimination thresholds in dolphins are of the same order. It follows from equation (2.42) that δI values of 0.5 to 1 dB appear at D of $p/3.5$ to $p/4.5$; a value of $p/4$ can be taken as a first approximation. Thus, the model predicts resolution $R \approx p/4$. Equivalent rectangular bandwidth (ERB) of the *roex* filter is equal to $4F_0/p$, and $Q_{ER} = p/4$; hence:

$$R \approx Q_{ER}. \tag{2.43}$$

Thus, the found FRP values of 7 to 48 indicate quantitatively equal Q_{ER} of the frequency-tuned filters. It is a little higher than the ERB estimates obtained in bottlenose dolphins by masking methods, but this difference is not an indication of data disagreement. In humans, FRP is also higher than predicted by peripheral filter tuning (Supin et al., 1994, 1998); it may be a result of sharpening the spectrum resolution at higher levels of the auditory system. On the other hand, the filter form and δI value used for computation are not known precisely; therefore, the result of computation should be taken as a very first approximation. At least, the estimates of Q_{ER} obtained by FRP measurements are not less than those obtained in masking experiments, thus confirming the very acute frequency tuning of the dolphin's hearing.

Attempts were undertaken to apply the evoked-potential method in FRP measurements. For this purpose, EP to phase reversals of rippled noise should be recorded. The major problem is, however, that the phase-reversal stimulus is not efficient enough to provoke ABR because replacing one rippled-noise version with another is not a quick transient; it is limited by narrow-band ripple filtering. Therefore, attempts to record ABR to phase-reversal stimuli were not successful. However, longer auditory cortical responses (ACR, see Section 2.2.3) were more sensitive to phase-reversal stimuli and were used to measure FRP in bottlenose dolphins (Supin and Popov, 1988, 1990).

Response examples obtained with the use of the rippled-noise phase-reversal stimuli are presented in Fig. 2.85. In these experiments, rippled noises of 1-oct bandwidth were used, and ripple spacing δ was constant

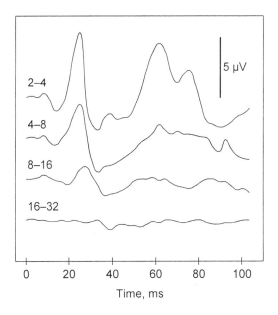

Figure 2.85. ACR evoked by phase reversals of rippled noise in a bottlenose dolphin *Tursiops truncatus*. Frequency band of the rippled noise was 45 to 90 kHz, intensity 120 dB re 1 μPa. Ranges of ripple density (from lower to upper boundary of the noise band) in relative units are indicated near records.

throughout the passband. Therefore, the ripple density $D = f/\delta$ was not constant but varied twofold from the lower to upper passband boundary. For example, at a passband of 45 to 90 kHz and ripple spacing of 10 kHz, the ripple density varied from 4.5 at the lower boundary to 9 at the upper boundary of the noise band. The wide noise bands were used since the noise of the narrower band was not effective enough even for ACR.

The records in Fig. 2.85 show that ACR amplitude depended on ripple density: the higher the ripple density (the more difficult the spectrum-pattern discrimination), the lower response amplitude, until responses disappeared when the spectrum pattern became unresolvable. Thus, the resolvable ripple density limits (FRP) were estimated as a ripple density at which response amplitude fell to zero; the zero amplitude point was found basing on amplitude-versus-ripple density regression line.

The result of measurements is presented in Fig. 2.86. Since ripple density varied across the noise band, there was some uncertainty as to which ripple density within the noise band is to be taken as an FRP estimate. The most conservative approach is to use the lowest values characteristic for the lower boundary of the noise band. These values are presented by plot 1 in Fig. 2.86. It shows that FRP increases with frequency until reached values of 28–30 relative units at frequencies of 64 to 128 kHz. However, it seems hardly possible that ACR to the phase-reversal stimulus was evoked by only a narrow fraction of the noise band near its lower boundary. If a higher fraction of the noise band contributed to ACR, FRP should be estimated basing on

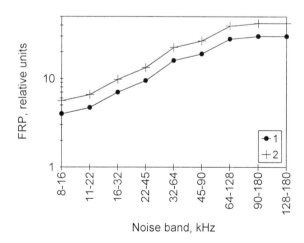

Figure 2.86. FRP as a function of noise band estimated by ACR to phase-reversal stimuli in a bottlenose dolphin *Tursiops truncatus.* 1 – estimates based on ripple density at the lower boundary of 1-oct noise band; 2 – estimates based on ripple density in the band center.

higher ripple density within the noise band above the lower boundary. As an arbitrary compromise, ripple density at the band center (a half-octave above the lower boundary) can be taken. In this case, FRP estimates are 1.4 times higher (plot 2 in Fig. 2.86) and reach 39 to 42 relative units. It is rather close to the data of psychophysical FRP measurements.

At lower frequencies, FRP estimates obtained by the evoked-potential method were markedly lower than psychophysical estimates. At a noise band of 8 to 16 kHz, the conservative FRP estimate (at the lower boundary of the noise band) was 4, correspondingly that for the band center (11.2 kHz) was 5.6. Psychophysical mean FRP estimate at this frequency was 13.9 (see Fig. 2.83). It is not clear yet either this difference is a result of interindividual variation or lesser ACR sensitivity to phase-reversal stimuli at lower frequencies.

Taking into account problems of EP recording to phase reversals in rippled noise, the results of evoked-potential measurements of FRP in dolphins can be considered as satisfactorily close to psychophysical data. In general, all these data confirm the very high frequency tuning of the dolphin's hearing.

2.7. SOUND-INTENSITY DISCRIMINATION

We are aware of the only psychophysical study on sound intensity discrimination in cetaceans (Burdin et al., 1973). This study was performed using low-rate (3 Hz) amplitude-modulated noise, and the animal was trained for go-response to an amplitude-modulated signal and no-go response to an unmodulated signal. The modulation depth threshold was evaluated as 5% (– 26 dB).

Only a limited body of evoked-potential data refer to this problem. The first attempt to measure AM sensitivity in bottlenose dolphins *Tursiops truncatus* was performed by Popov and Supin (1976a) using EP recorded in the auditory cerebral cortex through implanted electrodes. In that study, a single shift of tone or noise intensity was used as a stimulus to elicit a cortical EP; the rise-fall time was 10 ms for tones and less than 1 ms for wide-band noises. EP to such stimuli exhibited an amplitude dependence on stimulus value: the greater sound-intensity shift, the higher response amplitude. The minimal intensity shift that was able to evoke a just-detectable response was used as an estimate of the AM threshold.

It was found that at high tone intensities (above 110 dB re 1 μPa), threshold intensity shift was as small as 0.5 dB – about 12% of sound power or 6% of sound amplitude. At low sound intensities (below 100 dB re 1 μPa), the threshold (as expressed in the Weber's fraction) increased (Fig. 2.87). For

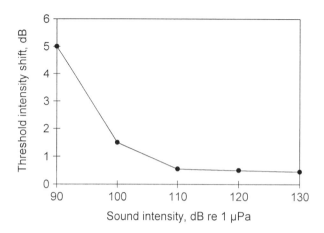

Figure 2.87. Thresholds of cortical EP to tone intensity shifts, as a function of sound intensity, in a bottlenose dolphin *Tursiops truncatus.*

noise stimuli, intensity-discrimination thresholds were higher than for tones: 2 to 3.5 dB. Higher noise thresholds could be associated with amplitude fluctuation intrinsic in noise, which masks the stimulus change of sound intensity.

Further data on sound intensity discrimination were obtained in experiments with EFR recording (Supin and Popov, 1995a). In that study, EFR was recorded at various values of modulation depth, from 1 to 0.01, to investigate linearity of the response. Although response amplitude diminished with decreasing the modulation depth, detectable EFR was recorded at modulation depth below 0.1 (Fig. 2.88).

Fourier analysis of the records revealed a detectable peaks corresponding to the modulation-rate frequency when modulation depth was as low as 0.02 to 0.03. Figure 2.89 demonstrates a well detectable 1000-Hz peak at modulation depths down to 0.05. Thus, the modulation-depth threshold can be taken as 0.02 to 0.03. It should be taken into account that the modulation depth m is deviation from the mean level; the maximal intensity difference is from $1-m$ to $1+m$ – that is, $2m$. Thus, the intensity discrimination threshold can be taken as 0.04 to 0.06 (4–6%), which constitutes 0.3 to 0.5 dB. It may be a conservative estimate since noticeable EFR can be revealed by Fourier analysis at modulation depths lower that 0.02.

Indeed, lower amplitude-modulation thresholds were found in a study of Supin and Popov (2000), which was designed to study frequency-modulation sensitivity but included also measurements of amplitude-modulation thresholds. In that study, threshold modulation depth varied from less than 0.01 at

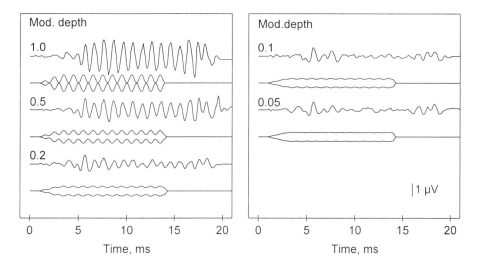

Figure 2.88. EFR evoked at various modulation depths in a bottlenose dolphin *Tursiops truncatus*. Tone carrier of 64 kHz, 60 dB above threshold. Modulation rate 1000 Hz, modulation depth is indicated near the curves. In each pair of records, the upper one – EFR record, the lower one – stimulus envelope.

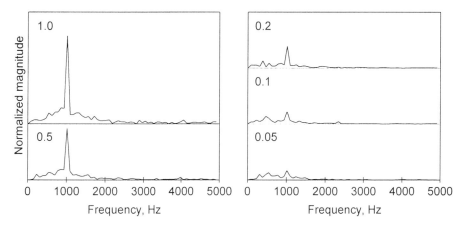

Figure 2.89. EFR frequency spectra in a bottlenose dolphin *Tursiops truncatus* at various modulation depths, as indicated. Tone carrier of 64 kHz, 60 dB above threshold. Modulation rate 1000 Hz.

a frequency of 64 kHz to 0.1 at 32 kHz (see Fig. 2.80). Again, taking that the intensity-discrimination threshold is twice as much as the threshold modulation depth, it should be estimated as 0.02 to 0.2 (0.17 to 1.7 dB), depending on frequency. Within this range, the lowest estimate (0.17 dB) may be considered as a very good intensity discrimination. For comparison, in humans this threshold varies from 0.5 to 1.6 dB, depending on frequency (Jesteadt et al., 1977); in many mammals, these thresholds are of an order of a few dB (Fay, 1988).

2.8. DIRECTIONAL SENSITIVITY, SPATIAL, AND BINAURAL HEARING

Directional sensitivity – the dependence of hearing sensitivity on direction to the sound source (azimuth) – is an important property of the auditory system that helps it to select the most important sound sources from the overall acoustic environment due to positioning of these animal's head in such a way that the sound-source of interest appears in the most sensitive sector. A narrow reception "beam" directed to a certain sound source allows the animal to minimize the perception of other interfering sounds.

Closely related to directional sensitivity is spatial hearing, which is an ability to localize a sound source. Spatial hearing can be based on spatial sensitivity: when a narrow receiving beam is directed in such a way as to gain the maximum sound power from a certain source, the beam direction indicates the source position. However, spatial hearing can be also based on more complicated mechanisms, particularly on binaural hearing.

2.8.1. Psychophysical Studies

A common method of measuring the directional sensitivity of a receiver is to move a sound source around the receiver along the arc and to measure the response as a function of the sound-source position. In psychophysical studies, hearing thresholds should be an equivalent of the response. However, no measurements of such kinds were performed in dolphins until now. When threshold measurements are made in unrestrained animals, it is very difficult to train the animal in such a way as to keep a constant head position and not to turn the head in the best-hearing direction. Therefore, instead of direct threshold sensitivity, masked hearing thresholds were measured as a function of the angular distance between the sources of the probe and masking sounds. In this case, it is natural for the animal to keep its head position directed toward the probe-sound source, but the probe-detection threshold is dependent on the masker-source position, as dictated by the directional sensitivity.

Using this approach, directional sensitivity of a bottlenose dolphin *Tursiops truncatus* was first measured by Zaytseva et al. (1975). They measured masked thresholds at angular distances of 0, 10, and 20 deg between the probe and noise transducers in the azimuth plane, at frequencies of 80 and 120 kHz. The 3-dB beamwidth found in their studies was as narrow as 8.2 deg at a frequency of 80 kHz.

These studies were extended in much more detail by Au and Moore (1984). In their experiments, a bottlenose dolphin was trained to hold a stationary position on a bite plate underwater. The probe and masker transducers were positioned along an arc centered on the animal's head. For the beam-pattern measurements in the azimuth plane, the animal took a normal position, its back upward, and the transducers were moved along a horizontal arc. For the beam-pattern measurements in the mid-sagittal (elevation) plane, the animal was trained to turn to its side, and the transducers moved along the same horizontal arc. A standard go/no-go paradigm was used to determine the animal's masked thresholds.

Two modes were used to vary the probe and masker transducer positions. For measurements in the azimuth plane, two masker-emitting transducers were stationary posed at ±20 deg relative to the midline, and the probe-emitting one was moved along the arc. It was done to prevent the dolphin, if it was capable of doing it, from steering internally the axis of its beam. For measurements in the elevation plane, no such precaution was taken, and mutual positions of the transducers was varied in an opposite manner: the probe-emitting transducer was posed stationary in the arc midpoint, and one masker-emitted transducer was moved along the arc.

Measurements carried out at three frequencies – 30, 60, and 120 kHz – have shown that the beam width is frequency-dependent, both in the azimuth and elevation planes: the higher frequency, the narrower the beam. In the azimuth plane, the 3-dB bandwidths were estimated as 59.1, 32.0, and 13.7 deg for frequencies of 30, 60, and 120 kHz, respectively. In the elevation plane, the corresponding values were 30.4, 22.7, and 17.0 deg. In the elevation plane, the beam was not symmetric; the sensitivity dropped off more rapidly at increasing angles above the animal's head than at angles below it.

Using these data, Au and Moore (1984) calculated the directivity index (DI) of the dolphin's receiving beam. DI is defined as a ratio of the power received by an omnidirectional receiver (P_0) to that received by a directional receiver (P_D) in isotropic noise field, expressed on dB-scale:

$$DI = 10\log(P_0 \ / \ P_D).$$ (2.44)

For frequencies of 30, 60, and 120 kHz, DI estimates were 10.4, 15.3, and 20.6 dB, respectively.

It should be noted, however, that the measurements based on the masked thresholds may give not the real pattern of the receiving beam. Their results depend both on the sound source position *relative to the head* (which gives the true receiving beam pattern) and on the *mutual position* of the probe-emitting and masker-emitting transducer. The latter reveals a capability of *discrimination* of sound-sources spatial positions rather than *sensitivity* as a function of the source position.

Some results of indirect measurement of the beam width are even more difficult for interpretation. Dubrovskiy (1990) reported results of measure-ment of spatial selectivity experiments designed in the following way. A dolphin faced two transducers symmetrically positioned relative to the mid-line. In each pair, one transducer was masker-emitting and the other was probe-emitting. Both masker and probe were simultaneously emitting short clicks. In each trail, both the masker and probe clicks were emitted from one side, whereas only the masker click was emitted from the other side. The dolphin was required to find the side of probe emission and swim to that di-rection. Masked thresholds were measured as a function of angular distance between the probe- and masker-emitted transducers. A very narrow receiving beam was obtained in those experiments: when the masker and probe trans-ducers were separated by 3.2 deg, the threshold decreased by 8 to 10 dB as compared to zero separation. It is a much more narrow function than that obtained by Au and Moore (1994).

It is unclear, however, whether these results really reflect spatial-selectivity properties of the dolphin's hearing. When the two speakers were spatially separated, short pulses emitted by them might arrive with a short delay relative to one another. As described above, in this case they fuse into a single click with a rippled frequency spectrum. It is a very effective cue to discriminate a two-pulse click (masker + probe) from a single-pulse one (masker only) that has a nonrippled spectrum.

True spatial-discrimination capability was also studied in a bottlenose dolphin *Tursiops truncatus* (Renaud and Popper, 1975). In that study, the dolphin faced two transducers positioned at equal angles relative to its mid-line. An acoustic signal was transmitted from one of these two transducers, and the dolphin was required to detect the emitting transducer and to touch the paddle at the same side. It was just the situation when an animal could be successfully trained in the two-alternative forced-choice paradigm. Varying the angle between the transducers according to the adaptive one-up-two-down procedure, threshold angles (the minimum audible angles, MAA) were determined both in the azimuth and elevation planes.

In the azimuth plane, discrimination thresholds were within a range of 2 to 2.5 deg at frequencies of 20 to 60 kHz; at both lower and higher frequen-cies, thresholds were a little higher: up to 3.6–3.8 deg. Paradoxically, rather similar thresholds were obtained for discrimination in the elevation plane: mainly from 2.5 to 3.5 deg at frequencies from 20 to 100 kHz (Fig. 2.90).

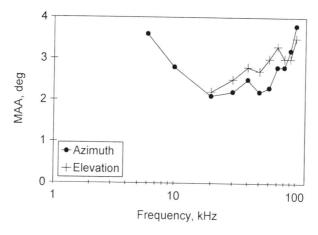

Figure 2.90. Minimal audible angle (MAA) in a bottlenose dolphin *Tursiops truncatus,* as a function of frequency, in the azimuth and elevation planes, as indicated. Plotted by table data in Renaud and Popper (1975).

The latter result was somewhat surprising since, according to a commonly adopted belief, the azimuth angular discrimination is based on a binaural hearing mechanism, which is not effective for the elevation angular discrimination. Thus, azimuth discrimination was expected to be markedly better. Au (1993) proposed an explanation addressing the fact that in experiments of Renaud and Popper, the dolphin was trained to bite a plate at the start position; this plate could shadow sounds coming from above or from below it, thus playing an important role in the angle discrimination process.

Directional selectivity as measured by Renaud and Popper (1975) depended on the azimuth of the transducer pair position. The values mentioned above are characteristic for zero azimuth – that is, when the two transducers were positioned symmetrically relative to the midline. At an azimuth of 15 deg, MAA slightly decreased (1.6 deg), and at an azimuth of 30 deg it increased again up to 5.3 deg.

Directional selectivity found by Renaud and Popper (1975) may be assessed as moderately acute. In humans, MAA varies within a limit of 1 to 3 deg (Mills, 1958).

2.8.2. Directional Sensitivity: Evoked-Potential Studies

Evoked-potential technique was helpful in studies of directional sensitivity. Contrary to psychophysical experiments, the animal can be restrained during evoked-potential recordings. It makes it possible to measure directly the hearing sensitivity as a function of the sound-source position relative to the head – that is, the true receiving beam pattern. For this purpose, ABR thresholds are to be measured at various sound-source azimuthal positions.

A usual problem in studying the directional sensitivity of hearing is the necessity to create an echoless acoustic field. Otherwise, it is difficult to attribute a found threshold to a certain sound-source position. However, an advantage of the ABR method is that even at the presence of echo, its effect on measurement results can be eliminated. Indeed, the duration of ABR is very short. If the distance from the animal to the pool walls and bottom is at least 2–3 m, the echo delay becomes longer than the ABR duration. Therefore, the response to a direct signal does not interfere with those of echoes. Moreover, instead of using the experimental pool 2–3 m deep, an alternative way us to place the animal in shallow water, 30–40 cm deep. In such shallow water, sounds spread as in a flat layer if the distance to the sound source is several times longer than the depth – that is, interference of echo from both water surface and bottom is minimal.

Using this technique, ABR thresholds were measured as a function of sound-source azimuth position in a few dolphin species: the bottlenose dolphin *Tursiops truncatus* (Popov and Supin, 1988, 1990b), tucuxi dolphin *Sotalia fluviatilis*, Amazon river dolphin *Inia geoffrensis* (Popov and Supin, 1990b), and beluga whale *Delphinapterus leucas* (Klishin et al., 2000). In an initial simple version, measurements were performed using wide-band clicks as stimuli. In each azimuth position of a transducer varied with 15-deg steps, stimulus intensity was varied, and ABR threshold was estimated using extrapolation of the amplitude-versus-intensity regression line. It should be noted that ABR was recorded using the standard electrode position on the head midline behind the blowhole (see Section 2.2.2); thus, responses to stimulation of the right and left ears were not separated. Therefore, it was reasonable to average thresholds at symmetrical left-side and right-side positions. The results are presented in Fig. 2.91 as threshold-versus-azimuth functions in three species.

The plots demonstrate a considerable dependence of the ABR thresholds on azimuth. The lowest were thresholds at zero azimuth – that is, when the transducer was disposed on the head midline. With the azimuth increase, thresholds rose up to 20–35 dB. The beam width at the 3-dB level varied from 12 deg (±6 deg) in the bottlenose dolphin *Tursiops truncatus* to 22.5 deg (±11.25 deg) in the Amazon river dolphin *Inia geoffrensis*.

In a bottlenose dolphin, ABR threshold-versus-azimuth function was estimated separately in different frequency bands. It was made using band-filtered clicks as stimuli (Popov and Supin, 1988; Popov et al., 1992). The spatial sensitivity was strongly dependent on the frequency band: it was

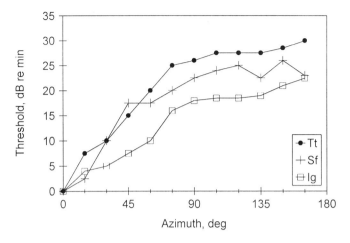

Figure 2.91. ABR thresholds as a function of the sound-source azimuthal position. Wide-band click stimulation. Thresholds are presented in dB relative to those at zero azimuth (head midline). Tt – *Tursiops truncatus*, Sf – *Sotalia fluviatilis*, Ig – *Inia geoffrensis*.

steepest at high-frequency stimuli and shallowest at low-frequency ones (Fig. 2.92; note that the ordinate scale in the figure is sensitivity – that is, thresholds plotted in the inverse order). At a high-frequency band (70–150 kHz), the difference between the highest and lowest sensitivity was more than 30 dB, whereas at lower frequencies (10–25 and 25–50 kHz) it was not more than 10 dB.

In a beluga whale *Delphinapterus leucas*, an attempt was also made to

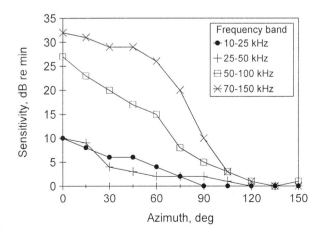

Figure 2.92. ABR sensitivity (inverted thresholds) in a bottlenose dolphin *Tursiops truncatus*, as a function of azimuth in various frequency bands, as indicated in the legend.

measure ABR thresholds dependence on azimuth separately at various frequencies (Klishin et al., 2000, Fig. 2.93). It was done using tone-pip stimuli of frequencies from 16 to 90 kHz (by half-octave steps). However, no regularity was noticed in the beam width dependence on frequency. Being averaged across frequencies, the 3-dB beam width was 15 deg (±7.5 deg).

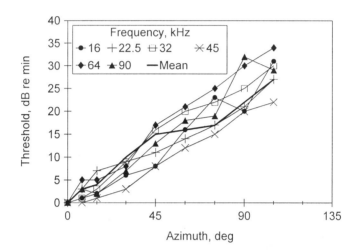

Figure 2.93. ABR thresholds as a function of the sound-source azimuth position in a beluga whale *Delphinapterus leucas.* Tone-pip stimulation. Frequencies are indicated in the legend; Mean – averaging across frequencies. Thresholds are presented in dB relative to those at zero azimuth (head midline).

In the experiments described above, contributions of the right and left ears to the recorded ABR were not separated, so the threshold-versus-azimuth function was adopted to be symmetrical by definition. However, it was found that the evoked-response technique allowed to record separately responses to stimulation of one or the other ear which made possible to measure directional sensitivity of one ear. These measurements were carried out in the Amazon river dolphin *Inia geoffrensis* (Popov and Supin, 1992).

To separate responses evoked by two ears, recording electrodes were positioned not at the dorsal but at the lateral head surface, near the auditory meatus (under water). In this case, ABR waveform differed from that obtained at the standard electrode position. In particular, the earliest response wave became more pronounced (Fig. 2.94). Its onset latency was as short as 1.4 to 1.5 ms including the 0.7-ms acoustic delay. When a sound source was placed close to the auditory meatus (to minimize the acoustic delay), the onset latency of this wave was 0.7 ms. The generation of such a short-latency response requires no more than one synaptic delay thereby suggesting that

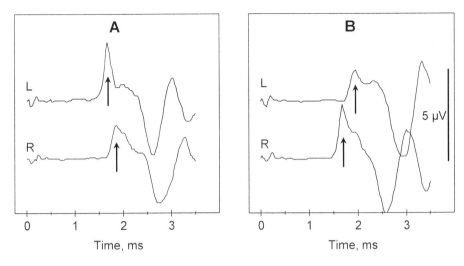

Figure 2.94. Evoked responses recorded from the lateral surface of the head of an Amazon river dolphin *Inia geoffrensis.* **A**. Responses to a click stimulus when the sound source was located 1 m from the head, 15 deg to the left from the longitudinal axis. **B**. The same as (**A**), but the sound source was located 15 deg to the right. L, R – records from the left or right side of the head (the auditory meatus region), respectively. Stimulus emission corresponds to the record beginning. Upward headed arrows indicate ANR wave.

the earliest wave manifests the activity of the auditory nerve. Thus, this wave was designated as the auditory nerve response (ANR).

In correspondence with interpretation of the first wave as ANR, it displayed clear monaural properties – that is, ANR recorded from the left and right ears behaved independently. When a sound source was positioned by the side of the longitudinal axis, the ipsilaterally recorded wave was of greater amplitude and shorter latency than the contralateral one, and no cross-contamination of one record by another was visible (Fig. 2.94). Note that later ABR waves recorded at two sides do not feature such independence: they differ in phase but rather similar in amplitude and duration.

Therefore, based on ANR amplitude and threshold, it was possible to measure directional sensitivity of one ear. At each azimuth position of the sound source, ANR of the right and left ears were recorded at a variety of stimulus intensities, and intensities providing a few criterion amplitude levels (from 0.5 to 3 µV) were found. Averaged data obtained at several criterion values are presented in Fig. 2.95 as sensitivity-versus-azimuth function. The data for the left and right ears were averaged in terms of ipsi- and contralateral sound-source positions.

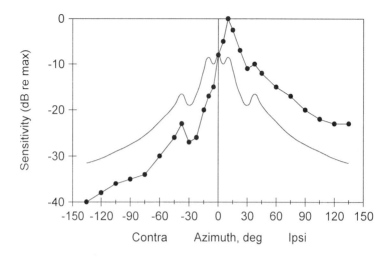

Figure 2.95. Effect of the sound-source azimuth position on ANR sensitivity in the Amazon river dolphin *Inia geoffrensis*. Zero azimuth corresponds to the frontal direction. Thin line is the average of ipsi- and contralateral branches of the experimental plot.

Two prominent features of the monaural directional-sensitivity function were: (1) it peaked not at zero azimuth but at 10 deg (as measured with 5-deg steps) ipsilateral to the recorded ear; (2) it was much more acute than binaural directional-sensitivity functions; its 3-dB width as estimated by interpolation between the nearest experimental points was 8.8 deg (from –3 to +5.8 deg). With azimuth deviation to any direction from the best-sensitivity point, the sensitivity decreased, but at the ipsilateral side sensitivity was always 15 to 20 dB better than at the symmetrical contralateral azimuth.

It is reasonable to assume that less acute and symmetric binaural directional-sensitivity function appears because of averaging of two monaural ones. Indeed, averaging of the right and left branches of the experimental plot in Fig. 2.95 results in a curve very similar to binaural directional-sensitivity functions.

2.8.3. Binaural Hearing: Evoked-Potential Studies

In mammals, the ability to localize sound sources is based mainly on binaural hearing. In terrestrial mammals, two main cues serve for this purpose: the interaural intensity difference (IID) and interaural time delay (ITD). IID is a result of sound shadowing by the head. If a sound source is positioned sidewise of the head midline, and the sound wavelength is smaller than the head dimension, the sound reaches the ipsilateral ear unobstructed, whereas its passage to the contralateral ear is shadowed by the head. The result is a difference of sound intensities at two ears (IID), which is a cue to localize the sound source. Apart from that, the distance of a sidewise-localized source to the ipsilateral ear is shorter than to contralateral one; the result is a time (or phase) difference between the sounds at the ipsilateral and contralateral ears (ITD).

It remained unclear whether these cues are as effective in aquatic mammals as in terrestrial ones. In air, sound shadowing by the head is possible due to large difference of acoustic impedance of air and body tissues. In water, the acoustic impedance of the media is rather close to that of many head tissues. Although the impedance of bones and air cavities within the head differs from that of water, it was not known whether the dolphin's head is really capable of sound shadowing large enough to provide significant IID.

On the other hand, sound speed in water is 4.5 times higher than in air. Therefore, at similar head dimensions, ITD in water is 4.5 times shorter than in air. Even it air, possible ITD values (hundreds of microseconds) are at the lower limit of time intervals resolvable by neuronal mechanisms. So it remained uncertain whether short ITD in water can be analyzed in the nervous system and used as a cue to localize sound-source positions.

A way to explore this problem is evoked-potential recording from monaural structures in the auditory system. Such recording may allow the assessment of the sensitivity and latency difference between two ears at various sound-source positions. Bullock et al. (1968) were the first who made an attempt to measure IID by recording evoked responses in the dolphin's inferior colliculus, both contralateral and ipsilateral to the stimulated ear. However, the responses in that study were not purely monaural; thus IID remained undetermined.

Supin et al. (1991) and Popov and Supin (1992) explored this problem in Amazon river dolphins *Inia geoffrensis* by the use of the recording of monaural evoked potentials, the auditory nerve response (ANR). As shown above, this kind of evoked potentials can be recorded separately from the left and right sides of the head and exhibits clearly monaural properties (see Fig. 2.94), thus making it possible to measure one-ear hearing sensitivity (see Fig. 2.95). As Fig. 2.95 shows, the one-ear sensitivity is markedly higher at the ipsilateral rather than at the contralateral side.

Using these data, IID can be found by subtraction of the two branches of the function plotted in Fig. 2.95. The resulting IID is plotted in Fig. 2.96 as a function of sound-source azimuth. First of all, it is noteworthy that IID reached values as high as 20 dB. IID rose with the azimuth increase up to 10–15 deg with the most steep rise near the midline. Within the azimuth range of ±5 deg, IID increased up to 10 dB, thus the slope was 2 dB/deg;

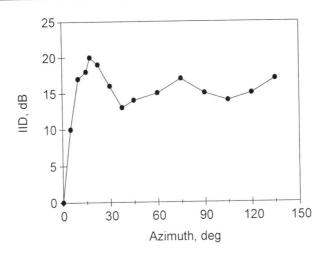

Figure 2.96. IID obtained by ANR recording in the Amazon river dolphin *Inia geoffrensis*, as a function of sound-source azimuth.

further azimuth increase from ±5 to ±10 deg resulted in more 7 dB IID increase (1.4 dB/deg). With further azimuth increase, IID fluctuated around a value of 15 dB. Thus, the data suggest that the dolphin's head does create significant shadowing to cause a large IID, with maximal IID appearing at a rather small azimuthal angles.

As to ITD, if exists, it should manifest itself in corresponding difference of latencies of the left- and right-ear responses. Independent recording of ANR of the right and left ears allows us also to measure their latencies and thus to find the interaural latency difference (ILD).

Figure 2.97 demonstrates that ILD did exist: the contralateral ANR latency was markedly shorter than the ipsilateral one. Figure 2.97(1) presents ILD as a function of the sound-source azimuth. ILD rose steeply with the azimuth increase, the ipsilateral latency being always shorter than the contralateral one. ILD reached its maximum of 200 to 250 μs at azimuth angles of 10 to 15 deg. The steepest ILD dependence on azimuth was near the midline: it was 175 μs at an angle of 5 deg – that is, 35 μs/deg; at an angle of 15 deg, ILD reached almost 250 μs. At greater angles, ILD fluctuated within a range of 175 to 275 μs.

The obtained ILD values are much larger than the expected acoustical ITD. Indeed, for a pair of receivers positioned in free field symmetrically to an axis,

$$ITD = d(\sin \alpha) / c,$$ (2.45)

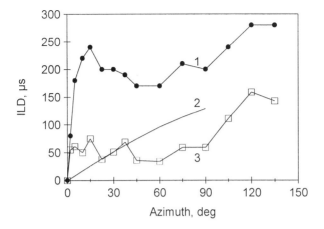

Figure 2.97. ILD obtained by ANR recording in the Amazon river dolphin *Inia geoffrensis*, as a function of sound-source azimuth. 1 – ILD in conditions of constant stimulus intensities, 2 – computed acoustic ITD, 3 – ILD in conditions of constant response amplitudes.

where d is the interreceiver distance, α is the incidence angle, and c is the sound velocity. In human psychoacoustics, a little more complicated formula is used that takes into account the diffraction of sound around the quasi-spherical head:

$$ITD = r(\sin \alpha + \alpha) / c, \qquad (2.46)$$

where $r = d/2$ is the head radius. At small α, the difference between the two equations is negligible. ITD computed according to the last equation for $c = 1530$ m/s and $d = 0.2$ m, is plotted in Fig. 2.97(2) as a function of azimuth. In particular, at $\alpha = 15$ deg, ITD is of 34 µs. It is almost an order of magnitude shorter than the experimentally found ILD.

Thus, ILD cannot be interpreted as a manifestation of ITD only. It is not surprising taking into account that the response latency to a sound stimulus consists of two parts: the acoustic delay and physiological latency. The latter includes the time of all physiological events resulting in the evoked-potential generation: peripheral transduction, nervous conduction, synaptic delays, and so on. ILD can be taken as manifesting acoustic ITD only when the physiological latency is constant. This is not the case, in particular, for the dolphin's ANR. Its latency is dependent on sound intensity (Fig. 2.98). With stimulus increase from 105 to 145 dB, the latency decreased by 350 µs with a slope close to 10 µs/dB. It was shown above that IID in dolphins is about 20 dB at a sound-source azimuth of 15 deg; this IID is expected to result in a

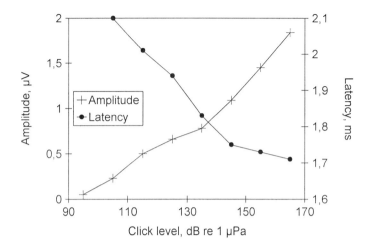

Figure 2.98. Effects of stimulus intensity on ANR amplitude (left ordinate scale) and latency (right scale) in the Amazon river dolphin *Inia geoffrensis.*

difference of physiological latencies of about 200 μs between the ipsi- and contralateral responses. This is very close to the ILD values really observed.

Therefore, it was reasonable to suppose that the found ILD dependence on azimuth reflects changes in physiological ANR latencies rather than ITD. Similarity of the form of IID-versus-azimuth and ILD-versus-azimuth functions – compare Fig. 2.96 and Fig. 2.97(1) – supports this hypothesis. To explore this suggestion, ILD was measured in conditions where stimulus intensities were adjusted in such a way as to provide responses of equal amplitudes from the ipsi- and contralateral ear. That is, the ipsilateral-ear response to a certain stimulus was compared in latency not with the contralateral-ear response to the same stimulus but with a response to the stimulus of 15 to 20 dB higher intensity that evoked a contralateral response of the same amplitude as the ipsilateral one. It was assumed that equalizing response amplitudes compensated the azimuth-dependent variation of sensitivity thus making the physiological ANR latency almost constant.

The computation has shown that at equal response amplitudes, ILD was much lower than at constant stimulus intensities (see Fig. 2.97(3)). At an azimuth of 10 to 15 deg, ILD was as low as 50 to 70 μs. These values are rather close to the computed acoustical ITD, thus confirming the suggestion that at real stimulation conditions, ILD appears mainly due to the IID and dependence of latency on intensity, not due to acoustic ITD.

So it can be suggested that IID rather than temporal disparity operates as a main cue for binaural hearing in dolphins. ILD also exists when a sound source is positioned sidewise, but it replicates more IID rather than the actual temporal acoustic disparity. This conclusion does not mean, however, that ILD is not a significant cue for binaural hearing. Recoding of IID to ILD may be a very effective way to measure IID with high precision by neurons sensitive to interaural delays. In brainstem auditory nuclei in dolphins, cell rows were found (Zook et al., 1988; Zook and DiCaprio, 1990) that run either perpendicular to the course of afferent fibers (in the ventral cochlear nucleus) or slanted at a consistent angle relative to these fibers (in the medial nucleus of the trapezoid body). The latter may function as delay-lines being a basis for measurement of latency difference.

The data on IID and ILD dependence on azimuth can be used to assess the angular selectivity of the dolphin's hearing. For this purpose, threshold values of IID and ILD should be known. These threshold estimates were obtained recently by Moore et al. (1995) in a bottlenose dolphin *Tursiops truncatus*. They used small transducers mounted in suction cups that allowed the stimulation of the right and left ears separately. With this technique, variation of the intensity and time difference between signals emitted by the two transducers made it possible to measure directly the IID and ITD thresholds. IID threshold (as estimated by a criterion of 75% performance in a two-alternative forced-choice procedure) were below 1 dB, and ITD threshold (as estimated by the same criterion) were 9 to 18 μs, depending on frequency. As shown above, in the frontal sector, IID dependence on azimuth is about 2 dB/deg and ILD dependence on azimuth is about 35 μs/deg. Thus, both 1-dB IID threshold and 18-μs ILD threshold correspond to angular resolution not worse than 0.5 deg. This estimate is several times better that those obtained in psychophysical studies in dolphins.

2.9. FREQUENCY-TEMPORAL AND FREQUENCY-SPATIAL INTERACTIONS

Until this point, analysis of temporal, frequency, and spatial stimulus features of sound signals in the dolphin's auditory system were considered separately. However, any real signal is characterized by all these features in conjunction. Interactions between temporal, frequency, and spatial stimulus characteristics in the auditory system may result in effects that are not predictable from data on analysis of each of the signal properties separately.

2.9.1. Temporal Interaction of Frequency-Colored Sound Pulses

Any natural sound is characterized by both temporal and spectrum patterns, so the temporal and frequency processing proceed in conjunction. One of models allowing the investigation of the combined frequency-temporal processing is double-click stimulation when two clicks (the conditioning and probe ones) have different frequency coloration.

A preliminary investigation of effects of paired stimuli with different frequency bands was undertaken by Klishin et al. (1991). They used stimuli as octave band-filtered clicks of various frequency bands, from 5–10 to 50–100 kHz, and recorded ABR to paired clicks with various ISI (both stimuli in a pair were of equal intensity). It was found that when the frequency bands of the two clicks in a pair were octave-different, the recovery time was about twice as shorte as when the clicks were of equal bandwidth.

Later, a more detailed study was undertaken to investigate the relationship between conditioning and probe stimuli when they differ by frequency band. For this purpose, paired stimuli as short tone pips were used. The pip envelope was always one 0.5-ms long cosine cycle. The frequency spectrum of such a signal is narrow enough: its bandwidth is 4 kHz at the –3-dB level and 7 kHz at the –15-dB level. The paired pips were presented with interstimulus intervals (ISI) from 0.5 to 10 ms, and ABR evoked by these stimuli were recorded. To extract the response to the probe stimulus from that to the double stimulus, the point-by-point subtraction procedure was used. The extracted probe response was compared with the response to the probe stimulus alone to present its amplitude in a normalized measure – that is, as a percentage of the control amplitude. To evaluate small probe-response amplitude, the match-filtering procedure was used.

Using these procedures, probe-response amplitude dependence on ISI was measured at a variety of stimulus parameters as follows: probe stimulus frequency was always 64 kHz; conditioning stimulus frequency was either equal to or above or below the probe one by 1/16, 1/8, 1/4, or 1/2 oct; probe stimulus levels were of 20, 30, 40, 50, and 60 dB above the response threshold (the threshold was about 60 dB re 1 μPa); conditioning stimulus levels were of the same values but not lower than the test stimulus level.

Recovery functions (ABR normalized amplitude as a function of ISI) obtained at various relations between intensities and frequencies of the conditioning and probe stimuli are presented in Fig. 2.99 and Fig. 2.100. In the figures, only one probe level (20 dB above threshold) and three conditioning stimulus levels (20, 40, and 60 dB – that is, equal to and 20 and 40 dB higher than the probe stimulus) are presented. To make the multiple plots better readable, they are separated into two figures: with conditioning stimulus fre-

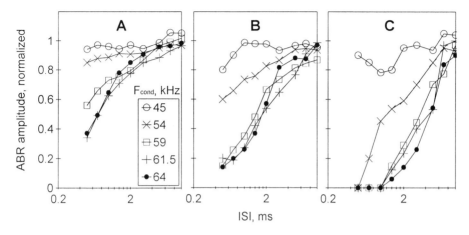

Figure 2.99. ABR recovery functions at different relations between conditioning and probe stimuli. Test stimulus: 64 kHz, 20 dB above threshold. Conditioning stimulus: from 45 to 64 kHz (as indicated in the legends), 20 dB (**A**), 40 dB (**B**), and 60 dB (**C**) above threshold.

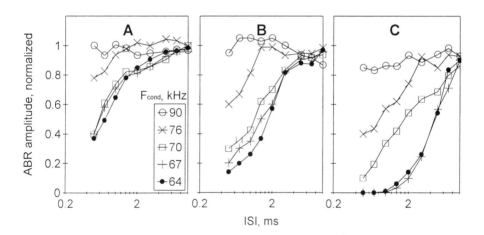

Figure 2.100. The same as Fig. 2.99, at conditioning stimulus frequencies from 64 to 90 kHz.

quencies below (Fig. 2.99) and above (Fig. 2.100) the test one. Both the figures demonstrate regularities as follows:

1. When conditioning and probe frequencies were equal or close, the probe-response suppression was rather weak and short at equal conditioning and probe intensities (Figs. 2.99A and 2.100A; both conditioning and probe intensities of 20 dB) and became deeper and longer with conditioning intensity increase above the probe one (Figs. 2.99B and C and 2.100B and C;

conditioning intensities of 40 and 60 dB, respectively, at the probe intensity of 20 dB).

2. With increasing the frequency difference between the conditioning and probe stimuli, the probe-response suppression became weaker; in particular, at a frequency difference as large as ±0.5 oct (conditioning frequencies of 45 kHz in Fig. 2.99 and 90 kHz in Fig. 2.100), the probe response was not markedly suppressed even at the shortest ISI (0.5 ms) and the highest difference between conditioning and test stimulus intensities (60 to 20 dB).

Thus, ABR suppression by a preceding (conditioning) stimulus is frequency-selective. This frequency selectivity can be characterized quantitatively by frequency-dependence curves which present the normalized response amplitude as a function of conditioning stimulus frequency at a certain ISI. Figures 2.101 and 2.102 present frequency-dependence curves at two representative ISIs – 1 and 5 ms, respectively – and at a variety of conditioning and test stimulus intensities. As expected, all the curves demonstrate that increase of the conditioning stimulus intensity resulted in

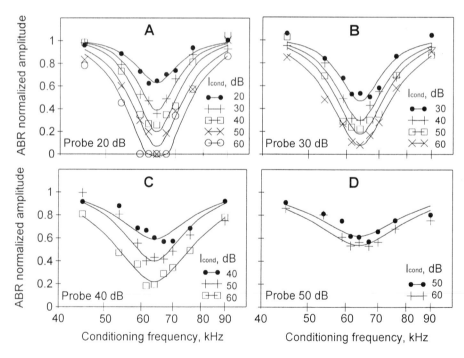

Figure 2.101. Frequency-dependence curves (normalized ABR amplitude as a function of the conditioning stimulus frequency) at ISI of 1 ms, in a bottlenose dolphin *Tursiops truncatus*. Probe stimulus: 64 kHz; from 20 to 50 dB above threshold, as indicated at the panels **A** to **D**. Conditioning stimulus levels (above threshold) are indicated in the legends. Pointing symbols – experimental data, smooth curves – approximation by *roex* functions.

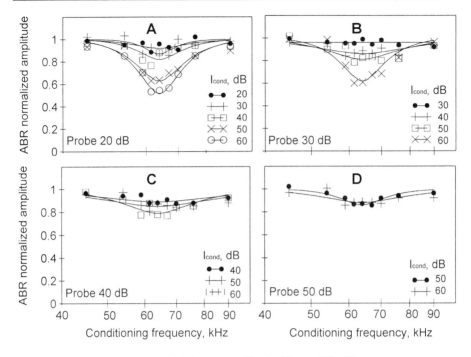

Figure 2.102. The same as Fig. 2.101 at an ISI of 5 ms.

deeper suppression of the test response. It is noteworthy, however, that this effect looks not like a shift of the curves downward as should appear if masking at all frequencies became deeper. Instead, there is an increase of the range between the curve base and tip – that is, the curves became sharper. Quantitatively, this effect was more prominent at shorter ISIs (Fig. 2.101) rather than at longer ones (Fig. 2.102), however qualitatively it was similar at all intervals.

To quantify the acuteness of the frequency-dependence curves, the data were approximated by inverted *roex* functions – that is, the function was described as:

$$A(g) = 1 - k(1 + pg)\exp(-pg),\tag{2.47}$$

where A is the normalized test response amplitude, g is the deviation of the conditioning stimulus frequency from the probe one (on the logarithmic scale), p is the parameter determining the acuteness of the curve (the higher p, the more narrow the curve), and k is the parameter determining the depth of suppression (the higher k, the deeper suppression). The equation implies by definition that suppression is maximal at equal frequencies of condition-

ing and test stimuli ($A = 1 - k$ at $g = 0$ and $A > 1 - k$ at $g > 0$) and disappears
at large differences between these frequencies ($A \to 1$ at $g \to \infty$). For each
curve, values of the parameters p and k were adjusted to minimize the mean-
square deviation of the curve from the experimental data. Idealized curves
obtained in such a way are presented in Figs. 2.101 and 2.102 for ISI of 1
and 5 ms, respectively.

Basing on the parameter p of the adjusted *roex* curves, their equivalent
rectangular bandwidth (ERB) can be found as $4/p$. Figure 2.103 presents
ERB of the frequency-dependence curves as a function of conditioning and
test stimulus intensities; the data are presented for a variety of ISI, from 0.5
to 5 ms. Although the base-to-tip range of the curves depended on ISI (compare Figs. 2.101 and 2.102), their ERB was rather similar at all intervals
(compare Figs. 2.103A–D). The general trend was that ERB was little dependent on the conditioning stimulus level (the plots are almost horizontal)
but much more dependent on the test stimulus level (the higher level, the
wider ERB). At low-level test stimuli (20 and 30 dB above threshold), ERB

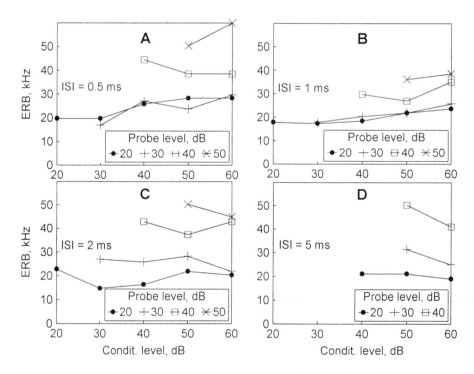

Figure 2.103. ERB of frequency-dependence curves as a function of conditioning and test
stimulus levels (the latter is indicated in the legend). Both probe and conditioning levels are
indicated in dB above threshold. The curves were obtained at ISI from 0.5 to 5 ms, as indicated at the panels **A** to **D**. Data from a bottlenose dolphin *Tursiops truncatus*.

was 18 to 24 kHz which corresponded to ±0.2 to ±0.26 oct. At higher-level test stimuli, the curves became wider, up to than ±0.5 oct at 50 dB.

The interaction between the conditioning and test stimuli can be considered as the forward masking – a phenomenon that is widely used to study frequency selectivity of hearing. However, an important difference between the frequency-dependence curves described herein (response amplitude as a function of conditioning stimulus frequency) and the standard tuning curves described above (response threshold as a function of masker frequency) should be stressed. Very rapid recovery of ABR in dolphins made it difficult to achieve deep suppression of the test response in forward-masking conditions. Therefore, the present study was not intended to obtain tuning curves that require deep (near-complete) masking of the test response, and true tuning curves were not obtained.

Nonetheless, interaction between stimuli of different frequencies is also of interest when this interaction is expressed in terms of response amplitude instead of thresholds. Being expressed in such a way, the data also demonstrate a prominent frequency selectivity: ERB of frequency-dependence curves was as narrow as ±0.2 oct at low probe levels. This ERB value is of an order of magnitude wider than that of the dolphin's tuning curves presented in terms of masking thresholds: that ERB was as narrow as 1/30 to 1/35 of the central frequency – ±0.021 to ±0.025 oct (Section 2.6.2). There is no contradiction between these data. Taking into account that several-times (that is, a few dB) change of stimulus power results in a rather small change of response amplitude, the latter yields much wider frequency-dependence curves than the former at the same degree of frequency selectivity.

From the point of view of frequency-temporal interactions, it is of importance that the frequency tuning and temporal resolution of the dolphin's auditory system provides a significant response magnitude very soon after the conditioning stimulus if the latter differs in frequency from the probe one.

2.9.2. Paradoxical Lateral Suppression

In the preceding section, frequency-dependent interaction was described in conditions of stimulation by paired sound pulses. It was also of interest to study similar interactions in conditions of rhythmically modulated sounds. An unexpected result if investigation of this problem was the finding of a paradoxical interaction between two sinusoidally amplitude-modulated (SAM) sounds of different carrier frequencies; it was found that under certain conditions, a weak sound deeply suppressed responses to much stronger sounds (Popov et al., 1997b, 1998).

In this study, interaction of two tone stimuli was investigated: the probe and masking ones. The probe stimulus was SAM tone burst, which evoked a complex response, as Fig. 2.104 exemplifies. Note that the probe stimulus (Pr record) did not have a shallow rise ramp, so the response waveform consisted of both a transient on-ABR and a subsequent quasi-sustained rhythmic response, EFR (in the presented case, the stimulus onset was 5.5 ms after the record beginning; thus, the on-response occupied a part of the time scale approximately from 7 to 11 ms).

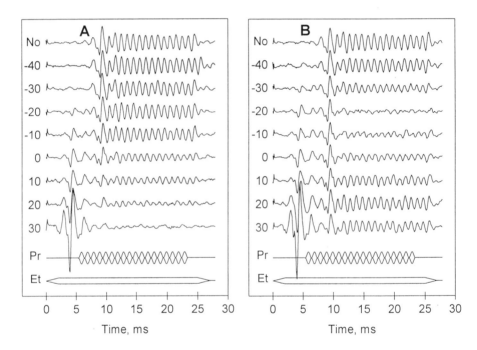

Figure 2.104. ABR and EFR suppression by extra-tones in a bottlenose dolphin *Tursiops truncatus.* Probe: SAM tone burst 76 kHz, 100 dB re 1 µPa (40 dB above response threshold), burst duration 18 ms, modulation rate 1000 Hz. Extra-tone: nonmodulated 25-ms tone burst with 2-ms rise-fall ramps. **A.** Extra-tone frequency 78 kHz. **B.** Extra-tone frequency 85 kHz. Extra-tone levels are shown near records in dB re probe level; No – no-extra-tone condition. Pr and Et – envelopes of the probe and extra-tone.

Both ABR and EFR were modified by the simultaneous presentation of an additional tone that differed in frequency from the probe. Below it is referred to as the extra-tone. The result depended, both quantitatively and qualitatively, on the frequency difference between the probe and extra-tone.

When the probe and extra-tone were little different in frequency (76 and 78 kHz, respectively, in Fig. 2.104A), low-intensity extra-tones (20 dB below the probe level and lower) influenced neither transient ABR nor sustained EFR. As the intensity of the extra-tone increased, both ABR and EFR were suppressed. Note that at high levels, the extra-tone onset itself evoked an ABR in spite of the shallow rise ramp; this response occupied the time scale from 2 to 5 ms, whereas the probe response subjected to suppression appeared after 7 ms. The observed suppression had typical features of regular masking as described above (Section 2.6.2).

A very different effect was observed when the extra-tone frequency was 5 to 20 kHz higher than that of the probe (76-kHz probe and 85-kHz extra-tone in Fig. 2.104B). In this case, suppression of the probe EFR appeared at very low extra-tone levels. A small but noticeable decrease in EFR amplitude occurred at an extra-tone level as low as 40 dB below the probe level. The extra-tone level increase resulted in a deeper EFR suppression, but the suppression appeared at still low extra-tone levels; when the extra-tone level was 20 dB below the probe, the suppression was maximal so as EFR disappeared almost completely. Further increase of the extra-tone level paradoxically resulted not in further response suppression but, instead, in releasing the probe EFR. When the extra-tone level was 20 dB above the probe, EFR amplitude recovered to almost the same value as in no-extra-tone conditions. At very high extra-tone levels (30 dB above the probe in Fig. 2.104B), EFR was slightly suppressed again. The latter suppression can be attributed to a regular masking effect that occurs at high masker levels when the probe and masker frequencies are different.

Note that contrary to the masking by the 78-kHz extra-tone (Fig. 2.104A), the low-level extra-tone of 85 kHz suppressed only the sustained EFR while the transient ABR remained unchanged (Fig. 2.104B).

Figure 2.104 exemplifies typical effects only at two extra-tone frequencies. Actually, at each of the probe frequencies, a wide variety of extra-tone frequencies and levels were tested. The result is presented in Fig. 2.105, which shows EFR amplitude as a function of both extra-tone frequency and level. There are two distinctly separated areas in the plot. The first, arbitrarily designated as area A, present suppression appearing at extra-tone intensities higher than that of the probe. This suppression was the most effective when the extra-tone frequency coincided with that of the probe. Obviously, this is a regular masking effect, and the iso-amplitude lines in this area represent the frequency-tuning curves, which reveal the frequency selectivity of the masking.

The other area of suppression (B) appeared at higher extra-tone frequencies and lower intensities than those of the probe. This suppression was the most effective at extra-tone frequencies 10 to 12 kHz above the probe fre-

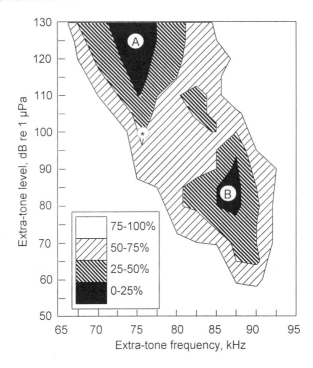

Figure 2.105. EFR amplitude dependence on extra-tone frequency and level in a bottlenose dolphin *Tursiops truncatus*. Asterisk shows the probe frequency and level (75 kHz, 100 dB re 1 μPa). EFR amplitude is shown by hatching according to the legend in % re no-extra-tone response. **A** and **B**. Areas of masking and lateral suppression, respectively.

quency and levels 15 to 20 dB below the probe level. This effect can be assessed as a kind of lateral suppression since it appeared when frequencies of the probe and extra-tone differed markedly. It is noteworthy that the lateral suppression was only observed at extra-tone frequency higher than that of the probe. No suppression was produced by low-level extra-tones of frequencies below the probe. The two areas of suppression, A (masking) and B (lateral suppression) were separated by a gap – an area of less effective suppression, which is evidence that the two suppression areas reflect two different mechanisms.

Measurements at a variety of probe frequencies have shown that the lateral suppression was deep only at rather high probe frequencies, mainly from 45 to 76 kHz. At these probe frequencies, the suppression reached 70–75% when the extra-tone was 8 to 12 kHz above the probe (Fig. 2.106). The effect decreased when the probe frequency was less than 40 kHz.

In principle, lateral suppression is a well-known effect in the auditory system (Evans, 1992). Inhibitory areas around the excitatory area of the

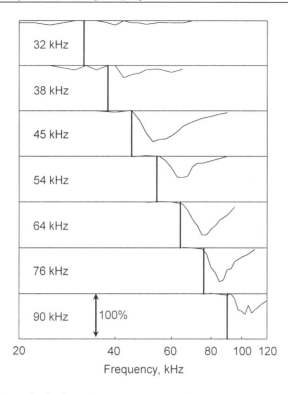

Figure 2.106. EFR amplitude dependence on extra-tone frequency at various probe frequencies in a bottlenose dolphin *Tursiops truncatus*. Probes: SAM tone bursts 40 dB above threshold, frequency as indicated. Extra-tones: nonmodulated tone bursts of varying frequency, 20 dB below the probe level. Abscissa – extra-tone frequency, ordinate – EFR normalized amplitude. Vertical solid lines indicate probe frequencies.

receptive field are a feature of many neurons in cochlear nuclei (e.g., Evans and Nelson, 1973; Young and Brownell, 1976; Voigt and Young, 1980; Young, 1985; Rhode and Greenberg, 1994) and inferior colliculus (e.g., Ehret and Merzenich, 1988). Psychophysical manifestations of lateral suppression in the auditory system have also been shown (Houtgast, 1974). However, the phenomenon revealed in dolphins differs from the known cases of lateral suppression in a few respects: (1) The suppression was observed when the extra-tone (suppresser) level was much lower than the probe level; (2) An increase in the extra-tone level over a certain value did not lead to a further increase in suppression but rather to its decrease; (3) Only the rhythmic response (EFR) was subjected to this kind of suppression whereas the transient ABR was not.

The latter property gives a key to understanding the nature of the phenomenon. It seems that extra-tones of low levels do not reduce the overall excitability in the responding auditory structures but reduce their ability to reproduce high stimulation rates. To test the validity of this explanation, the dependence of lateral suppression on probe modulation rate was studied. However, when SAM are used as probes, the tested rate range is restricted since rates lower than 300–400 Hz are not very effective for producing EFR. Therefore, another type of probe stimulus was used to investigate the lateral suppression at a variety of probe rates. It was a train of cosine-enveloped tone pips of constant duration (1 ms) and of variable rate (Fig. 2.107). Contrary to SAM sounds, the rise-fall time of the pips was independent of the modulation rate, so the efficiency of this kind of stimulus did not decrease at low presentation rates.

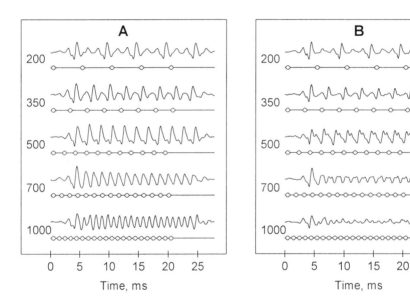

Figure 2.107. Lateral suppression at various rates of probe pips in a bottlenose dolphin *Tursiops truncatus.* Probes are trains of cosine-enveloped pips of 64 kHz carrier

Figure 2.107 demonstrates responses to pip trains at rates from 200 s^{-1} to 1000 s^{-1}. In the absence of extra-tone (A), all rates produced robust responses. The 200 s^{-1} stimulus evoked separated ABRs; at 1000 s^{-1}, the 1-ms pips fused into a SAM signal, and ABRs fused into a sustained EFR. In the presence of the extra-tone (B), the responses to low pip rates were little influenced, but suppression increased with rate, so at a rate of 1000 s^{-1} the suppression became the most effective.

These data confirm the suggestion that the paradoxical lateral suppression is the reduction of ability of evoked response to reproduce high stimulation rates. This explanation makes less paradoxical the fact that a weak sound suppresses the response to a much stronger one. Actually responses are suppressed by the preceding pulses of the probe itself while the presence or absence of extra-tone enables or disables this suppression.

The functional significance of the paradoxical lateral suppression is not clear yet. Anyway, the suppression makes the dolphin's auditory system more sensitive to pulsation of weak sound of higher frequency rather than to pulsation of stronger sounds but of lower frequency. It was demonstrated directly in experiments with interaction of two carriers of different frequencies and levels when both of them were amplitude-modulated and thus capable of producing EFR. In order to distinguish responses to one and another carrier, they were modulated by different rates: each of them was "labeled" with its own modulation rate, so that the EFR frequency showed which of them evoked the response.

Figure 2.108 exemplifies interaction of two carriers of different frequencies and levels: St1 of 76 kHz 100 dB and St2 of 85 kHz 80 dB re 1 μPa. Carrier St1 was modulated by a rate of 1000 Hz, and carrier St2 by a rate of 600 Hz. Presented separately, St1 evoked a robust EFR at a rate of 1000 Hz (Fig. 2.108A (1)), while St2, although of 20-dB lower level, also evoked a well defined EFR at a rate of 600 Hz (Fig. 2.108A (2)). The record 3 shows the result of the two carriers overlapping. During the action of St1 alone (the

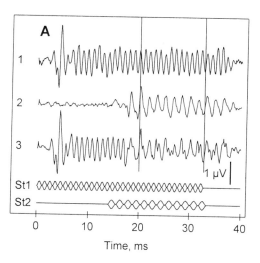
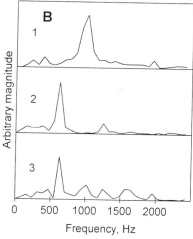

Figure 2.108. Lateral suppression between two SAM carriers in a bottlenose dolphin *Tursiops truncatus.* **A**. Original records. **B**. Fourier transforms of the temporal window delimited by vertical lines. 1 – EFR to carrier 1 of 76 kHz, 100 dB re 1 μPa, modulation rate 1000 Hz. 2 – EFR to carrier 2 of 85 kHz, 80 dB re 1 μPa, modulation rate 600 Hz. 3 – EFR to both carriers superimposed. St1 and St2 – envelopes of carrier 1 and 2, respectively (the envelope records show only their waveforms, not amplitude relations).

initial part of the record), the 1000-Hz EFR was observed. As soon as St2 was added (the later part of the record), EFR of 1000 Hz was suppressed and replaced by the 600-Hz EFR. To evaluate the effect quantitatively, the 12.8-ms long fragments of all the records were Fourier-transformed to obtain their frequency spectra. The spectra show (Fig. 2.108B) that the EFR to St1 alone had a prominent peak at 1000 Hz (1) and EFR to St2 alone had a peak at 600 Hz (2). EFR to the combination of the both carriers (3) had a prominent peak at 600 Hz, which was of almost the same magnitude as that of St2 alone whereas the peak at 1000 Hz was several times less than that to St1 alone. Thus, contribution of the weaker sound to the combined response was more substantial than that of the stronger sound.

The described experiment demonstrates directly that due to the paradoxical lateral suppression, a weak sound can dominate over a stronger one instead of being masked by that strong sound. It is not known exactly yet what an advantage the dolphin gains from this effect. Some hypotheses can be suggested as described below (Section 2.12.5).

Phenomena similar to the paradoxical lateral suppression in dolphins have not been described yet in any other animals. It is noteworthy, however, that the phenomenon was found at stimulus parameters that are not accessible to hearing of most other animals: carrier frequencies above 30–40 kHz, modulation rate of a few hundred Hz. So it remains unknown whether the paradoxical lateral suppression is absent in other mammals and humans or it exists at sound parameters accessible to their hearing.

2.9.3. Interaction of Directional and Frequency Sensitivity

It was shown above that (1) the dolphin's auditory system features a significant directional selectivity and (2) the directional sensitivity is frequency-dependent: the higher sound frequency, the more acute directional selectivity (Section 2.8). Thus, the perceived spectrum of a stimulus may differ depending on the sound-source direction thus being indicative of the direction.

Spectral cues for spatial sound localization are well known. In humans, spectral cues are the most important for assessment of sound-source elevation (Roffler and Butler, 1968a, 1968b; Butler, 1969; Blauert 1969/1970; Herbank and Wright, 1974; Musicant and Butler, 1984), although they participate also in determination of the azimuth (Belendiuk and Butler, 1977; Butler and Flannery, 1980; Flannery and Butler, 1981). The spectral cues thus could be expected to participate in spatial sound localization in dolphins as well.

This problem was explored using the recording of the auditory nerve response (ANR) in the Amazon river dolphin *Inia geoffrensis* (Supin and

Popov, 1993). In this study, tonal pips of varying frequency, from 20 to 100 kHz, were emitted through a transducer positioned at various azimuthal angles, from 135 deg to the left to 135 deg to the right of the head longitudinal axis. ANR evoked by such stimuli were recorded separately from the right and left ears.

As described above (Sections 2.8.2, 2.8.3), ANR displays clear monaural properties – that is, the responses recorded simultaneously from the left and right ears behave independently. When a sound source departs from the longitudinal axis, the ipsilaterally recorded ANR is of greater amplitude and of shorter latency than the contralateral one. Figure 2.109 demonstrates that this effect markedly depends on frequency: at higher sound frequency (85 kHz in Fig. 2.109A) the difference between the left-side and right-side records is much more than at lower frequency (20 kHz in Fig. 2.109B).

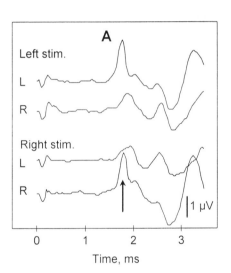
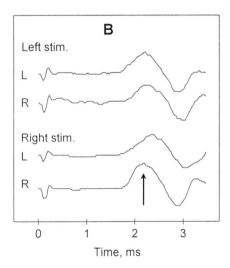

Figure 2.109. ANR recorded in an Amazon river dolphin *Inia geoffrensis* to pips of different frequencies. **A**. Pip frequency 85 kHz. **B**. Pip frequency 20 kHz. L and R – records from the left and right sides of the head, respectively. Upper pair of records (Left stim.) – the sound source is located 15 deg to the left from the longitudinal head axis; lower pair (Right stim.) – the sound source is located 15 deg to the right. Upward headed arrows indicate ANR peak in the lower trace.

This effect was used to investigate the frequency dependence of interaural relationships. For this purpose, at each sound direction, ANR to stimuli of various frequencies and intensities were recorded, and their thresholds were measured at a few criterion amplitudes. The results were plotted as sensitivity (inverse threshold) dependence on azimuth, keeping sound frequency as a parameter (Fig. 2.110). All the plots feature the highest sensitivity at azimuth of 5 to 10 deg ipsilateral to the recording side, and higher sensitivity at ipsilateral directions as compared with symmetrical contralateral ones. However, the acuteness of the functions depended on frequency. The most sharp increase in sensitivity was observed at the highest of the tested frequencies (100 kHz), and the least increase in sensitivity was at the lowest frequency (30 kHz); the intermediate frequency (55 kHz) resulted in a function of an intermediate shape.

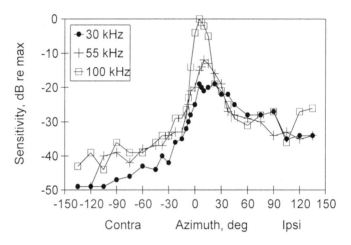

Figure 2.110. ANR sensitivity as a function of azimuth at various sound frequencies (30 to 100 kHz) in an Amazon river dolphin *Inia geoffrensis*. Sensitivity is plotted relative to the peak at the 100-kHz frequency.

Different sensitivity dependence on azimuth resulted in a difference of interaural intensity difference (IID) dependence on azimuth. The IID-vs-azimuth functions were derived by subtraction of the left from the right branch of the plots presented in Fig. 2.110. Figure 2.111 presents the IID dependence on azimuth at a few frequencies. Common features of all the functions were the steep rise at small azimuthal angles and subsequent small decay or fluctuations around a certain value. The highest frequency (100 kHz) resulted in the most steep increase of IID with its peak at a minimal azimuth. The lower frequencies (55 and 30 kHz) resulted in a less steep IID increase with their peaks at progressively greater azimuths.

From these results it follows that spectral patterns of sounds are perceived differently depending on the sound-source direction. It is demonstrated by sensitivity-versus-frequency functions (spectral sensitivity) for

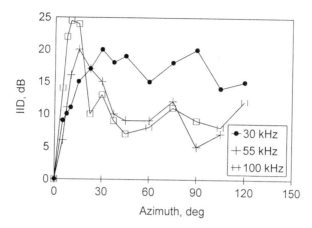

Figure 2.111. IID as a function of azimuth at different sound frequencies (30–100 kHz) in an Amazon river dolphin *Inia geoffrensis*.

different sound directions (Fig. 2.112). At zero azimuth, sensitivity to higher frequencies is much higher than to lower ones (the difference exceeds 20 dB); with the azimuth increase, the spectral sensitivity function flattens.

Different spectral sensitivity at different sound directions suggests that a perceived spectral pattern of a broadband sound depends on the sound direction. It may be an additional cue for a dolphin to localize the sound source. Of course, the perceived spectral pattern cannot indicate unambiguously the sound direction because the perceived spectrum depends also on the stimulus

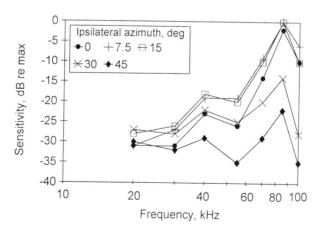

Figure 2.112. Sensitivity-vs-frequency functions (inverse thresholds) at various sound directions in an Amazon river dolphin *Inia geoffrensis*.

spectrum. However, spectral cues can provide more exact spatial information when inputs from two ears are compared: the difference between two ears does not depend on the stimulus spectrum (Grinnell and Grinnell, 1965; Batteau, 1967; Fuzessery and Pollak, 1985). This mechanism may function in dolphins as well. It was shown that not only sensitivity-versus-frequency function but also IID-versus-frequency function varied depending on the sound direction (Fig. 2.113). At zero azimuth, IID is equal to zero by definition, but at the azimuth of the highest monaural sensitivity, IID of high-frequency sounds much exceeds that of low-frequency ones (7.5 deg in Fig. 2.113). With the azimuth increase, the IID-vs-frequency function flattens (15°) and then inverts in slope (30–45 deg). This effect can be referred to as an interaural spectrum difference.

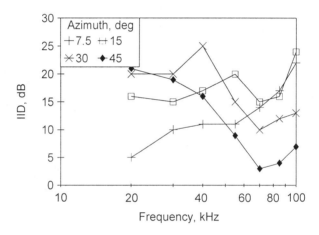

Figure 2.113. IID-versus-frequency functions at various sound directions in an Amazon river dolphin *Inia geoffrensis.*

It is quite reasonable to suggest that the mechanism based on the interaural spectrum difference may function in dolphins as a cue to determine the sound direction, in addition to the IID-based mechanism.

2.10. SOUND-CONDUCTION PATHWAYS

All hypotheses of functioning of the ear in cetaceans imply delivering the sounds to the middle ear (see Section 2.1.1). Since a classic path of sound conduction through the external auditory canal does not function in cetaceans, it remains debatable how sounds are delivered to the middle ear. Search for sound-conduction pathways in these animals was a topic of many investigations. The best known hypothesis implicates the lower jaw as such a pathway. Norris (1968, 1969, 1980) was the first who paid attention to a fatty body filling the mandibular channel. Acoustic impedance of the fat channel is close to that of sea water (Varanasi and Malins, 1971, 1972). The distal end of the fatty body contacts the outer surface of the lower jaw through a thin bony plate; its proximal end is close to the tympanic bulla. It was supposed that sounds enter the fat channel through the thin bony plate, and that this channel functions as a specific pathway conducting sounds to the middle ear which is posed just near the rear edge of the lower jaw. The region of the lower-jaw surface where sounds enter the fatty body was referred to as the *acoustic window*.

This so-called mandibular hypothesis was supported by a number of various data. First of all, it was found that intracranially recorded evoked responses (Bullock et al., 1968) and cochlear action potentials (McCormick et al., 1970, 1980) have the lowest threshold when a sound source is placed on or near the lower jaw. Later similar study but in more detail and with the use of more sophisticated stimulation technique (a specially designed transducer in a suction cup) was carried out by Møhl et al. (1999), who noninvasively recorded ABR in a beached animal to contact acoustic stimulation of various points of the head surface. They have also found the highest sensitivity at the middle of the lower jaw surface.

It should be noted, however, that stimulation performed by a sound source contacting to the head surface may yield results substantially differ from those in a free acoustic field in natural conditions. Efficiency of sound penetration into head tissues from a surface-contacting transducer may depend on tissue properties in a different manner than in free acoustic field. In particular, local application of sound pressure to a certain point may produce transversal waves contrary to longitudinal waves in homogenous media. Indeed, some results in the study of Møhl et al. (1999) were unexpected, in particular, the very large increase of the response delay with increasing the distance from the lowest-latency point (the latter was found far behind the highest-sensitivity point). These delays were much longer than predicted by sound velocity in water and head tissues. Thus, the acoustic situation in conditions of contact stimulation was not understood.

An attempt to investigate the role of the lower jaw in natural conditions was made by Brill (1988, 1991) and Brill et al. (1988). They placed a sound-shielding neoprene hood on the lower jaw of a dolphin and investigated its echo-location performance. Closed-cell neoprene markedly attenuates sounds in water; so it was expected that shielding of the lower jaw by the hood must impair the echolocation capabilities of the dolphin if it perceives

echo-sounds through the mandibular pathway. As a control, a hood con-
structed of gasless neoprene was used that did not block acoustic signals. It
was found that the hood impaired echolocation performance; this result was
considered as evidence in favor of the mandibular hypothesis.

However, interpretation of these data is problematic. First, the hood could
shield not only the lower jaw. Inevitably, it should also distort the acoustic
field in the vicinity of other regions of the head. Second, there may be no
direct connection between sound attenuation and behavioral performance;
both experimental data and everyday experience show that lowering of
sound (as well as illumination, and so on) level within a certain range may
not prevent recognition of auditory or visual images. In this respect, it should
be noted that impairment of echolocation performance in experiments of
Brill manifested itself more in increase of false-alarm percentage rather than
in decrease of fit percentage, as demonstrated by ROC-format presentation
of the data in Brill et al. (1988). The percentage of hits with the use of the
control (sound-transparent) and sound-shielding hoods was almost equal,
and the difference in action of the these two hoods was only an increase of
false-alarm percentage.

Popov and Supin (1990d) approached this problem in a different manner:
not measuring the sound-conducting effectiveness of various parts of the
head surface but determining the spatial position of the acoustic window by
measuring the acoustic delays at different position of sound sources in
evoked-potential experiments.

The general idea of the study is illustrated in Fig. 2.114. Suppose that the
positions of sound sources S_1 and S_2 are known and that the position of a
receiving point R is to be found. The acoustical delay depends on the mutual
position of the sound source and receiver. If the acoustic delays from the
sound sources S_1 and S_2 were measured and converted to the distances L_1
and L_2, then the position of the receiving point R could be computed unambiguously. However, in evoked-potential experiments, response latencies are
measured instead of acoustic delay. The latency is composed of the acoustic
delay and the physiological latency. The latter is not known in advance, so
the acoustic delay can not be evaluated. However, the delay *difference* can
be measured even if the physiological latency is unknown but constant.

Suppose the response latencies of the receiving point R to signals S_1 and
S_2 (Fig. 2.114) were measured and their difference was $\delta T_{1,2}$. Assuming that
the physiological response latency is constant, the latency difference can be
adopted as the acoustic-delay difference. The sound velocity c being known,
the difference $\delta L_{1,2}$ between distances L_1 (from S_1 to R) and L_2 (from S_2 to R)
can be found as

$$L_1 - L_2 = \delta L_{1,2} = c \cdot \delta T_{1,2}. \tag{2.48}$$

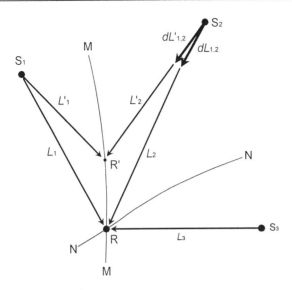

Figure 2.114. Determination of the position of a receiver by the difference between acoustic delays. S_1, S_2, S_3 – sound sources, R, R' – receivers, L_1–L_3 – distances from the sources S_1–S_3 to the receiver R, L'_1, L'_2 – the distances to the receiver R', $dL(1,2)$ – the difference between L_1 and L_2, $dL'(1,2)$ – the difference between L'_1 and L'_2.

This equation does not define an exact position of the receiver. For example, the receiving point R' result in the same distance difference $dL'_{1,2}$ as the point R. Thus, the receiving point is localized on a MM line defined by the equation. However, one coordinate of the point (the horizontal one in Fig. 2.114) can be obtained in such a way with a satisfactory approximation. The use of one more sound source (S_3 in Fig. 2.114) allows us to draw another line basing on the distance difference $dL_{2,3}$ (NN in Fig. 2.114) which the receiving-point position also belongs to. The intersection point of the MM and NN lines gives the exact position of the point R.

The nature of a receiving point found in such a way depends on organization of the sound-conducting system. If there is a punctiform receiver in free field, the measurement will indicate the position of this receiver properly. If there is a specific acoustic window through which sounds are transmitted to the receiver properly, the position of this window will be found, since the acoustic delays depend on the window position.

These measurements were performed by Popov and Supin (1990d) using ABR recording in three dolphin species: the Amazon river dolphin *Inia geofrensis*, bottlenose dolphin *Tursiops truncatus*, and tucuxi dolphin *Sotalia fluviatilis*. ABR were evoked by short wide-band clicks, and their peak latencies were measured with sampling intervals as short as 20 µs. The latencies were measured at sound-source positions at a constant 1-m distance from a voluntary chosen reference point, the melon tip, at azimuthal angles from 0 to ±165 deg from the head midline.

An example of transducer positions around the dolphin's head is presented in Fig. 2.115. The positions were at azimuth angles of 0 (S_1), 60 (S_2), and 120 deg (S_3) relative to the head midline. Responses obtained at these three transducer positions are presented in Fig. 2.116, and the peak latencies of the first positive wave are marked by cursors. When recording ABRs, it was taken into account that hearing sensitivity depends on the sound-source position, and this dependence may influence the ABR physiological latency. In order to compensate this effect, the stimulus intensity was chosen in such a way as to evoke responses of equal amplitudes at all transducer positions. Therefore, both ABR waveform and amplitude was similar in all records, but their latencies were markedly different: the longest latency was at the transducer position S_1, and the shortest latency at the position S_3. It is an indication that the "acoustic window" is at a longer distance from the position S_1 than from S_3. In this particular experiment, latency measurements resulted in $dT_{1,3} = 250$ μs ($dL_{1,3} = 375$ mm) and $dT_{2,3} = 140$ μs ($dL_{2,3} = 210$ mm). Based on these data, the computation resulted in receiver-point coordinates of $x = 196$ mm and $y = 84$ mm from the reference point (the melon tip); this is a point W in Fig. 2.115.

In each of the studied species (the Amazon, tucuxi, and bottlenose dol-

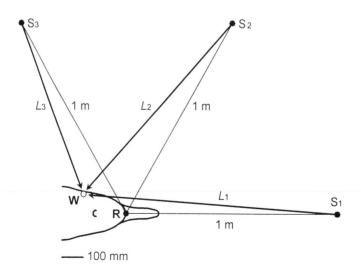

Figure 2.115. Positions of the sound sources relative to the head of an Amazon river dolphin *Inia geoffrensis* (dorsal view). R – reference point (the melon tip), S_1–S_3 – sound-source positions, L_1–L_3 – distances from the sources to an acoustic window.

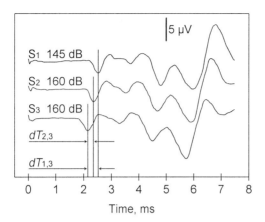

Figure 2.116. ABR recorded in an Amazon river dolphin *Inia geoffrensis* as evoked by stimulation at three transducer positions S_1, S_2, and S_3, as shown in Fig. 2.115. Vertical lines are temporal cursors indicating the first wave peak latencies; $dT_{1,3}$ and $dT_{2,3}$ – the difference between peak latencies of responses to stimuli S_1–S_3 and S_2–s_3, respectively. Transducer positions and stimulus intensities (dB re 1 µPa) are indicated near the records.

phins) measurements as described above were repeated 50 and more times. The results of all measurement are summarized in Fig. 2.117 as histograms showing distributions of estimates of the longitudinal coordinate of the receiving point in each species. The means of the longitudinal coordinate of the receiving point were found to be as follows: 190 mm for the Amazon river dolphin, 307 mm for the bottlenose dolphin, and 183 mm for the tucuxi dolphin.

To compare these results with the basic head and cranial measurements, Fig. 2.118 presents the side view of the dolphin heads. Additionally, the side view of partially dissected heads of dead Amazon and tucuxi dolphins with the intact melon and exposed lower jaw, basal, and occipital parts of the skull are shown. The points indicated by above-presented results – about 19 cm from the melon tip in the Amazon river dolphin, 31 cm in the bottlenose dolphin, and 18 cm in the tucuxi dolphin – are near the tympanic bulla (at the intact head surface this place is marked by the auditory meatus) and far behind of any part of the lower jaw.

Thus, the measurements have shown the presence of a receiving point (a kind of acoustic window) in the vicinity of the middle ear (the tympanic bulla). These results cannot be explained if a unique acoustic window is located at the lower jaw – that is, if sounds can reach the middle ear only through the lower-jaw pathway. In this case, acoustic delays would depend only on the distance from the sound source to the mandibular window (the

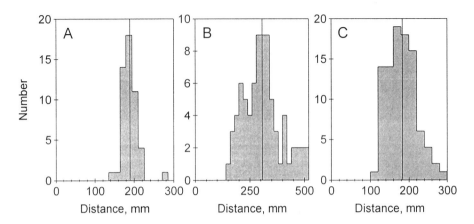

Figure 2.117. Distributions of distances between the melon tip and the acoustic window (along the longitudinal direction) in three dolphin species. **A**. The Amazon river dolphin *Inia geoffrensis*. **B**. The bottlenose dolphin *Tursiops truncatus*. **C**. The tucuxi dolphin *Sotalia fluviatilis*. Abscissa – distance; ordinate – number of measurements. Vertical lines indicate means of the distributions. Number of measurements: **A** – 50; **B** – 78; **C** – 91. Mean ± SD values: **A** – 190 ± 20 mm; **B**: 307 ± 87 mm; **C** – 183 ± 36 mm.

transmission time from the mandibular window to the bulla should be equal for all sounds), so delay-based calculation should indicate the receiving point at the window. This was not the case. The results presented above can be explained only if the direct sound transmission to the middle ear or a "window" just near the middle ear is possible.

On the other hand, the presented result do not suggest that sounds are transmitted *only* directly to the bulla. These results do not exclude a possibility of additional ways of specific sound transmission, especially from the frontal or near-frontal directions where the auditory sensitivity is maximal. Significant dependence of sensitivity on the sound direction (Sections 2.8.1–2.8.2) indicates by itself some specificity of sound delivery to the receiver.

To explore a possibility of multichannel sound transmission, Popov et al. (1992) used the delay-difference technique to investigate in more detail the sound transmission from various directions. The main idea of the study is illustrated in Fig. 2.119. When a sound source is moved around an arbitrary reference point O, the distance between the source and a receiving point R depends on azimuthal angle and on the distance *d* between the receiving and reference points (Fig. 2.119A). Accordingly, the acoustic delay varies. If the recorded responses is evoked through a single receiving point, the delay dependence on angle looks like the curve in Fig. 2.119B. The amplitude of the curve is equal to ±*d*, thus indicating the distance between the reference and receiving point, and the position of the curve peaks indicates the receiving-

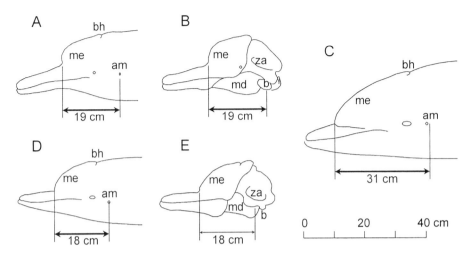

Figure 2.118. The side-view of intact and partially dissected dolphin heads. **A** and **B**. Amazon river dolphins **C**. A bottlenose dolphin. **D** and **E**. Tucuxi dolphins. **A, C,** and **D** – intact heads of experimental animals, **B** and **E** – partially dissected heads of dead animals. Double-headed arrows indicate the mean distance from the reference point (the melon tip) to the found receiving point. bh – blowhole, me – melon, am – auditory meatus, b – bulla, za – zygomatic arc, md – mandible.

point azimuth. Beyond that, fitting an experimental curve to the theoretical one indicates that the receiving system can be approximated by a single receiving point.

Measurements according to the described paradigm were performed in bottlenose dolphins *Tursiops truncatus* using click stimuli and ABR recording. As well as in the preceding experiments, at each azimuthal position of the transducer, the sound intensity was adjusted in such a way as to compensate azimuth-dependent variation of sensitivity and evoke responses of equal amplitudes. In these conditions, physiological latency of the recorded ABR was assumed to be equal, and variations of its latency were attributed to changes of the acoustic delay. A typical dependence of ABR latency on sound-source azimuthal position is presented in Fig. 2.120. For comparison, the calculated acoustic delay from the sound source to the auditory meatus-bulla region is shown. The plot demonstrates that the experimental curve did not coincide with the computed function. It did not coincide either with idealized functions computed for any other single receiving points. Within an azimuth range from 70–80 to 150 deg, the experimental curve did correspond well to the theoretical one if a receiving point was supposed to be localized near the bulla. However, within an azimuth range from 0 to 70–80 deg, the latencies were markedly shorter than predicted. The difference be-

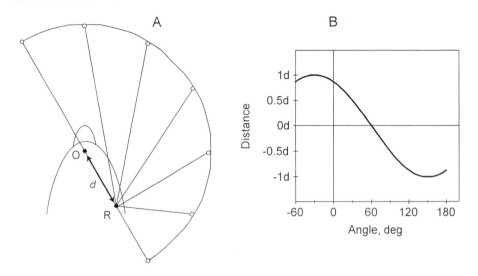

Figure 2.119. Dependence of acoustic delay on position of a sound source relative to a receiver. **A**. Positions of a sound source on an arc centered on a reference point O; R – receiver, *d* – distance from the reference point to the receiver. **B**. A typical dependence of distance between the sound source and receiver on a sound-source azimuthal position.

tween the experimental and theoretical latencies exceeded 100 μs, which corresponded to a distance more than 15 cm in water.

Markedly shorter response latency at frontal sound-source positions can be explained in two ways: either by the presence of a rapid sound-conduction pathway from the frontal part of the head to the bulla or by a change of the physiological latency in spite of equalized response amplitudes. The first explanation seems not very probable. In the bottlenose dolphin, the distance from the mandibular "acoustic window" to the bulla is less than 25 cm (see the head dimensions in Fig. 2.118). In water, sound transmission to such distance takes around 0.15 ms. Thus, to save 0.1 ms at this way, the sound velocity should be around three times higher than in water. There is no indication of such velocity of sound transmission in any of body tissues.

The second possibility seems more realistic, particularly, due to changes of perceived stimulus spectrum depending on the sound-source position. As shown above (Section 2.9.3), at frontal sound-source positions, the sensitivity increase is larger for higher rather than for lower frequencies. With the use of a wide-band stimulus, this effect may lead to a modification of the perceived stimulus spectrum: the proportion of higher frequencies increase when the sound source is positioned frontally. In turn, the ABR latency depends to some extent on the stimulus spectrum: the latency of responses to

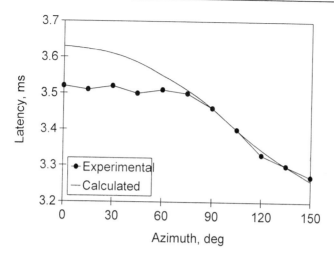

Figure 2.120. ABR latency dependence on the sound-source azimuthal position in a bottle-nose dolphin *Tursiops truncatus* (experimental data). The calculated curve is based on the acoustic delay in free field for a single sound-receiving point near the bulla and auditory meatus. Zero azimuth corresponds to the frontal direction.

high-frequency stimuli is shorter than that of responses to low-frequency stimuli. As a result, the response latency becomes shorter at frontal sound-source positions.

Summarizing all the data presented above, it seems that sounds from all directions can reach the bulla region in a way other than through the mandibular window, probably by a short way through the head tissues near the bulla. With this respect, it is worthwhile to note that magnetic resonance imaging recently revealed a second laterally positioned fat body with a density equal to that of the mandibular fat body (Ketten, 1994).

On the other hand, sound-conducting tissues in the frontal part of the head (maybe, the mandibular pathway) seem to conduct sounds, particularly high-frequency ones, with less attenuation than other head tissues. This results in a higher hearing sensitivity to frontally located sound sources. Thus, a multichannel sound-conduction system was suggested (Popov et al., 1992). A similar idea was expressed that the anterior (mandibular) channel may be specialized for capturing high-frequency echo signals, while the lateral channel may capture lower-frequency communication signals (Ketten, 1997).

2.11. CENTRAL REPRESENTATION OF THE AUDITORY SYSTEM

Anatomical studies of the cetacean brain revealed the same general structure of auditory pathways as in terrestrial mammals – that is, the presence of the dorsal and ventral cochlear nuclei, the trapezoid body, the nucleus of lateral lemnisc, the inferior colliculus, and the medial geniculate body. A distinctive feature of the auditory nerve centers in odontocetes is their hypertrophy. Relative volume of the bulbar auditory nuclei, the inferior colliculus, and medial geniculate body is several times greater than volume of analogous nuclei in many other mammals (Kruger, 1959; Zvorykin, 1963; Pilleri, 1964; Osen and Jansen, 1965; Pilleri and Gihr, 1968). For example, in the common dolphin, the superior olive in dolphins is of almost 150 times higher volume than in humans, the nucleus of lateral lemnisc is of about 200 times higher volume, and so on (Zvorykin, 1963).

None of morphological studies succeeded to identify the auditory projection area in the cerebral cortex. Cytoarchitecture of various regions of the cerebral cortex in cetaceans was investigated by a number of authors (Riese, 1925; Rose, 1926; Langworthy, 1931; Grünthal, 1942; Zvorykin, 1963; Kesarev, 1969, 1970; Kesarev et al., 1977; Garey and Revishchin, 1990). In spite of large diversity in descriptions of cytoarchitectonic patterns, all the authors pointed out that the cerebral cortex of cetaceans was agranular: it did not contain a significant amount of granular cells in the layer IV. In most other mammals, the well-developed granular structure of the layer IV is a characteristic feature of projection sensory areas and is used as a cue for identification of these areas. Absence of this feature in the cetacean cerebral cortex deprived investigators of the main cytoarchitectonic cue. As to experimental morphological investigations that make it possible to identify directly a projection cortical area (such as labeling by axonal transport), they are limited in cetaceans because of technical difficulties and ethic restrictions.

Therefore, localization of the auditory projection area in the dolphin cerebral cortex was not determined until a series of evoked-potential studies was performed in 70-th (Ladygina and Supin, 1970, 1974; Sokolov et al., 1972). In these studies, EP to sound stimuli (clicks) were recorded in harbor porpoises *Phocoena phocoena* and bottlenose dolphins *Tursiops truncatus* through electrodes implanted into the cerebral cortex. Numerous electrode locations covered the lateral, suprasylvian, ectosylvian, post- and precruciate gyri (Fig. 2.121). EP to sound stimuli were detected only in the suprasylvian gyrus. The area of EP recording covered a major part of this gyrus, from the coronar sulcus at its rostral end to the caudal end of the suprasylvian sulcus.

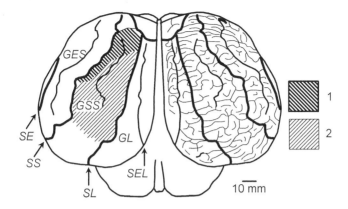

Figure 2.121. Localization of the auditory projection area in the cerebral cortex of the bottle-nose dolphin *Tursiops truncatus.* Dorsal view of the dolphin brain. On the right hemisphere, the pattern of sulci and gyri is replicated. On the left hemisphere, only first-order sulci and gyri are shown: *SE – sulcus ectosylvius, SS – s. suprasylvius, SL – s. lateralis, SEL – s. endolateralis, GES – gyrus ectosylvius, GSS – g. suprasylvius, GL – g.* lateralis. 1 – area of short-latency EP (type I in Fig. 2.122); 2 – area of long-latency EP (type II in Fig. 2.122).

No responses to sound stimuli were recorded in any cortical regions outside this area. Thus, the suprasylvian gyrus can be regarded as the cortical auditory projection area.

Within this area, subdivisions were found where different EP types were recorded (Ladygina et al., 1978). These EP types were arbitrarily designated as type I and type II. Type I responses were recorded in the frontal and lateral parts of the auditory projection area (field 1 in Fig. 2.121). These EP were of rather short onset latency (7–8 ms) and short duration (around 30 ms, Fig. 2.122A). Depending on the electrode position, peak latencies of the positive and negative waves varied within the 30-ms response duration (records 1–3 in Fig. 2.122A), but the onset latency and overall duration remained almost constant, indicating that all these variations belong to one and the same response type.

Type II responses were recorded in the major remaining part of the auditory projection area (field 2 in Fig. 2.121). These EP were of somewhat longer onset latency (13–16 ms), and their duration was also longer (45–50 ms) (Fig. 2.122B). Similarly to type-I responses, peak latencies of positive and negative waves of type-II responses varied depending on the electrode position, but the onset latency and overall duration were almost constant for all responses of this type.

It can be supposed that the area producing type-I responses is a primary projection area receiving afferent signals from the previous level of the auditory system (the medial geniculate body) by a short, direct way and through

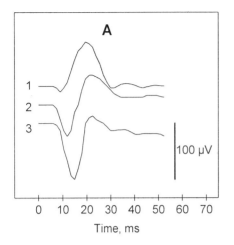
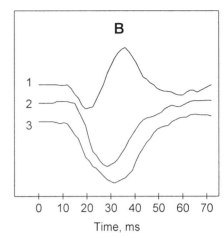

Figure 2.122. EP types recorded in different parts of the auditory area of the cerebral cortex in a bottlenose dolphin *Tursiops truncatus*. **A**. EP of shorter latency and duration (type I). **B**. EP of longer latency and duration (type II). 1–3 – variations of EP waveform depending on the electrode position relative to the cortex layers.

pathways with high conduction velocity. The area generating type-II responses receives afferent signals through longer or slower pathways. Longer response latency in this area indicates either lower conduction velocity in afferent pathways or the presence of additional synaptic delays.

The position of the auditory projection area in the cerebral cortex was partially confirmed by experiments with retrograde axonal transport in harbor porpoises *Phocoena phocoena* (Garey and Revishchin, 1990). Tracer injections into the suprasylvian gyrus resulted in labeling of neurons in the medial geniculate body (MGB), the thalamic relay station of the auditory system. Apart of that, labeling of MGB neurons was observed after tracer injections into the ectosylvian gyrus; it remains uncertain yet whether the latter result was a sequence of tracer diffusion or revealed some auditory projections that did not manifest themselves in evoked-potential experiments.

No attempts were made to investigate in more detail the tonotopic organization of the cortical auditory projection. Such investigation would require numerous EP recordings from a very large number of cortical points to find the characteristic frequency of each point, basing on its responses to tonal stimuli. The number of investigated cortical points should be several times larger than in the study described above. Implantation of an extremely large number of electrodes would be harmful for the animal and contradicted to ethic restrictions.

The position of the auditory projection area in the dolphin cerebral cortex markedly differs form that in most of other investigated mammals. Apart from neurological data on cortical areas in humans, localization of sensory projection areas of the cerebral cortex, particularly, the auditory projection area, was experimentally studied in a number of mammal orders: monotremates (Lende, 1964; Allison and Goff, 1972), marsupials (Lende, 1963), insectivores (Lende and Sadler, 1967), rodents (Krieg, 1946; LeMessurier, 1948; Lende and Woolsey, 1956), lagomorphs (Woolsey, 1947), carnivores (Rose and Woolsey, 1949), and primates (Diamond et al., 1970). According to all these data, the auditory projection area, including all its subdivisions, is located in the temporal lobe of the cerebral cortex. In carnivores, it corresponds mainly to the ectosylvian gurus. In dolphins, no signs of auditory responses were found in the ectosylvian gyrus, whereas the suprasylvian gyrus exhibited obvious auditory responses. Thus, the auditory projection area in dolphins (maybe, in all cetaceans) is markedly shifted to the medial direction as compared to other mammals.

Apart of the unusual position, a remarkable feature of the auditory cortical area in dolphins is its large size; it occupies a major part of the dorsal surface of the cerebral hemisphere. This feature correlates with all other data, indicating the high development of the auditory system in odontocetes.

2.12. IMPLEMENTS TO ECHOLOCATION

Echolocation is a remarkable peculiarity of the odontocete's auditory system. Many of the specific features of the odontocete's auditory system described above may serve specific demands of echolocation. Below we consider briefly the possible connections of some of these features with the performance of the sonar.

2.12.1. Hearing Frequency Range

Association of the extremely wide frequency range of odontocetes with functioning of their sonar system is almost self-evident. There are several reasons that the biosonar of odontocetes evolved in such a way as to use very high sound frequencies. The first one is that the spatial resolution of a sonar depends directly on the wavelength that is used. Objects reflect sounds (as well as any other wave energy) effectively only if their size is more than the wavelength. Therefore, high spatial resolution requires the use of short wavelengths. Short wavelengths correspond to high sound frequencies, particularly at high sound velocity in water. For example, frequencies of 1 to 2 kHz (the region of the best hearing in humans) at the sound velocity of 1530 m/s correspond to wavelengths of 1.53 to 0.76 m. A sonar with such spatial resolution would be hardly usable for dolphins. A frequency of 20 kHz (which is close to the upper limit of hearing of humans and many other mammals) corresponds to the wavelength of 7.6 cm, which does not provide very high resolution either. Dolphins use locating pulses with spectra peaking mostly around 100 kHz (Au, 1980, 1993; Kamminga, 1988). This frequency range provides a wavelength and spatial resolution of around 1.5 cm, which allows a dolphin to discriminate rather small details of objects.

Another reason to use high sound frequencies for echolocation is the necessity to detect the echo shortly after emitting the locating pulse. To avoid overlapping the emitted and echo signals, they should be markedly shorter than the echo delay. Because of high sound velocity in water, this delay may be very short at a short distance to the ensonified target. For example, at a distance of 1 m, the two-way delay is as short as 1.3 ms. The possibility of echolocation at distances shorter than 1 m, which results in echo delays shorter than 1 ms, must not be ruled out. For the emitted and echo signals separated so shortly to discriminate, they must be no longer than hundreds microseconds, preferably tens of microseconds. Indeed, echolocation pulses emitted by dolphins are just of the same order of duration: wide-band locating clicks of the bottlenose dolphin and many other dolphin species last tens microseconds (Au, 1980, 1993), and narrow-band signals used by a few species, such as the harbor porpoise (Kamminga and Wiersma, 1981; Kamminga, 1988), last around a hundred microseconds . A pulse lasting a few tens of microseconds and containing a few wave cycles as well as a narrow-band pulse lasting around a hundred microseconds and containing around 10 cycles have a frequency spectrum as wide as tens to hundreds kHz.

The ability of the dolphin's auditory system to perceive sound frequencies more than 100 kHz fits the necessity to perceive echo signals just within this frequency range.

2.12.2. Frequency Tuning and Temporal Resolution

The very acute frequency tuning and extremely high temporal resolution of the dolphin's hearing are features as remarkable as the wide frequency range. No doubt, these features are very helpful for echolocation orientation. Acute frequency tuning allows dolphins to discriminate very fine details of frequency spectra of echo signals that depict sound "portraits" of objects, thus making possible the precise discrimination of objects by echolocation. High temporal resolution is necessary to discriminate echo from preceding locating pulses when the echo delay is as short as milliseconds or fractions of a millisecond.

One question is how the dolphin's auditory system has achieved such frequency and temporal resolution. Although direct microelectrode recordings from the dolphin's auditory nerve and brainstem nuclei have never been performed, there is no reason to suppose that their nerve spikes, synaptic potentials, and so forth, differ markedly from that in other mammals. Thus, one and the same neuronal basis provides several times better frequency tuning and many times better temporal resolution in dolphins than in many other mammals.

The model described above suggests a mode of temporal processing that allows the temporal resolution to be as high as it was really observed in dolphins. This model is based on a nonlinear transform of an input signal into the neuronal response, without involving any inhibitory process to separate successive acoustical events. The absence of inhibitory separation provides the temporal resolution comparable with duration of a single nerve spike (hundreds microseconds), and this is possible since the nonlinear transform itself is sufficient for providing temporal separation of signals.

However, another question may arise: if high temporal resolution is available using the simplest neuronal mechanisms, why is the temporal resolution as high as it is in dolphins but absent in other mammals? It seems that the combination of acute frequency tuning and high temporal resolution in dolphins is associated with the wide frequency range of their hearing. Indeed, both frequency tuning and temporal resolution in the auditory system are limited by passbands of peripheral auditory filters. Acute frequency tuning requires narrow passbands, whereas high temporal resolution requires wide passbands. Therefore, increasing the temporal resolution is possible only at the cost of decreased frequency tuning, and vice versa.

This contradiction, however, becomes less dramatic at high frequencies because the filter center frequency F, passband B, and tuning Q are related as: $Q = F/B$. Thus, at high frequency F, auditory filters can combine high tuning Q with wide passband B, which renders transfer of rapid sound modulations. For example, at a frequency of 100 kHz, the Q_{ER} as high as 30 corresponds to the bandwidth as wide as 3.3 kHz (± 1.65 kHz). A filter of such passband is capable to transfer amplitude modulations of the same rate – that is, 1.65 kHz, which is a very high temporal resolution. If a 1-kHz centered filter had the same frequency tuning, its temporal resolution limit would be as low as 16.5 Hz. Thus, at high frequencies, acute frequency tuning can be combined with high temporal resolution. This possibility is realized in the dolphin's auditory system.

2.12.3. Recovery Functions as a Basis of Invariant Perception of Echo Signals

Temporal-resolution mechanisms in the dolphin's auditory system not only allow dolphin to discriminate an echo signal after an emitted locating pulse but also may be a basis for echo perception invariantly of variation of some parameters. At first, a feature of double-click-evoked ABR attracts attention – namely, that the probe-response amplitude remains unchanged when intensities of both conditioning and test clicks change proportionally (Section 2.5.3). This constancy of a probe-response amplitude arises since an increase (decrease) of the probe response because of increase (decrease) of the probe stimulus level is almost exactly compensated by the increased (decreased) response suppression by a proportionally changed conditioning stimulus.

Suppose that two stimuli, the conditioning and test clicks, are the emitted and echo signals during echolocation. Indeed, performing echolocation, a dolphin probably hears its own locating signal and, after a short interval, an echo signal. Dolphins are able to change the intensity of emitted locating clicks within a wide range, as dictated by the distance to a target, target strength, and other conditions of echolocation. When they do so, the echo intensity changes proportionally to the emitted signal intensity. If the echo delay is within a range when the regularity pointed above takes place, the response to the echo signals may be of constant magnitude irrespective of the emitted-click intensity.

Another feature of the response recovery functions concerns the found relations between conditioning and test click levels and ABR recovery time. It is not known yet how loudly a dolphin hears its own locating signal, but it seems fairly probable that the intensive locating pulse is heard more loudly than the weak echo from a distant target. This situation is similar to that in experiments with paired clicks when the conditioning stimulus is of higher intensity than the probe one. During echolocation, both the delay between the emitted and echo signal and the echo intensity are dependent on the distance to the target. As to the delay, its dependence on distance is directly proportional. As to the echo-signal intensity, it can be assessed as follows. The intensity of a locating signal when it reaches a target may be assumed as determined by the law specifying spreading of spherical waves – that is, its sound pressure is proportional to $1/r$, where r is the distance to a target. It can be also taken that spreading of the echo signal from the target toward the animal is determined by the same law. Then the sound pressure of the echo signal returning to the animal will be proportional to $1/r^2$. This means that a 10-fold increase (decrease) in the distance results in the 10-fold increase (decrease) of the echo delay and 40-dB decrease (increase) of the echo-sound level. It is remarkable that the experiments with paired clicks showed that just the same relation between the two variables (40-dB intensity difference per 10-fold delay change) provides a constant value of the response to the test click. Thus, it may be concluded that within a limit of keeping this relation true, responses of the dolphin's auditory system to the echo signal have constant magnitude irrespective of the distance to the target. It is because attenuation of the echo signal with the distance increase is compensated completely by recovery of the auditory system reactivity at longer delay after the locating signal.

So the peculiarities of response recovery after a preceding stimulus may maintain a constant echo-response magnitude irrespective of both the distance to a target and the emitted-click intensity. Of course, at the moment, it is not more than a hypothesis. However, the character of reactivity recovery that perfectly compensates the echo attenuation with distance hardly can be an incidental coincidence. Quite probably, it reflects an adaptation of the dolphin's auditory system to the echolocation function.

2.12.4. Rippled Spectrum Resolution and Echolocation

Normark (1961), Johnson and Titlebaum (1976), and Johnson (1980) hypothesized that discrimination of rippled spectra plays an important role in determination of the distance to a target. If a dolphins hears both its own emitted click and the echo, then the spectrum of this combined signal has a rippled pattern with ripple spacing equal to $1/\tau$, where τ is the echo delay. The ripple structure of the spectrum may be a cue to determine the distance.

However, measurements of frequency resolving power (FRP) in dolphins did not support this hypothesis. The highest FRP estimates of around 50 at high frequencies (around 100 kHz) correspond to the ripple frequency spacing of around 2 kHz, that is the delay of 500 µs. In experiments of Au and Pawloski (1989), wide-band rippled-noise stimuli of constant ripple spacing better simulated the situation when rippled spectrum pattern appears because of summation of two coherent signals; these experiments also demonstrated ripple-density resolution limit at delays not longer than 500 µs. On the other hand, to be analyzed as a signal with common spectrum, the emitted pulse and echo should merge into a single image, which probably is possible within the integration time of the auditory system; all measurements of the temporal integration in dolphins also indicated the integration time less than 500 µs. Thus, all the available data are indicative that the use of rippled spectrum analysis is hardly possible at echo delays longer than 500 µs. This corresponds to two-way length of 0.75 m – that is, the distance to a target should be shorter than 0.375 m. Obviously, the distance range of 20 to 30 cm is not the most usable for the dolphin's sonar.

It seems much more probable that analysis of rippled spectra is used by dolphins to discern various within-echo target attributes (Moore et al., 1984). Indeed, as a rule, echo arises with multiple sound reflections within a target; delays between these reflections may be of an order of tens to hundreds microseconds (Au, 1993). Multiple reflections with different delays result in a rippled pattern of the spectrum of the complex echo with ripple spacing of an order of tens of kHz. This is just the range of ripple spacing that can be discriminated by the auditory system of dolphins. The complex rippled pattern of the spectrum can be a valuable cue for discrimination of different targets by their echo signals.

2.12.5. Frequency-Temporal Interactions

Experiments with double-pulse stimulation when two pulses (pips) were of different frequencies can be considered as simulating some events during echolocation. Indeed, the emitted signal and echo do not coincide in their frequency spectra. There are several causes of their spectral differences. First, when a dolphin hears both the emitted click and echo, they probably reach the middle ear through different pathways, which can have different frequency responses. Second, the echo spectrum becomes different from that of the ensonifying pulse due to frequency characteristics of the target. Third, the frequency sensitivity of the dolphin's hearing depends on the sound-source (target) position relative to the head (near the longitudinal axis, the hearing is more sensitive to higher frequencies as compared to lower ones; with the azimuth increase, this difference diminishes). All these reasons suggest that if a dolphin hears both its own emitted click and the echo, it hears them as differently frequency-colored. Therefore, experiments with double pulses differing in frequency seem to be a simulation of this process.

As shown above (Section 2.9.1), masking of a stimulus by a preceding one diminished markedly when the two pulses differed in frequency. If the difference was as large as half-octave, the masking was negligible at any ISI even if the conditioning pulse was of 40 dB higher intensity than the probe one. Thus, it can be expected the response to an echo releases from suppression by the emitted pulse if spectral compositions of the emitted pulse and the echo differ markedly; this releasing may exist even if the emitted pulse is heard much more loudly than the echo.

Does this suggestion contradict another one – namely, that recovery of echo perception after suppression by the emitted pulse plays an important role in maintenance of invariant perception? It probably does not. Even if the echo spectrum markedly differs from that of the emitted pulse, this difference never can be complete. Assuming that the sound propagation and reflection are linear, the echo can contain only those frequencies that exist in the emitted pulse; only the ratio of spectral components can be changed. Therefore, the spectra of the emitted pulse and the echo inevitably overlap though do not coincide completely; they contain both coinciding and different components. Thus, both regularities found for double pulses of equal spectral composition (Section 2.5.3) and those found for double pulses of different spectral composition (Section 2.9.1) have to manifest themselves in interactions between the emitted and echo signals during echolocation.

Another kind of frequency-temporal interaction between the emitted and echo signals can appear due to the paradoxical lateral suppression described in Section 2.9.2. Indeed, due to increased hearing sensitivity in the frontal direction, the weight of high frequencies in perceived echo may be higher than in the emitted pulse. On the other hand, the echo intensity is probably much lower than that of the emitted pulse as it heard by the dolphin. Lower intensity and higher frequency of the echo – these are just the conditions for paradoxical lateral suppression of the response to the emitted signal of higher intensity and lower frequency. It may be an effective mechanism to release the response to the echo signal from masking by a more intensive emitted signal. The same mechanism may result in releasing responses to echo signals from masking by sounds of other origins (external noise, reverberation, and so on) if the perceived echo is of lower intensity and higher frequency than the masking sound.

It should be noted, however, that this kind of lateral suppression influences only the ability of the auditory system to reproduce high stimulation rates. Thus, it can be expected to act only in conditions of a very high rate of location pulses.

2.12.6. Spatial Resolution

Steep dependence of IID and ILD on the sound-source azimuthal position may provide very acute localization of a located target. Both the transmission and high-frequency receiving beams of the dolphin's sonar are about 10 deg at the −3 dB level and around 20 deg at the −10 dB level (Au, 1993). However, binaural hearing of dolphins may provide much more precise location of sound-source position. As shown above, comparison of IID-versus-azimuth and ILD-versus-azimuth functions with IID and ILD thresholds suggests a possibility to localize a sound-source position with an accuracy of about 0.5 deg. Interaural spectral difference may be an additional cue to localize the azimuthal target position.

2.13. SUMMARY

Hearing of cetaceans is deeply adapted to underwater conditions. The outer ear has lost its sound-conducting functions (the pinna is absent, the auditory canal is closed), and the tympanic membrane does not function as a sound-pressure-sensitive device. The ear ossicles are present but organized in a manner different from ossicles in terrestrial mammals. The malleus is connected not to the tympanic membrane but to a thin wall of the middle-ear cavity (the tympanic plate), which is supposed to substitute the tympanic membrane as a sound-pressure-sensitive device matching the high-impedance aquatic medium. From the malleus-incus union, vibrations are transferred to the inner ear in a conventional way, through the stapes and oval window. In the inner ear of odontocetes, the basilar membrane is very narrow and thick at the base and wide and thin at the apex; this yields high-frequency tuning of the proximal part of the membrane. The brain auditory centers are extremely developed.

A variety of EP types can be recorded in odontocetes both intracranially and non-invasively from the head surface. Intracranially recorded EPs were obtained from various levels of the auditory system, from the brain stem to the cerebral cortex. Among noninvasively recorded EP, the most prominent one is the auditory brainstem response, ABR. Due to the high development of auditory brain centers, ABR amplitude in dolphins is several times greater than that in other mammals and humans in similar recording conditions. Another type of noninvasively recorded EP is the auditory cortical response, ACR. Specific versions of EP are their rhythmic sequences obtained either at sinusoidally amplitude-modulated (SAM) sounds (the envelope-following response, EFR) or rhythmic click sequences (the rate-following response, RFR). In dolphins, they are mostly composed of rhythmic sequences of ABR, which can follow stimulation rates higher than 1000 Hz.

The contribution of different frequency bands to ABR generation is unequal; the contribution of high frequencies is larger than that of low frequencies. Nevertheless, the contribution of all frequency bands is significant; thus, ABR can be used to study hearing abilities and mechanisms in the whole frequency range of hearing.

EP, particularly ABR, can be used for both above-threshold and threshold measurements. The latter task requires evaluation of low, near-zero, response amplitudes. ABR amplitude in dolphins could be measured with a satisfactory precision even at near-threshold stimuli. For precise measurements of low-amplitude responses, a more sophisticated method was used – namely, the match filtering using the standard ABR waveform as a filter. Amplitude of rhythmic responses (EFR and RFR) could be measured precisely by their

Fourier transform and evaluation of the fundamental magnitude of the frequency spectrum. Response (ABR, EFR, or RFR) thresholds were estimated either using a certain criterion amplitude or by extrapolation to zero a regression line approximating response-amplitude dependence on stimulus intensity.

Using the methods of EP recording and evaluation, a number of hearing characteristics were studied in dolphins. First of all, the hearing sensitivity and frequency range were measured. Psychophysical studies on a number of odontocete species have shown than all of these species feature a wide frequency range (more than 100 kHz) and high hearing sensitivity (40–60 dB re 1 µPa) at frequencies of a few tens of kHz. Similar results were obtained in a variety of odontocete species with the use of EP methods. The use of prolonged SAM stimuli evoking EFR had some advantage since it allowed EP thresholds to specified in terms of sound pressure, similarly to psychophysical studies. The efficiency of short tone pips evoking ABR depends on their both sound pressure and duration.

Temporal resolution of the dolphin's hearing was investigated in a few psychophysical studies that yielded integration time estimates from hundreds milliseconds in temporal-summation experiments to hundreds microseconds in forward- and backward-masking tests. Evoked-potential methods allowed several experimental paradigms to be applied to explore this problem. Temporal-summation experiments have shown that ABR amplitude and threshold depend on the stimulus duration within a range not longer than 0.5 ms. The double-click test has shown that at equal intensities of the two clicks in a pair, the second (probe) ABR appears at interstimulus intervals (ISI) as short as 200 to 300 µs and its complete recovery takes from 2 ms to more than 10 ms, depending on stimulus intensity. If the first (conditioning) response exceeds the probe one in intensity, the recovery time enlarges. Gap-in-noise-detection experiments have revealed ABR to gaps as short as 100 µs. From these data, the temporal transfer function of the auditory system was derived using a model based on a nonlinear (quasi-logarithmic) transform of signals. The derived temporal integrating transfer function, as presented in the sound-intensity domain, features a rapid initial decay (around 300 µs to a –3-dB level) and a subsequent longer decay with a slope of 35 to 40 dB per decade. Equivalent rectangular duration of this function is around 300 µs. This model is satisfactory to explain a wide variety of experimental data. Rhythmic SAM stimulation and EFR recording allowed to assess the temporal resolution in terms of the modulation transfer function (MTF). MTF in dolphins looked like a low-pass filter with a cut-off frequency of around 1700 Hz at a level of 0.1. A rhythmic-click test yielded the same cut-off frequency when a confounding effect of long-term rate-dependent adaptation was eliminated. This cut-off frequency corresponds well to the integration time of around 300 µs.

Frequency tuning of hearing was studied psychophysically using the critical ratio, critical band, and tone-tone masking paradigms. However, the results were uncertain: critical-ratio and critical-band data displayed a large scatter, whereas tuning curves obtained by tone-tone masking were distorted by beats between the probe and masker. The tone-tone masking paradigm in conjunction with ABR or EFR recording yielded tuning curves free of that distortion. EFR had some advantage as a probe because of strictly limited spectral width of SAM stimuli. The obtained tuning curves featured very sharp tuning with Q_{ER} (quality estimated by the equivalent rectangular bandwidth) up to 35 and Q_{10} (quality estimated by the curve width at a 10-dB level) up to 18 at high frequencies. Measurements with the use of notched noise as a masker gave similar results (Q_{ER} of around 35). This is several times better acuteness that in humans. Correspondingly, frequency discrimination limens are low in dolphins: psychophysical studies gave thresholds down to 0.2%; evoked-potential studies, down to 0.1%. Psychophysical measurements of the frequency resolving power (FRP, the ability do discriminate rippled spectrum structure) showed the ripple-density resolution limits up to 50 relative units, which is several times better than in humans.

Data on sound-intensity discrimination in cetacean are limited. Both psychophysical and evoked-potential data indicated thresholds of an order of a few percent (less than 0.5 dB).

Directional sensitivity of hearing in odontocetes was studied psychophysically as masked threshold dependence on the angular distance between the probe and masker sound sources. The beam-width estimates varied markedly. The minimum audible angle (the angular difference between distinguishable sound sources) was estimated as more than 3 deg. EP recording allowed the spatial sensitivity to be measured directly as threshold dependence on azimuth. The beam width obtained in such a way varied from ±6 to ±11 deg in different odontocete species. Unilateral EP recordings indicated the best sensitivity of each ear to a sound-source position of around 10 deg from the midline. Separated EP recordings from the right and left ear revealed interaural intensity difference (IID) as large as 20 dB; IID reached this level at an azimuth as small as 10 to 15 deg. Near the midline, IID dependence on azimuth was as steep as 2 dB/deg. Interaural latency difference (ILD) was revealed also, but it more reflected the difference in physiological latencies rather than interaural time delay (ITD). These interaural differences may provide angular selectivity of around 0.5 deg.

Some mechanisms of complex interactions of temporal, spectral, and spatial resolution were also studied in cetaceans. ABR to double frequency-colored sound pulses were used as a model of the frequency-temporal interaction. With increase of spectral difference between two pulses, the second (probe) ABR released of suppression; at a frequency difference as large as ±0.5 oct, the probe response was not suppressed even at a short ISI and a high-intensity conditioning stimulus. When the probe stimulus was a high-rate rhythmic sound (SAM or a pip train), the evoked EFR was deeply suppressed by another sound of 8 to 12 kHz higher frequency and 15 to 20 dB lower intensity (a paradoxical lateral suppression effect). Interaction of spectral and spatial analysis manifested itself in different sensitivity-versus-frequency functions at different sound-source directions; at small azimuth, higher sensitivity to high frequencies was much more prominent than at large azimuth. It results, in particular, in interaural spectral difference, which may be an additional cue for sound-source localization.

Pathways of sound conduction to the middle ear are not completely understood in cetaceans. The most popular is a mandibular hypothesis that implicates the lower jaw as a sound-conducting pathway. However, computation based on acoustic delays (as revealed by ABR latencies) revealed an "acoustic window" just near the middle ear, thus suggesting a possibility of direct sound conduction to the middle ear through head tissues. More detailed studies indicated multichannel sound conduction to the middle ear, different channels being of different spectral sensitivity.

Representation of the auditory system in the cerebral cortex of dolphins was found by EP recording. The position of the auditory projection area in the dolphin's cerebral cortex differs markedly from that in other mammals. It occupies a large part of the suprasylvian gyrus (a parietal region), whereas in other mammals it is located in the temporal lobe.

Implements of revealed peculiarities of the auditory system to echolocation are discussed. The wide frequency range of hearing is dictated by the need to use short-wave (high-frequency) sounds that yield high spatial resolution of the sonar. The need to hear a weak echo after much stronger emitted signals dictates both the use of short (thus, high-frequency) sonar signal and high temporal resolution of the auditory system. Operating in the high-frequency range makes it possible to combine the high temporal resolution and the acute frequency tuning; at lower frequencies this combination would be prohibited because of inverse relation between the frequency tuning and bandwidth of peripheral auditory filters. The temporal transfer function of the dolphin's auditory system not only helps to distinguish the emitted and echo pulses but also may provide echo perception invariantly of the emitted-sound intensity and the distance to a target; it is possible because the response dependence on echo intensity (which, in turn, depends on the emitted sound intensity and the distance to a target) is compensated by the recovery after the emitted pulse. Discrimination of fine spectrum patterns, which depends on FRP, is more important for target recognition than for estimation of the distance. Frequency-temporal interaction helps to extract those features of the echo spectrum that differ from those of the emitted pulse and depend mostly on target properties. The paradoxical lateral suppression gives an advantage to a high-rate weak echo over emitted sonar signals and extraneous sounds. Binaural mechanisms (steep IID and ILD dependence on azimuth and interaural spectral difference) may help to localize target positions with high precision.

Chapter 3
HEARING IN PINNIPEDS AND SIRENIANS

Pinnipeds and sirenians have not been subjects of as extensive investigation of hearing abilities and mechanisms as cetaceans have been. The studies performed on these animals did not reveal extreme hearing capabilities or echolocation. Nevertheless, their hearing system presents a great opportunity for understanding the various ways they have adapted to an aquatic environment.

3.1. HEARING IN PINNIPEDS

From the point of view of bioacoustics, pinnipeds (*Pinnipedia*) are of great interest because of their amphibious hearing. Being actually not completely aquatic but semiaquatic animals, pinnipeds spend their life both in water (during foraging and long migrations) and on land (during breeding seasons). So both underwater and aerial orientation, including acoustic orientation and communication, are of importance for them. Many findings have shown that both aerial and underwater hearing are of comparable importance for pinnipeds. Airborne vocalizations by pinnipeds play an important role in social function, including the delineation of territory, advertisements of dominance status, and female attendance. Some underwater vocalizations are also related to social interactions, particularly among breeding males (see Schusterman, 1978, for a review).

Contrary to cetaceans, vocal repertoire of pinnipeds is low frequency (see Schusterman et al., 2000, for a review). They are not able to emit ultrasonic signals.

Among three pinniped families – true seals (*Phocidae*), eared seals (*Otariidae*), and walruses (*Odobenidae*) – hearing of otariids and phocids have been the subject of more or less detailed investigations.

No physiological studies (neither evoked-potential nor other kinds) of the hearing abilities of pinnipeds have been carried out until now, except data on the localization of the auditory sensory projection in the cerebral cortex. Only anatomical and psychophysical data are available concerning their hearing abilities. Nevertheless, the data have revealed many features of the pinniped's auditory system.

3.1.1. Ear Anatomy

Detailed anatomical investigation of the pinniped ear are presented by Repenning (1972) and Ramprashad et al. (1972). It can be stated that the pinniped ear demonstrates adaptation for both aerial and underwater hearing.

Phocids lack an external ear pinna. Their external auditory canal is long, narrow, and filled with cerumen and hairs. It is supported by cartilage throughout most of its length and is flexible. Muscular attachments allow closure of the meatal opening under water. It remains unclear whether the canal remains air-filled when closed under water. In contrast to phocids, otariids have a pinna, which is markedly reduced as compared to the pinna of terrestrial carnivores.

The middle-ear ossicles are larger and denseer that those of terrestrial mammals. In otariids, the ossicles are less massive than in phocids. The ossicles are loosely attached to the wall of the middle-ear cavity and surrounded by a highly vascularized *corpus cavernosum*, which presents also in the auditory meatus (Odend'hal and Poulter, 1966; Møhl, 1967b, 1968b). The round window is partly or entirely shielded from the middle ear cavity in both phocids and otariids. In the elephant seal, the round window opens at the junction of the bulla, and the mastoid to the exterior of the skull (Repenning, 1972).

3.1.2. Hearing Sensitivity and Frequency Range

To date, underwater and aerial audiograms or individual threshold measurements are available in many pinniped species of both *Otariidae* and *Phocidae* families: the California sea lion *Zalophus californianus* (Schusterman et al., 1972), the northern fur seal *Callorhinus ursinus* (Moore and Schusterman, 1987; Babushina et al., 1991) – *Otariidae*, the harbor seal *Phoca vi-*

tulina (Møhl, 1968a; Terhune, 1988, 1989, 1991; Terhune and Turnbull, 1995), the harp seal *Pagophilus groenlandicus* (Terhune and Ronald, 1971, 1972), the ringed seal *Pusa hispida* (Terhune and Ronald, 1975b), the Hawaiian monk seal *Monachus schauinslandi* (Thomas et al., 1990) – *Phocidae*.

All the data were obtained by the psychophysical method, mainly using various versions of the go/no-go experimental paradigm with training the animal to minimize the false-alarm probability. In some studies (e.g., Schusterman et al., 1972; Schusterman, 1974), the animal was trained to respond by vocalization instead of movement, but it is unlikely that this change of the experimental paradigm markedly affected the results. In a few cases, the animal was successfully trained to push one paddle (key) in the presence of a stimulus and press another paddle (key) in the absence of a stimulus instead of no-go response (Møhl, 1968a; Terhune and Ronald, 1971; Terhune, 1988, 1989, 1991; Terhune and Turnbull, 1995 – that is, an element of the two-alternative forced-choice procedure was introduced into the experimental paradigm. Actually, however, the latter version is more close to the go/no-go paradigm since only one stimulus interval was used in each trial, so the animal had to respond either yes or no (the principle of the go/no-go paradigm) instead of answering the question "which of the intervals or places?" (the principle of the forced-choice paradigm). In particular, a preference of responding either yes or no (preference for one of the keys) was not eliminated in this procedure, so the animals were trained to reach a low false-alarm level. Thus, all the measurements were made mostly with similar experimental approaches. Therefore, all the data seem comparable in spite of insignificant methodological differences.

Some of the obtained results are presented in Fig. 3.1 (an otariid species) and Fig. 3.2 (a few phocid species). The exemplified audiograms demonstrate that the underwater hearing of most pinnipeds features frequency ranges varying within a limit of a few tens of kHz; however, in some cases the range exceeded 100 kHz. The frequency range of aerial hearing is somewhat narrower; nevertheless, thresholds were measurable at frequencies as high as 32 kHz.

Comparison of the underwater and aerial sensitivity of hearing is of special interest. The best underwater-hearing thresholds in different pinniped species varied from 63 to 80 dB re 1 μPa, which corresponds to an intensity of 1.2×10^{-12} to 6.5×10^{-11} W/m^2. These values are of the same order of magnitude as in many odontocetes though obviously higher than the best odontocetes' hearing thresholds. Aerial thresholds are at least 20 to 40 dB higher than the standard hearing threshold of 20 μPa; thus, the sensitivity is not very good.

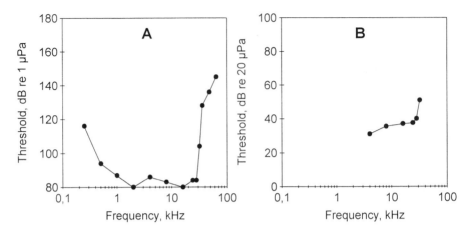

Figure 3.1. Underwater (**A**) and aerial (**B**) audiograms of an otariid species, the California sea lion *Zalophus californianus*. Plotted by table data in Schusterman et al. (1972) and Schusterman (1974).

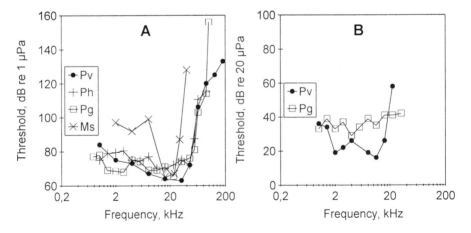

Figure 3.2. Underwater (**A**) and aerial (**B**) audiograms of a few phocid species. Pv – the harbor seal *Phoca vitulina*, Ph – the ringed seal *Pusa hispida*, Pg – the harp seal *Pagophilus groenlandicus*, Ms – the Hawaiian monk seal *Monachus schauinslandi*. Plotted by table data in Møhl (1968a) (the harbor seal), Terhune and Ronald (1971, 1972) (the harp seal), Terhune and Ronald (1975b) (the ringed seal), and Thomas et al. (1990) (the monk seal).

For better comparison, Katsak and Schusterman (1995, 1998) measured underwater and aerial hearing thresholds in the same subject. Measurements were carried out in a variety of species: one otariid (a California sea lion *Zalophus californianus*) and two phocids (a harbor seal *Phoca vitulina* and northern elephant seal *Mirounga angustirostris*). In each subject, both aerial and underwater thresholds were measured using the go/no-go procedure. The results are presented in Fig. 3.3 as thresholds (in terms of sound pressure) versus frequency. As commonly used, aerial thresholds were presented in dB re 20 μPa, whereas underwater thresholds were presented in dB re 1 μPa. For better comparison, panels A and B in Fig. 3.3 are shifted relative one another by 26 dB (the dB-difference between 1 and 20 μP), so the equal heights of experimental points relative to the panel frames correspond to equal sound pressures.

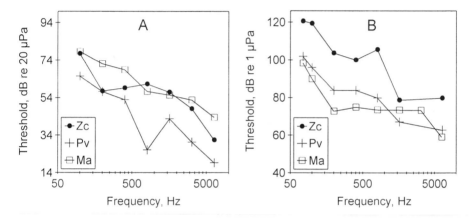

Figure 3.3. Aerial (**A**) and underwater (**B**) sound-detection thresholds for three pinniped species. Zc – *Zalophus californianus*, Pv – *Phoca vitulina*, Ma – *Mirounga angustirostris*. Plotted by table data presented in Katsak and Schusterman (1998).

The comparison shows that in all of the tested species, aerial and underwater hearing thresholds, being expressed in terms of sound pressure, were of the same order: aerial thresholds were either a few dB lower (in *Zalophus* and *Phoca*) or a few dB higher (in *Mirounga*) than underwater thresholds. However, it should be taken in consideration that the same sound pressure in the low-impedance aerial medium corresponds to 35.5-dB higher intensity (sound power flux density) than in the high-impedance aquatic medium. Thus, being expressed in terms of sound intensity, the underwater hearing thresholds in all the tested pinnipeds were 15 to 35 dB lower (the sensitivity better) than aerial ones.

Katsak and Schusterman (1998) interpreted the similarity of aerial and underwater sound-pressure thresholds as evidence that just sound pressure is the parameter determining sound perception in pinnipeds. It seems quite probable. However, it indicates that the pinniped's ear is better adapted to underwater rather than aerial hearing. Indeed, underwater thresholds as low as 57 to 80 dB re 1 µPa (at 6400 Hz) are comparable to those of cetaceans and indicate a good sensitivity. On the contrary, thresholds as high as 19 to 43 dB (at 6400 Hz) above the *standard aerial hearing threshold* of 20 µPa indicate rather bad sensitivity. At low frequencies, the sensitivity is even worse: thresholds are of 60 to 70 dB above the standard threshold level.

Very little is known about hearing abilities of walruses. Kastelein et al. (1993, 1996) have made an attempt to study aerial hearing sensitivity in walruses. The first of these studies (Kastelein et al., 1993) was a preliminary study of sensitivity of wild Atlantic walruses *Odobenus rosmarus rosmarus* based on observation of behavioral responses (opening the eyelids, rolling the eyes, and raising the heads) to band-filtered noise signals. All the found thresholds were around the level of ambient noise, indicating that there were masked rather than absolute hearing thresholds. In the latter study (Kastelein et al., 1996), much more detailed experimental measurements were carried out in a Pacific walrus *Odobenus rosmarus divergens* using the go/no-go paradigm and a staircase stimulus-presentation procedure. Two kinds of measurements were made: using headphones reducing the background ambient noise by 20 to 36 dB (depending on frequency) and in free aerial field without reducing the ambient noise, which was higher than 50 dB re 20 µPa in one-third-octave bands. Both kinds of measurements resulted in very high hearing thresholds. In the headphone-stimulation conditions, the best pure-tone thresholds (at 2 kHz) were about 60 dB re 20 µPa (60 dB higher than the standard human hearing threshold). In the free-field conditions, the best thresholds to modulated tones and band-filtered noise were 53 to 54 dB re 20 µPa – that is, above the one-third-octave ambient noise level. The authors themselves did not assess these results as an estimate of hearing sensitivity. The free-field thresholds were probably masked thresholds, whereas high headphone thresholds did not agree with the ability of walruses to follow orally given commands. The authors supposed that the outer ear canal could be closed off by the auricular muscle due to the presence of headphones, thus giving abnormally high thresholds.

No attempts were made to study the underwater hearing sensitivity of walruses. Thus, the hearing abilities of walruses remain to be investigated.

3.1.3. Temporal Processing

Hearing abilities associated with temporal processing were studied to a very limited extent. Terhune (1988) investigated the temporal summation of hearing in a harbor seal *Phoca vitulina*. He measured hearing thresholds to tone pulses of various duration within a wide frequency range, from 1 to 64 kHz. It was found that when pulses were shorter than approximately 400 cycles, the thresholds increased linearly with the logarithm of the number of cycles, independently of frequency within a range of 4 to 32 kHz. This was an indication of temporal summation. Calculation of the pulse total energy has shown that within this range, thresholds were almost independent of pulse duration when expressed in terms of energy (not power) flux density – that is, complete energy summation took place. At duration longer than 400 cycles, thresholds reached their minimum as expressed in terms of sound power or sound pressure.

Thus, no constant (independent of frequency) integration time was found in that study. This conclusion is not in agreement with numerous psychophysical data obtained in other mammals, first of all, in humans, which showed a significant role of a central integrator that dictated an integration time little dependent on frequency (Shailer and Moore, 1985, 1987; Moore and Glasberg, 1988; Grose et al., 1989; Plack and Moore, 1990; Eddins et al., 1992; Moore et al., 1993). Various paradigms of measurement of the integration time in odontocetes have also shown the existence of a certain characteristic integration time that was little dependent on frequency. Taking into account the small body of data concerning temporal resolution of the pinniped's hearing, this question calls for further investigation.

3.1.4. Frequency Tuning

3.1.4.1. Critical Ratio and Critical Band Measurements

The frequency-tuning capabilities of pinnipeds were not a matter of investigations as detailed as those in dolphins. Nevertheless, Terhune and Ronald (1971, 1975a), Turnbull and Terhune (1990) presented results of critical-ratio measurements in a few phocid species. Measurements in a harp seal *Pagophilus groenlandicus* (Terhune and Ronald, 1971) were made in conditions of aerial hearing; measurements in ringed seals *Pusa hispida* (Terhune and Ronald, 1975a) were made underwater; measurements in a harbor seal *Phoca vitulina* (Turnbull and Terhune, 1990) were made both in air and underwater. All the studies were carried out using a procedure when the animal was trained to push one or another key depending on the stimulus presence or absence. Thresholds measured in such a way in conditions of masking broadband noise were presented in terms of critical ratio (Fig. 3.4).

Measurements in all species and in both hearing conditions (aerial and underwater) gave qualitatively similar results. Critical ratio increased with the probe frequency (except low-frequency aerial data in the harbor seal), as expected for frequency-proportional bandwidth of the auditory filters. How-

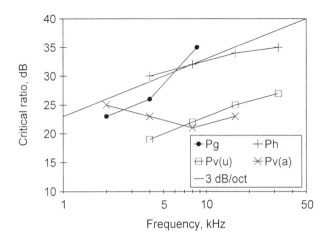

Figure 3.4. Critical ratio as a function of frequency in a few phocid species. Pg – *Pagophilus groenlandicus* (aerial hearing), Ph – *Pusa hispida* (underwater hearing), Pv(u) – *Phoca vitulina* (underwater hearing), Pv(a) – *Phoca vitulina* (aerial hearing). Oblique straight line shows a slope of 3 dB/oct. Plotted by table data in: Terhune and Ronald (1971) (the harp seal), (1975a) (the ringed seal), Turnbull and Terhune (1990) (the harbor seal).

ever, a noticeable data scatter was observed, so the slope of the critical ratio-versus-frequency function markedly deviated from the slope of 3 dB/oct as should be for frequency-proportional filters. In the harp seal, the critical ratio estimates varied from 23 dB at 1 kHz (assuming the signal-to-noise ratio at the masked threshold being 1, this corresponds to $Q_{ER} = 5.0$) to 35 dB at 8.6 kHz (under the same assumption, this corresponds to $Q_{ER} = 2.7$). In ringed seals, critical ratio varied from 30 dB at 4 kHz to 35 dB at 32 kHz; under the same assumption, this corresponds to Q_{ER} from 4 to 10, respectively. In a harbor seal, underwater critical ratios were from 19 dB at 4 kHz to 27 dB at 32 kHz (which corresponded to Q_{ER} from 50 to 64), and in-air critical ratios were from 25 dB at 2 kHz to 23 dB at 16 kHz (which corresponded to Q_{ER} from 6.3 to 80).

Thus, the scatter of estimates of frequency tuning is extremely large, more than an order of magnitude. It should be stressed again, however, that the critical ratio is a very poor estimate of frequency tuning since real values of the signal-to-noise ratio at the masked threshold are not known, and uncertainty in this ratio of only a few dB results in uncertainty of the frequency-tuning estimates by several times. A moderate scatter of critical-ratio data also result in a very large variation of frequency-tuning estimates. In the studies mentioned above, standard deviation of data in the critical-ratio domain varied from a few dB to around ±10 dB, sometimes up to ±16 dB (data for 16 kHz in air by Turnbull and Terhune, 1990). Note that a ±10 dB deviation corresponds to variation of frequency tuning estimates by two orders of magnitude, from 0.1 to 10 of the mean value. The ±16 dB deviation results in variation of frequency tuning estimates from 1/40 to 40 of the mean value (1600 times). This shows that the critical ratio cannot be used to assess the frequency tuning with more or less satisfactory precision.

Turnbull and Terhune (1990) have also made an attempt to obtain psychophysical tuning curves in a harbor seal *Phoca vitulina*. Although they called this study as a measurement of the critical bandwidth, actually it was a kind of inverted tuning-curve paradigm: tone-probe thresholds were measured around a tone masker of a constant frequency and level. The authors claimed that the bandwidths were all under 2.25 kHz when estimated at 23 dB below the masker level and became proportionately narrow at higher frequencies. However, analysis of the presented data suggests that true tuning curves hardly were obtained in that study. The obtained curves featured very high probe thresholds when the probe and masker frequency coincided and fell steeply as soon as the probe frequency deviated by one step (500 Hz for underwater measurements and 250 Hz for aerial measurements) from the masker frequency; thus, the constant "width" of the curves really was the doubled step of probe frequency variation. Coincident probe and masker frequencies resulted in beats that produced unpredictable effects on the task performance. Should these points be neglected, the remaining points do not allow the real shape and width of the tuning curves to be estimated. Thus, the real frequency tuning in any of pinnipeds remains unknown.

3.1.4.2. Frequency Discrimination Limens

More definite data were obtained in measurements of frequency-discrimination limens. Although the frequency-discrimination limen is not a direct estimate of the auditory filter bandwidth and quality, it is an important indirect indication of the frequency-tuning acuity, assuming that the lower the limen, the better tuning (see Section 2.6.4).

Frequency-discrimination limens were measured in one otariid species, the California sea lion *Zalophus californianus* (Schusterman and Moore, 1978) and two phocid species, the harbor seal *Phoca vitulina* (Møhl, 1967a) and ringed seal *Pusa hispida* (Terhune and Ronald, 1976). All measurements were performed in conditions of underwater hearing and with the use of various versions of the go/no-go paradigm. These were the same modifications of the standard paradigm as in previously mentioned studies of the same authors – namely, Schusterman and Moore (1978) used vocalization instead of movement as yes-response; Møhl (1967a) and Terhune and Ronald (1976) used pushing of a key instead of staying the animal as no-response. There was an important difference between the studies. Schuster-man and Moore (1978) presented two 1-s signals, either both of equal fre-quencies or the second signal was of lower frequency than the first one. The animal was required to give a yes-response (vocalization) if the signals dif-fered in frequency. This procedure probably involved a short-term memory to compare frequencies of two successive signals, so it is expected to give rather conservative estimates. On the contrary, in studies of Møhl (1967a) and Terhune and Ronald (1976), the animal was required to discriminate a frequency-modulated signal from a constant-frequency one, responding as yes or no, respectively. In this case, performance depended on the ability to detect frequency modulation – that is, to compare frequencies that immedi-ately replaced one another. This procedure addressed mostly to true sensory rather than to memory abilities. In spite of this difference, all the studies provided comparable results (Fig. 3.5).

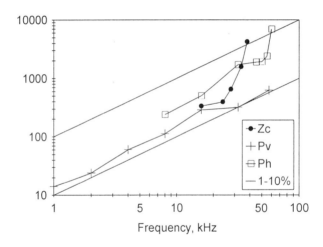

Figure 3.5. Frequency-discrimination limens in pinnipeds. Zc – *Zalophus californianus*, Pv – *Phoca vitulina*, Ph – *Pusa hispida*. Oblique straight lines delimit a region of 1 to 10% of the carrier frequency. Plotted by table data in Schusterman and Moore (1978) (*Zalophus califor-nianus*), Møhl (1967a) (*Phoca vitulina*), and Terhune and Ronald (1976) (*Pusa hispida*).

In all the tested subjects, frequency-discrimination limens were found within a range from 1 to 10% of the standard frequency. These values are to be considered as indicating rather moderate frequency discrimination. For comparison, in humans, frequency-discrimination thresholds are within a range from 0.2 to 1% (Weir et al., 1976) – an order of magnitude better than in pinnipeds. In dolphins, frequency discrimination limens were found to be below 0.1% (see Section 2.6.4). Thus, together with rather high critical ratios, the frequency-discrimination data are indicative of nonacute frequency tuning of the pinniped's hearing.

It is noteworthy that in the study of Schusterman and Moore (1978), the limens were found not higher than in other studies, although they used the procedure that is expected to give more conservative estimates. Difference in methods probably did not markedly influenced the results.

3.1.5. Intensity Discrimination

We aware of only one attempt to measure intensity-difference threshold in a pinniped – namely, in a California sea lion *Zalophus californianus* (Moore and Schusterman, 1976). The animal was trained in a version of the go/no-go paradigm with the use of a vocalization response instead of movement. In each trial, a single-tone pulse was presented to the subject. The animal was required to compare the tone intensity with an "internal standard" maintained by feedback: the positive response was rewarded if the stimulus was of higher level than the standard and was not rewarded if the stimulus was of lower level. Measurements were made at the only frequency of 16 kHz and resulted in an intensity-discrimination threshold of 3.2 dB – that is, the sound pressure excess of about 45% above the standard was detectable. It is not the best threshold among mammals; for comparison, in humans, intensity-discrimination thresholds are around 1 dB (Jesteadt et al., 1977), in the domestic cat they are mainly 1.5 to 2.5 dB, depending on frequency (Elliot and McGee, 1965), in the rhesus monkey 1 to 2.5 dB except for low sensation levels (Clopton, 1972). In dolphins, intensity-discrimination thresholds are also much lower than that obtained in the California sea lion (see Section 2.7). However, many other mammals feature intensity-discrimination thresholds of the same order as the California sea lion – for example, many rodents (see Fay, 1988, for a review). So the presented threshold estimate in the California sea lion does not seem unrealistic. It should be noted, however, that the procedure used seems to provide rather conservative results since the animal's performance depended not only on its sensory capabilities but also on its ability to use the "internal standard" stored in memory.

3.1.6. Directional Hearing

Minimal audible angle (MAA) – the minimum distinguishable angle between two sound sources – is a demonstrative measure of directional sensitivity of hearing. MAA were measured in a few pinniped species. Measurements in a harbor seal *Phoca vitulina* were performed using an experiment design that was natural for spatial-hearing studies (Terhune, 1974). The animal was trained to respond by touching the left or right of two paddles (keys), according to the sound-source position on the left or on the right of the center. Measurements were made in air at two tone frequencies, 1 and 8 kHz, and at a few center frequencies of band-filtered noise, from 2 to 16 kHz. MAA was found as 4.6 and 5.5 deg at 1 and 8-kHz tone frequencies. For one-third-octave band-filtered noise, MAA varied from 6.6 deg at 2-kHz center frequency to 1.2 deg at 16-kHz center frequency. The lowest MAA obtained at 4-kHz center frequency constituted 0.7 deg; however, this point was off the general trend, so it remains unclear yet whether MAA in the harbor seal is really as low as this estimate.

A similar technique applied to underwater measurements in a California sea lion *Zalophus californianus* resulted in larger MAA estimates: 15 deg at a frequency of 3.5 kHz and 10 deg at 6 kHz (Gentry, 1967). In another study on a California sea lion in conditions of aerial hearing (Moore and Au, 1975), MAA estimates varied in a very wide range, from 3.5 deg at 1 kHz to 42 deg at 4 kHz and nonmeasurable MAA (no discrimination at any angle) at 2 kHz. Since variations of MAA did not feature any regular dependence on frequency, this variation seems to be a result of a significant data scatter. It should be taken into consideration that in the latter study, the animal was trained in a different way than in other directional-hearing studies; namely, it had to push one paddle when the sound source was at zero azimuth and another paddle when the sound source was at nonzero azimuth. Perhaps this task was less natural and therefore more difficult for the animal than the task to choose a right or left object according to the right/left sound-source position.

In general, the sensitivity of pinnipeds to sound-source direction can be assessed as not very acute. It is a little worse than in humans (MAA of 1 to 3 deg depending on frequency, Mills, 1958), though comparable.

3.1.7. Auditory Representation in the Cerebral Cortex

The pattern of gyri and sulci in the cerebral cortex of pinnipeds is very similar to that in carnivores. Therefore, localization of the auditory projection area was not as debatable as in cetaceans; it was expected to be localized in the same manner as in carnivores.

Nevertheless, attempts were made to find the exact position of the auditory projection area in the cerebral cortex in pinnipeds. Alderson et al. (1960) succeeded in recording cortical EP to auditory stimuli in the harbor seal *Phoca vitulina* and found them in a temporal area of the cortex, similarly to carnivores.

Further study was undertaken by Ladygina and Popov (1986) in the northern fur seal *Callorhinus ursinus (Otariidae)* and the Caspian seal *Pusa caspica (Phocidae)*. EP to short sound clicks were recorded from the cortical surface through small holes drilled in the skull of deeply anaesthetized animals. Both in the northern fur seal and Caspian seal, EP to sound stimuli appeared in a rather small area between the sylvian and posterior suprasylvian gyri (Fig. 3.6). There was some differentiation within this area: short-latency (primary) EPs were recorded in a smaller area adjacent to the sylvian sulcus and in the depth of this sulcus. In the remaining part of the area, EP of longer latencies were recorded.

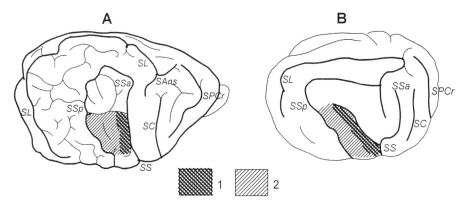

Figure 3.6. Localization of the auditory projection area in the cerebral cortex of the northern fur seal *Callorhinus ursinus* (**A**) and the Caspian seal *Pusa caspica* (**B**). Lateral view of the right hemisphere. *SL– sulcus lateralis, SS – s. sylvius, SSa – s. suprasylvius anterior, SSp – s. suprasylvius posterior, SC – s. cruciatus, SPCr – s. precruciatus, SAns – s. ansatus.* Hatched area is the auditory projection area; 1 – primary projection area (short-latency EP), 2 – non-primary area (EP of longer latency).

These data show that in pinnipeds, both otariids and phocids, the auditory projection area is not as domination among other sensory projection areas as it appears in dolphins.

No attempts were made to determine the tono-topic projection within the cortical auditory area.

3.1.8. Hearing Adaptation to Amphibious Lifestyle

All the studies mentioned above show that pinnipeds, both otariids and phocids, do not possess hearing as perfect as that of cetaceans (odontocetes). Their hearing is good enough but does not feature extremely high sensitivity, frequency range, frequency, and spatial resolution.

The ability of pinnipeds to hear somewhat better under water than in air should be emphasized. This feature of their hearing is intriguing since pinnipeds evolved from terrestrial carnivores, whose outer and middle ears were adapted to aerial hearing. Without significant modification, such an ear should not be effective in water because of the impedance mismatch.

Repenning (1972) supposed that bone sound conduction in pinnipeds is effective enough because of modifications of the skull bones; the bone conduction could be responsible for underwater hearing. Other authors (Møhl, 1968b; Moore and Schusterman, 1987) and also Repenning (1972) suggested that during diving, changes in the middle ear occur that result in the impedance matching to the aquatic medium. According to the hypothesis, a crucial role was ascribed to the cavernous tissue *corpus cavernosum* in the middle ear and the external auditory canal. It was supposed that when a pinniped dives, the cavernous tissue engorges with blood. This may be an active mechanism, accompanying changes in blood circulation that take place during diving, or a passive response to pressure changes at depth. The swelling of cavernous tissue collapses the lumen of the external auditory canal as well as the middle-ear cavity. The result of these changes of the cavity volume in the external auditory canal and the middle ear may be an increase of the ear input impedance that thus closely matches that of surrounding water. The input impedance increase reduces the sound reflection from the tympanic membrane. Under high pressure, both sides of the tympanic membrane may contact fluid (tissue of the meatus and *corpus cavernosum* in the middle ear) allowing sound to be transmitted efficiently because of minimized impedance difference across the tympanic membrane. This mechanism of input-impedance matching, if exists, allows underwater acoustic energy to be transmitted to the inner ear in a "conventional" way – that is, via the tympanic route and the ossicular chain.

The presence of echolocation in odontocetes and its absence in pinnipeds suggests itself as either a cause or a consequence of differences between the hearing abilities of these two orders of aquatic mammals. An idea was claimed that pinnipeds were capable of echolocation and some their sounds were sonar signals (Poulter, 1963; Renouf and Davis, 1982). Poulter (1963, 1966, 1967) argued for the presence of echolocation in the California sea lion *Zalophus californianus* and other otariids basied on the observation of captive animals approaching food under a variety of conditions and suggested that for this purpose they used vocalizations ranging in frequency from 5 to 13 kHz. However, this idea was not confirmed by numerous later studies and was a subject of criticism (Evans and Haugen, 1963; Schusterman, 1981; Wartzok et al., 1984; Schusterman et al., 2000). Most of vocalization of the California sea lion, as well as other pinnipeds, are lower than 4 kHz in frequency and are associated with social interactions rather than foraging (e.g., Schevill et al., 1963; Schusterman, 1967). Thus, no satisfactory evidence of the presence of echolocation in pinnipeds is present to date. Schusterman et al. (2000) argued that an advanced echolocation system did not evolved in pinnipeds primarily because of constrains imposed by the amphibious functioning of the auditory system; as a result of these constrains, pinnipeds have not developed truly aquatic high-frequency and high-sensitivity hearing required for underwater echolocation.

It should be also taken into consideration that the auditory system of cetaceans evolved in aquatic medium (in particular, becoming adjusted to echolocation) during a much longer time (50 to 60 million years) than that of pinnipeds (about 25 million years) which remain close relatives of some terrestrial carnivores (Tedford, 1976). The pinniped auditory system evolved from that of terrestrial carnivores being modified in such a way as to permit good underwater hearing and yet save a satisfactory aerial hearing, thus keeping a compromise between different requirements of these two media.

3.2. HEARING IN SIRENIANS

Hearing abilities of sirenians are of great interest, especially in comparison with other aquatic mammals, because being completely aquatic mammals, they differ from other both completely aquatic (cetaceans) and semi-aquatic (pinnipeds) mammals in phylogeny and lifestyle. All sirenians are obligate herbivores, whereas most other marine mammals are predators. Therefore, sirenians need quite different sensory abilities as compared to other marine mammals.

There are no data on echolocation in sirenians. However, manatees produce underwater vocalizations within a sound frequency range of up to 10 kHz (Schevill and Watkins, 1965; Evans and Herald, 1970; Sonoda and Takemura, 1973). Manatees have normally developed ears (Harrison and King, 1965) and normally developed brain stem acoustic centers (Verhaart, 1972). These data suggest that manatee have normally developed hearing.

However, since sirenians are not as widely available for investigations as many cetaceans, there were only a few studies of their hearing. Nevertheless, the results of these studies are worth mentioning.

3.2.1. Ear Morphology

Earlier studies of the manatee ear (Harrison and King, 1965) have shown that manatees have the normally developed middle ear with rather large ear ossicles. However, the pinna is absent, and the external auditory meatus is occluded in its outer part; the external auditory canal reaches the rather large tympanic membrane.

After earlier descriptions of the tympano-periotic bones of sirenians (Robineau, 1969; Fleischer, 1978), a detailed description of the ear in the West Indian manatee *Trichechus manatus* was presented by Ketten (1992c). Like cetaceans, manatees have no pinnae. The external auditory meatus is a surface dimple, leading to a narrow external auditory canal filled with discarded epithelial cells. It is unclear whether the canal participates in sound transmission to the middle ear.

The tympano-periotic complex consists of a smaller tympanic and larger periotic parts. Unlike the cetacean tympano-periotic complex, which is external to the skull, the manatee's tympano-periotic complex is attached to the inner wall of the cranium and occupies a substantial portion of the posterior brain case. Such intracranial position of periotics should result in relatively small interaural time delay (ITD).

The tympanic membrane forms the lateral wall of the middle-ear cavity; ventrally, the cavity is closed by the soft tissues of the throat. The cavity is large and lined with a thick, vascularized fibrous sheet. It is likely that the cavity is air-filled. The tympanic membrane is laterally convex; however, it is structurally similar to membranes of terrestrial mammals and has little in common with the tough membranes of cetaceans.

The ossicular chain in manatees is massive, nearly straight. The stapes and incus lie on a medial-lateral line posterior to the malleus, so that the two major axes of rotation of the ossicles form an angle of approximately 140 deg (Ketten, 1992c). The stapes is columnar; its footplate is medially convex and is attached to the oval window. The hooklike incus has two arms, one extending superiorly and attached to the periotic, and the other ending in a flat, articular plate that abuts the stapes head. Three facets of the anterior surface of the incus articulate with the malleus. The malleus is a thick ovoid with a flattened ridge to which the tympanic membrane attaches. Thus, it seems that in sirenians, unlike cetaceans, the traditional way of sound conduction through the tympanic membrane and then through the ossicles to the oval window plays an important role.

The cochlea morphometrically follows the conventional mammal pattern. It is a multiturn spiral. There is little base-to-apex differentiation in the basilar membrane. Measurements by Ketten (1992c) have shown the membrane thickness from 7 μm at the basal point to 5 μm at the apex, the membrane width from 200 μm at the basal point to 600 μm at the apex – that is, the apical membrane is only 1.5-fold thinner and 3-fold wider than the basal end (contrary to approximately 5-fold decrease of thickness and more than 10-fold increase of width in odontocetes). There is no outer osseous spiral lamina. These properties of the basilar membrane indicate much narrower frequency range in manatees than in odontocetes.

3.2.2. Psychophysical Audiogram

A psychophysical underwater audiogram was obtained in the West Indian manatee *Trichechus manatus* by Gerstein et al. (1999). Two animals were trained to wait for a signal at a start position (inside a stationing hoop) and then to push one of two paddles depending on the sound stimulus presence or absence in the trial. Thus, elements of a two-alternative forced-choice paradigm were present in the test procedure, although each trial contained only one stimulus interval (a feature of the go/no-go paradigm). Tone stimuli of variable intensity were present in an up-down (staircase) adaptive manner, and the threshold was calculated as a mean value of sound intensities at reversal points.

The obtained audiograms (Fig. 3.7) were of an U-shaped form typical of many mammals. They demonstrated quite good sensitivity of the manatee's hearing; thresholds were as low as 50 dB re 1 μPa; this is of the same order as the best thresholds found in cetaceans and better than underwater thresholds in pinnipeds. On the other hand, the hearing frequency range of the

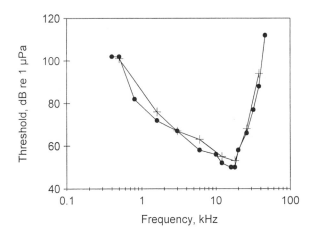

Figure 3.7. Underwater audiograms of two West Indian manatees *Trichechus manatus.* Plotted by table data presented in Gerstein et al. (1999).

manatee was much narrower than that of cetaceans; the lowest thresholds appeared at frequencies of 16 to 18 kHz, and the range of high sensitivity was assessed as 6 to 20 kHz. The upper frequency limit was around 40 kHz (the highest frequency that yielded an animal's responses was 46 kHz).

In the same study, thresholds to sound stimuli in an infrasonic frequency range (15–200 Hz) were also measured in the manatees. These stimuli were detected by the animals, though with very high thresholds. The authors reasonably suggested that sensitivity to high-intensity sound in this frequency region is probably vibrotactile rather than auditory. Therefore, these data are not presented in Fig. 3.7.

3.2.3. Evoked-Potential Data

3.2.3.1. Evoked-Response Features

Bullock et al. (1980, 1982) were the first who succeeded to record auditory evoked potentials (EP) in manatees. EP to acoustic stimulation were recorded noninvasively from the animal's head. EP of rather long latency and duration (tens of milliseconds) were obtained; presumably, they were generated by the auditory cortex. Irrespective of the EP origin, it was of importance to show that auditory EP are capable to be provoked by sound stimuli. In the West Indian manatee *Trichechus manatus*, the largest EP were obtained at sound frequencies of 1.0 to 1.5 kHz, and noticeable EP at frequencies up to 35 kHz. In the Amazonian manatee *Trichechus inunguis*, the largest EP were obtained at frequencies around 3 kHz. However, during those experiments, the animal was in air and was stimulated by airborne sounds. For a completely aquatic animal, the aerial sound stimulation hardly can be assessed as adequate. Therefore, none of hearing characteristics were measured in that study with a satisfactory precision.

An attempt to use EP for measurement of hearing characteristics in a manatee was made in a study by Klishin et al. (1990). That study was carried out on a young Amazonian (Brazilian) manatee *Trichechus inunguis*. During experimentation, the animal was kept in a water-filled bath, and stimulating sounds were presented under water, thus the stimulation was more adequate for the aquatic animal.

Evoked potentials recorded in such conditions consisted of two main parts (Fig. 3.8A): the initial fast wave complex (1) and later slow wave complex (2). The initial complex consisted of a sequence of short waves, each lasting about 1 ms (Fig. 3.8B); with subtraction of the acoustic delay constituting 0.5 ms, its onset latency was as short as 0.9–1 ms. This response was

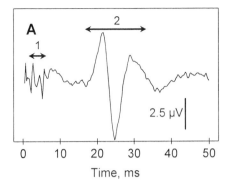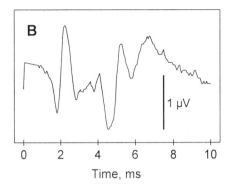

Figure 3.8. EP waveforms recorded from the head surface of a Amazonian manatee *Trichechus inunguis* in response to a sound click. **A**. Slow-sweep record showing both a rapid ABR (indicated by the double-headed arrow 1) and a slow ACR (indicated by the arrow 2). **B**. Fast-sweep record at higher gain demonstrating ABR in more detail.

considered as ABR. The later complex consisted of slower waves lasting 5 to 10 ms. It resembles the EP recorded by Bullock et al. (1980) though was somewhat less prolonged; it probably was a cortical response.

It is obvious that ABR in the manatee markedly differed from that in dolphins. It was longer, at its amplitude never exceeded a value of an order of 1 μV, which is an order of magnitude lower than in most of dolphins. Perhaps, it is because of less developed brainstem auditory centers in the manatee as compared to those in odontocetes. Lower ABR amplitude may depend also on some peculiarities of the brain and head anatomy and position of field-generating dipoles.

For measurement of hearing characteristics in more detail, ABR was used as more consistent. Its amplitude almost linearly depended on stimulus intensity (Fig. 3.9) which made it possible to find threshold by extrapolation of the regression line to its intercept with the zero-amplitude level.

3.2.3.2. ABR Audiogram

Using threshold measurements as described above, ABR threshold were determined in an Amazonian manatee at frequencies from 5 to 60 kHz. The resulting audiogram is presented in Fig. 3.10. It features the maximum sensitivity in the same frequency region as the psychophysical audiogram described above: the lowest thresholds were at 10 to 20 kHz. Absolute estimates of the best thresholds were 85 dB re 1 μPa.

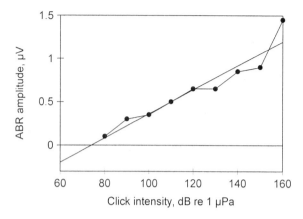

Figure 3.9. Dependence of ABR amplitude on sound click intensity in an Amazonian manatee *Trichechus inunguis*. Dots – experimental data, oblique straight line – their linear approximation (regression line). The regression line intercepts the abscissa axis at 74 dB.

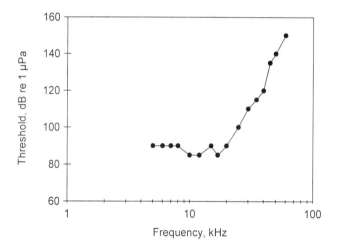

Figure 3.10. ABR audiogram of an Amazonian manatee *Trichechus inunguis*.

The found ABR thresholds were markedly higher than psychophysically found thresholds (down to 50 dB). It does not seem very probable that this difference reflects different hearing abilities of the two investigated species (*Trichechus manatus* and *T. inunguis*). Of course, some difference may exist, but it is not large. Some individual variation of hearing sensitivity cannot be excluded either. However, it seems more probable that the difference is a result of different kind of stimuli in psychophysical and evoked-potential experiments. In the evoked-potential study, short tone pips were used: 0.5 ms with a rise-fall time of 0.1 to 0.2 ms. The ABR thresholds were presented in peak sound pressure of short tone pips; these values can be higher than thresholds to long tones used in psychophysical experiments.

Although absolute threshold estimates differed in psychophysical and evoked-potential experiments, the general pattern of the audiograms seems rather similar. They show that the frequency region of the best sensitivity is narrower in manatees than in odontocetes.

3.2.3.3. Temporal Resolution: Rhythmic Click Test

To estimate temporal resolution of the auditory system, the rate-following response (RFR) was recorded in an Amazonian manatee at various rates of rhythmic clicks. A simplest stimulation mode was used – namely, steady-state rhythmic-click sequences. In such stimulation conditions, ABR waveform and amplitude were slightly changed at click rates up to 20 s^{-1}. Further increase of the rate resulted in a gradual decrease of the amplitude and in response waveform closer to a sinusoid.

Figure 3.11 presents RFR amplitude as a function of the rate of rhythmic stimulation. The shape of the plots was similar at stimulus intensities of 100 and 140 dB re 1 μPa, except for a difference in response amplitudes. The general trend is that RFR amplitude diminished when the stimulation rate increased above 20 s^{-1} though the responses were able to follow rhythmic stimulation at rates up to 500 s^{-1}.

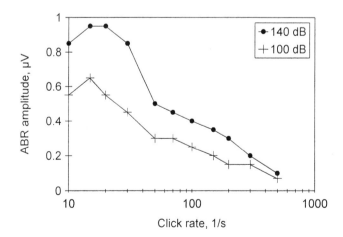

Figure 3.11. Dependence of RFR amplitude on rhythmic click rate at two stimulus intensities (140 and 100 dB re 1 μPa peak sound pressure) in an Amazonian manatee *Trichechus inunguis*.

Thus, the ability of RFR in the manatee to follow a rhythmic stimulation rate is not as high as in dolphins in similar stimulation conditions – that is, with the use of steady-state sequences. As shown above (Section 2.5.7), stimuli of such kind do not reveal the true temporal resolution of the auditory system because of influence of rate-dependent long-term adaptation. Nevertheless, comparison of data obtained in various animals with the use of one and the same test seems correct. This comparison shows that the auditory system of the manatee is able to reproduce several times less rates than that of odontocetes.

All the results presented above indicate that manatees do not have such exclusively high auditory abilities as those of odontocetes.

3.3. SUMMARY

In pinnipeds, the outer ear is modified, but the middle and inner ears have a number of features characteristic of terrestrial mammals. Sound transmission is through the tympanic membrane and middle-ear ossicles. Psychophysically measured hearing frequency ranges were of a few tens of kHz. Underwater hearing thresholds were found to be 63 to 80 dB re 1 µPa, and aerial hearing thresholds 20 to 40 dB re 20 µPa. Thus, expressed in the sound-pressure measure, the underwater and aerial thresholds were of the same order of magnitude. However, in terms of the power flux density, aerial thresholds are much higher than underwater ones. Temporal-summation data are very limited and have shown not constant but frequency-depended integration time. Attempts to measure frequency tuning by the critical-ratio and critical-band methods did not yield definite results. Underwater frequency discrimination limens in different pinniped species were found to be from 1 to 10%; intensity discrimination thresholds more than 3 dB. The minimal audible angle (MAA) in air was from a few deg to about 1 deg; in water, MAA estimates varied depending on frequency from 10 deg to a nonmeasurable angle (no discrimination at any angle). Representation of the auditory system in the cerebral cortex does not differ from that in terrestrial carnivores. Supposedly, the moderate hearing abilities and the absence of echolocation in pinnipeds are connected with the amphibious functioning of the auditory system, which did not result in evolving a truly aquatic high-frequency and high-sensitivity hearing required for underwater echolocation.

In sirenians, the middle-ear structure indicates the possibility of sound transmission in a conventional way, through the tympanic membrane and the middle-ear ossicles, though the outer ear is markedly modified. In the inner ear, the basilar membrane does not feature properties characteristic of ultrasonic hearing. Nevertheless, psychophysical audiograms showed a good sensitivity (thresholds as low as 50 dB re 1 μPa) and rather wide frequency range (more than 40 kHz). The evoked-potential (ABR) audiogram featured a similar frequency range though somewhat higher thresholds. In the rhythmic click test, ABR followed a stimulation rate several times lower than that in odontocetes.

Chapter 4
VISION IN AQUATIC MAMMALS

The visual system of aquatic mammals is of special interest because of its mostly amphibious mode of action. Although truely aquatic mammals (cetaceans, sirenians) spend their entire lives in water, their aerial mode of breathing confines them to the superficial water layer. Semiaquatic mammals (pinnipeds) spend a significant proportion of time on land. Therefore, the visual system in most of aquatic mammals functions both in air and water, though these two media have very different optical features and thus impose different requirements on the visual system.

4.1. VISION IN CETACEANS

4.1.1. Eye Morphology

4.1.1.1. Eye Anatomy and Optics

Ocular anatomy and optics in cetaceans are significantly different from those in terrestrial mammals. The structure of the eyecup and the refractive structures of the eye in cetaceans are determined primarily by the optical properties of the aquatic medium (the refractive index of water is much higher than that of air) and by a number of other factors: low temperature and low luminosity deep in water, strong light scatter by particles (plankton and others) suspended in water, and so on.

Characteristic examples of eye structure in cetaceans are presented in Fig. 4.1. Prominent features typical for all cetaceans are a thick sclera (particularly so in whales, Fig. 4.1B), a significantly thickened cornea, a highly developed choroid, and massive ocular muscles. These peculiarities probably serve to protect the eye from underwater cooling and mechanical damage.

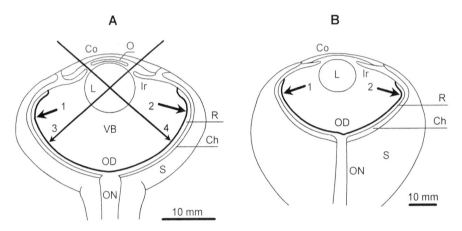

Figure 4.1. Diagrammatic drawings of the structure of cetacean eyes. **A**. The eye of the bottlenose dolphin *Tursiops truncatus*. **B**. The eye of the gray whale *Eschrichtius gibbosus*. Co – cornea, O – operculum, L – lens, Ir – iris, VB – vitreous body, R – retina, ON – optic nerve, Ch – choroid, S – sclera. Arrows 1 and 2 delimit the part of the retina that can adequately be approximated by a segment of a spherical surface concentric with the lens. Arrows 3 and 4 show the directions of light rays passing through the edges of the pupil and the center of the lens to the areas of maximum ganglion cell density (see below).

The shape of the eyeball is markedly altered as compared to terrestrial mammals. While the eyecup in terrestrial mammals is typically almost spherical, there is substantial flattening of the anterior segment in whales and dolphins, resulting in a hemispherical shape of the eyecup. The flattening of the anterior segment manifests itself in a small anterior chamber of the eye. Apart of that, the eyeball is somewhat flattened in the dorso-ventral direction as compared to the naso-temporal (horizontal) one, so from the front the eye looks as a slightly stretched ellipse. But as a first approximation, the eyecup shape is close enough to a hemisphere. More precisely speaking, the shape approximates not a complete hemisphere but a segment of a sphere of a large arc of about 150 deg, as delimited by arrows in Fig. 4.1.

In terrestrial mammals, the convex outer surface of the cornea is the major refractive element of the eye, as it separates media with significantly different refraction indices: air (the refraction index of 1) and cornea (the refraction index of around 1.35). In cetaceans, the refraction index of the cornea was found to be somewhat larger than in terrestrial mammals, about 1.37 (Dawson et al., 1972; Kröger and Kirschfeld, 1994). Nevertheless, even this rather high refractive index is little different from that of water (1.33–1.34). Therefore, the outer cornea surface provides little contribution to the overall refraction power of the cetacean eye, and the main refraction necessary to focus an image on the retina is provided by the lens (Sivak, 1980). This is why the lens is almost spherical in most cetaceans or, as in a few mysticetes and the beluga whale, of a slightly elliptical shape (see Fig. 4.1). The large curvature of the lens surface provides a sufficiently high refractive power to focus images on the retina despite the very weak refraction at the cornea surface in water. This optics is similar to that in fish, which is quite understandable as a common sequence of the optical properties of water.

A strongly convex (spherical) lens made of homogeneous material is known to have very strong spherical aberration. The cetacean lens is to a large extent free of this disadvantage. Elimination of spherical aberration is achieved due to heterogeneous lens structure: the outer layers have a lower refractive index than the inner nucleus (Rivamonte, 1976; Kröger and Kirschfeld, 1993).

An important feature of the cetacean eye is that the centers of the lens and of the spherical segment of the eyecup almost coincide. This produces a centrally symmetric optic system, so that light rays coming from any direction are focused almost identically on the retina. In this feature, the cetacean eye is different from the eyes of terrestrial mammals, which have clearly pronounced axial symmetry.

The spherical shape of the lens in cetaceans led to loss of the accommodatory mechanism typical of terrestrial mammals – that is, change of the lens shape by the ciliary muscle. The ciliary muscle is poorly developed in all dolphins and is absent from most whales (Waller, 1984; West et al., 1991), suggesting that accommodation can not be achieved by the change of the lens shape (Dral, 1972). This is supported by ophthalmoscopic observations (Dawson et al., 1987b) which revealed no significant accommodative changes.

It has, however, been suggested that accommodation in cetaceans does occur by a mechanism different from that operating in terrestrial mammals. According to this suggestion, accommodation in cetaceans, like in fish, is produced by axial displacement of the lens due to changes in intraocular pressure. The intraocular pressure in dolphins is rather high, 65 to 72 mm Hg in the harbor porpoise (Kröger and Kirschfeld, 1989) and 25 to 33 mm Hg in the bottlenose dolphin (Dawson et al., 1992). Intraocular pressure can be changed within a certain limit due to contraction/relaxation of the retractor/protractor muscles, which produce axial displacement of the eye in the orbit. It has been suggested that intraocular pressure increases when the eye is pulled back into the orbit; the increased intraocular pressure, in turn, results in shifting the lens forward. The lens moves backward when the eye is moved forward, thus producing a decrease of the intraocular pressure.

The cornea in cetaceans is much thicker than in humans and most terrestrial mammals, and its thickness is not uniform. The cornea is thinner in the center and thicker at the periphery, and it forms a characteristic limb at its edges attached to the sclera. This is characteristic of both odontocetes and mysticetes (Pütter, 1903; Rochon-Duvigneaud, 1939; Mann, 1946; Dawson, 1980; Pardue et al., 1993; van der Pol et al., 1995).

Although the role of the cornea in light refraction in the cetacean eye is not as crucial as in terrestrial mammals, it does contribute to refraction. The outer corneal surface has lower curvature than the inner surface, so the cornea has a shape of a divergent lens (Kröger and Kirschfeld, 1992, 1993, 1994; van der Pol et al., 1995). Under water, the divergent refraction by the cornea is small since the media on its both sides (water outside and anterior chamber liquid inside) have refractive indices not greatly different from that of the cornea itself. Nevertheless, even under water, refraction at the cornea cannot be neglected completely. Its refractive index varies from 1.37 in the central part to 1.53 in the peripheral thickened part, whereas the refractive index of water is 1.33 to 1.34. Thus, the cornea really acts as a weak divergent lens. Experimental studies with the use of laser interferometry have shown that the total refraction of the cornea and the lens make the cetacean eye just emmetropic under water (Kröger and Kirschfeld, 1994).

Adaptation of the cetacean eye to underwater vision also concerns the structure of the iris and pupil. The cetacean eye functions in conditions of large and rapid changes of illumination when the animal dives from the surface, where illumination is high, into the depth, where illumination level is greatly attenuated, and vice versa. This requires the pupil to react to a wide range of changes in illumination. An unusual pupil shape in cetaceans fits this requirement. The upper part of the iris in cetaceans has a characteristic protuberance, the operculum (Fig. 4.2). In conditions of low illumination, the operculum is contracted (raised), so that the pupil, similarly to other mammals, is a round or oval hole with a horizontal diameter of around 10 mm (Fig. 4.2A). When illumination increases, the operculum moves down, turning the pupil into an U-shaped slit (Fig. 4.2B). At high illumination, the operculum moves down so far that the slit becomes almost completely obstructed, leaving only two narrow holes at the nasal and temporal ends of the slit (Fig. 4.2C) (Dawson et al., 1979; Herman et al., 1975). Our observations have shown that this pupil shape is characteristic of many dolphin species, including the bottlenose dolphin *Tursiops truncatus*, harbor porpoise *Pho-*

A   B   C

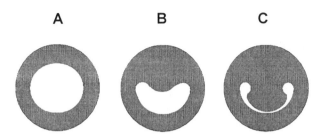

Figure 4.2. Shape of the pupil in the bottlenose dolphin *Tursiops truncatus.* **A**. Dilatated pupil. **B**. Partially constricted pupil. **C**. Completely constricted pupil.

coena phocoena, common dolphin *Delphinus delphis*, tucuxi dolphin *Sotalia fluviatilis*, and also the gray whale *Eschrichtius gibbosus*. The only found exception was the Amazon river dolphin *Inia geoffrensis*, which, like most terrestrial mammals, has a round pupil.

While it is clear how the refractive system of the cetacean eye functions under water and why the eye is emmetropic, its operation in air remains poorly understood. The problem is that in air, refraction by the lens is supplemented by refraction at the convex outer surface of the cornea. The significant difference between the refractive indices of air and the cornea makes this surface a powerful convergent unit. Its refractive power was estimated as large as 20 D (diopters) (Dral, 1972; Dawson et al., 1987b). Thus, the cetacean eye should be catastrophically myopic in air. This expectation, however, does not agree with a great many observations and experimental data (see below) demonstrating good visual acuity of dolphins not only in water but also in air.

Several attempts were made to find flattened regions of the cornea in dolphins or at least regions with minimal curvature. A flat corneal surface would not produce additional refraction in air, so such regions would function as emmetropic windows for clear vision both in air and water (the effect of a diving mask). Detailed keratoscopic studies of corneal surface in bottlenose dolphins revealed that, indeed, there were "spoon"-shaped regions near the nasal and temporal margins of the cornea, with markedly lower curvature than in the central region (Dawson et al., 1987b). No other regions with lower refraction, as hypothesized for the ventronasal region of the cornea (Dral, 1972) were found in the bottlenose dolphin.

Aerial myopia could be partially compensated by accommodatory axial displacement of the lens, which, as noted above, results from changes in intraocular pressure, which are in turn is caused by retraction/protraction movements of the eye. For aerial vision, the dolphin eye can move axially forward; it has been suggested that this produces a significant backward movement of the lens and therefore reduction of myopia. Additionally, it has been shown that changes in intraocular pressure in dolphins produce significant change in the curvature of the cornea. Under water, where the eye is retracted into the orbit and intraocular pressure is higher, the corneal surface is of more convex shape, while in air, when the eye is protracted from the orbit, intraocular pressure decreases, the cornea flattens, reducing myopia (Kröger and Kirschfeld, 1989).

There is, however, another model that explains the functioning of the cetacean eye both in water and air, which does not involve an accommodatory mechanism (Rivamonte, 1976). This model suggests that myopia in air does not arise because vision in water and vision in air use different parts of the lens with different refraction indices. According to the hypothesis, vision in water depends on the central core of the lens with a high refractive index, whereas vision in air exploits only the peripheral part of the lens with a lower refractive index; the latter compensates for additional refraction at the corneal surface in air.

An additional ability for correction of aerial myopia can be provided by constriction of the pupil. Above water, diurnal luminosity is high and evokes strong constriction of the pupil, which to a large extent corrects errors of refraction, including aerial myopia. Apart from myopia, constriction of the pupil above water reduces astigmatism arising from the nonuniform curvature of the cornea in the vertical and horizontal directions, as well as refraction errors arising from spherical aberrations in the lens.

As shown below, the characteristic shape of the constricted pupil in cetaceans (as two pin-holes) correlates with the retinal topography, which also plays an important role in the perception of well-focused optic images.

One more adaptation of the cetacean eye to underwater vision in conditions of low luminosity is a highly developed reflective layer, the tapetum. The tapetum is present in all cetaceans, and it is especially well developed in balene whales. In cetaceans, reflective cells containing guanine crystals are located parallel to the retinal surface and reflect mainly the blue-green part of the spectrum, corresponding well to the absorption maximum of the photosensitive pigment in the retinal receptors (rods), which is 490 nm. This increases visual sensitivity in scotopic conditions (Madsen and Herman, 1980). The structure of the tapetum and its properties have been described in a number of cetaceans (Dawson, 1980; Young et al., 1988). In all cetaceans, the tapetum covers at least the upper two-thirds of the fundus (Dawson, 1980), and in some whales it covers the entire fundus (Waller, 1984). Such complete coverage of the fundus by the tapetum in cetaceans is unique among mammals; in terrestrial mammals, the tapetum does not usually extend below the horizontal equator of the eye (Prince, 1956).

4.1.1.2. Eye Movements

The question of eye movements in cetaceans remained under discussion for a long time. Only since it has become possible to keep dolphins and small whales in captivity has the question been clarified. It was found that all dolphins and whales have quite mobile eyes. Morphological studies have shown that the oculomotor muscles are well developed in dolphins and whales, allowing eye movements in both the horizontal and vertical directions (Hosokawa, 1951). In addition, cetaceans have powerful retractor/protractor muscles that are not known in terrestrial mammals. Along with the four straight and two oblique muscles, the retractor/protractor produce forward/backward (axial) movements of the eyes (Yablokov et al., 1972). In air, the bottlenose dolphin *Tursiops truncatus* is capable to move its eyes forward some 10 to 15 mm and pull them back. This eye property may be used for binocular examination of objects (Dral, 1972; Dawson, 1980). The only kind of eye movements not observed in dolphins and whales is torsion movements. However, experimental data indicated that in general, eye mobility in the bottlenose dolphin is lesser than in humans, and the movements are slower (Dawson et al., 1981).

Another intriguing property of oculomotor activity in dolphins that distinguishes them from terrestrial mammals is the ability to move the left and right eyes independently. Independent eye movements in both horizontal and vertical directions have been described (McCormick, 1969; Dawson, 1980). The retraction/protraction movements of the left and right eyes can also be independent (Dawson et al., 1981). Quantitative measurements of coordinated activity of the two eyes in laboratory conditions have confirmed that the correlation coefficient between movements of two eyes is very low (Dawson et al., 1981).

Apart from independent eye movements, cetaceans feature independent pupil reflexes of the left and right eyes and independent eyelid movements. At first it was demonstrated in dolphins during sleep when one eye could be open whereas the other one closed (Lilly, 1964). In general, the bottlenose dolphin demonstrates long periods of swimming with only one eye open, the left and right eyes alternating (Lilly, 1964; Bateson, 1974). A similar observation was described for a gray whale (Lyamin et al., 2000).

4.1.1.3. Retinal Morphology

The laminar structure of the retina was investigated in a variety of cetacean species: in the bottlenose dolphin *Tursiops truncatus* (Perez et al., 1972; Dawson and Perez, 1973; Dral, 1977; Dawson, 1980), the common dolphin

Delphinus delphis (Dral, 1983), the Dall's porpoise *Phocenoides dalli* (Mu-rayama et al., 1992a, 1995), the beluga whale *Delphinapterus leucas* and finwhale *Balaenoptera physalus* (Pütter, 1903; Pilleri and Wandeler, 1964), and Minke whale *B. acutorostrata* (Murayama et al., 1992a, 1992b). The laminar structure of the retina in all cetaceans is qualitatively similar to that in terrestrial mammals. It consists of the receptor layer, outer nuclear layer, outer plexiform layer, inner nuclear layer, inner plexiform layer, ganglion cell layer, and, finally, the optic-fiber layer. This laminar structure is similar in all mammals, including cetaceans. Even in a cetacean species with strongly reduced visual system, the Ganges river dolphin *Platanista gangetica* (there is almost no lens and there are no oculomotor muscles in its eye), the laminar structure of the retina shows no radical changes and contains all the layers (Dral and Beumer, 1974; Purves and Pilleri, 1974).

Being qualitatively similar to other mammals, laminar structure of the retina in cetaceans markedly differs quantitatively. The cetacean retina is much thicker than that of terrestrial mammals, ranging from 370 to 425 µm (Dral, 1977; Dawson et al., 1982, 1983; Murayama et al., 1995). For comparison, the thickness of the retina in diurnal mammals lacking a foveal area is 112 to 240 µm in the dog, 115 to 220 µm in the cat, and 110 to 220 µm in the horse and cow (Prince et al., 1960).

The most detailed descriptions of the retina are available for the bottle-nose dolphin *Tursiops truncatus* (Perez et al., 1972; Dawson and Perez, 1973; Dral, 1975a, 1975b, 1977; Dawson, 1980; Dawson et al., 1982).

The receptor layer of the retina in all cetaceans thus far studied consists predominantly of rods. The question of the existence of cones remained debatable for some time. An early comparative study of the retina in a number of cetaceans, both odontocetes (*Delphinus delphis, Phocoena phocoena, Physeter catodon*) and mysticetes (*Megaptera noaveangliae, Balaenoptera borealis, B. sibbaldi, B. musculus*) have led to a conclusion that cones in these species are either absent or very few in number (Rochon-Duvigneaud, 1939). The first observation that the cetacean retina contains two types of photoreceptors was reported by Mann (1946). Subsequently, cones were described in the finwhale *Balaenoptera physalus* and beluga whale *Delphinapterus leucas* (Pilleri and Wandeler, 1964). The first study of the laminar structure of the retina in the bottlenose dolphin using Golgi preparations (Perez et al., 1972) as well as subsequent communications (Dawson and Perez, 1973; Dral and Beumer, 1974) reported different receptor profiles in the receptor layer and different sizes of photoreceptor endings in the outer plexiform layer, which are indicative of different receptor types. Later, tangential sections of the retina in the bottlenose dolphin provided the first description of a receptor mosaic, in which a small number of large conelike units were embedded among numerous small round units. This observations, along with electron microscopic studies, provided better evidence for existence of different receptor types in the bottlenose dolphin retina (Dawson, 1980). Two types of receptors were also seen by electron microscopy in the retina of the Amazon river dolphin *Inia geoffrensis* (Waller, 1982). Recently, similar data were obtained by light microscopy in the Dall's porpoise *Phocoenoides dalli* (Murayama et al., 1995). However, precise identification of the observed profiles has not been done in that study.

Recent studies with the use of immunochemical identification of visual pigments have shown that the cone opsin does exist in receptors of the retina in the bottlenose dolphin; thus, the retina does contain cone receptors. However, contrary to the majority of terrestrial mammals, which have two types of cones with different pigments providing color vision (short-wave sensitive S-opsin and middle-to-long-wave sensitive L-opsin), only L-opsin containing cones were found in the dolphin retina (Fasick et al., 1998; Peichl and Berhmann, 1999.

A detailed description of the structure of the outer nuclear layer, the outer plexiform layer, the inner nuclear layer, and the inner plexiform layer, based on Golgi preparations of the bottlenose dolphin retina, was reported by Perez et al. (1972), and a similar description was reported on other cetacean species (Dawson, 1980). According to these descriptions, amacrine, bipolar, and horizontal cells are generally similar to those in terrestrial mammals. However, detailed analysis of morphology of different retinal cell types has not yet been performed.

The inner plexiform and ganglion cell layers of the retina demonstrate the most prominent difference between terrestrial mammals and cetaceans. Particularly, the ganglion cell layer in cetaceans differs from that of terrestrial mammals. Ganglion cells have been described in most detail using Golgi preparations in the common dolphin *Delphinus delphis* and the bottlenose dolphin *Tursiops truncatus* (Shibkova, 1969; Perez et al., 1972; Dawson et al., 1982). The ganglion layer consists of a single row of large, sparsely distributed neurons separated by wide intercellular spaces. This cell pattern was also easily visible in Nissl-stained retinal transversal sections from the beluga whale *Delphinapterus leucas* (Pütter, 1903), the finwhale *Balaenoptera physalus* (Pilleri and Wandeler, 1964), and the Dall's porpoise *Phocoenoides dalli* (Murayama et al., 1995) as well as in retinal wholemounts from the same species (Dral, 1977, 1983; Mass and Supin, 1986, 1995a, 1995b; Murayama et al., 1995), the gray whale *Eschrichtius gibbosus* (Mass, 1996), and the Minke whale *Balaenoptera acutorostrata* (Murayama et al., 1992a, 1992b).

Both Golgi preparations and Nissl-stained wholemounts revealed very large neuron bodies with a clear membrane, large amount of cytoplasm, a clearly visible nucleus of up to 15 µm in diameter, and light nucleolus of 4 to

5 µm in diameter. Cell bodies contained easily visible, intensely stained Nissl granules. All large neurons (particularly giant neurons of the whale retina) showed several axon/dendrite bases.

A remarkable feature of the cetacean retina is the presence of extremely large, giant ganglion cells. Most ganglion cells in the cetacean retina are of rather large size, but the giant cell size reaches 75–80 µm and more. These large neurons have been described in many studies of a number of dolphin species: in the bottlenose dolphin *Tursiops truncatus* (Perez et al., 1972; Dawson et al., 1982; Mass and Supin, 1995a, 1995b), the common dolphin *Delphinus delphis* (Dral, 1983), the harbor porpoise *Phocoena phocoena* (Mass and Supin, 1986), the Dall's porpoise *Phocoenoides dalli* (Murayama et al., 1995), the Chinese river dolphin *Lipotes vexillifer* (Gao and Zhou, 1987), and a few mysticete species (Pilleri and Wandeler, 1964; Murayama et al., 1992a, 1992b; Mass, 1996) (Fig. 4.3). In the retina of the finwhale *Balaenoptera physalus* (Pilleri and Wandeler, 1964) and the bottlenose dolphin *Delphinus delphis* (Dawson and Perez, 1973), giant ganglion cells sized more than 100 µm were described; however, these observations were not confirmed by further studies.

It should be noted that the term *giant* as applied to terrestrial mammals

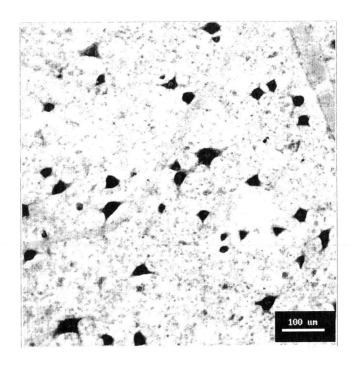

Figure 4.3. Ganglion cells in a retinal wholemount from a gray whale *Eschrichtius gibbosus.*

implies ganglion cells up to 15–35 μm including parts of dendrites (Fukuda and Stone, 1974; Hebel and Hollander, 1979; Hughes, 1981). In cetaceans, sizes of just cell bodies can exceed 75 μm.

In some cetacean species, ganglion cells do not reach such large sizes. In the Amazon river dolphin, the largest ganglion cells did not exceed 40–42 μm (Waller, 1982; Mass and Supin, 1989); in the retina of the Ganges river dolphin *Platanista gangetica*, ganglion cells not larger than 20 μm were found (Dral and Beumer, 1974). It should be noted, however, that even these cells are markedly larger than ganglion cells in many terrestrial mammals.

The smallest ganglion cells found in cetaceans by all authors cited above were not less than 10–12 μm.

More complete quantitative characterization of ganglion cell sizes were presented for the bottlenose dolphin *Tursiops truncatus* (Mass and Supin, 1995a, 1995b). Figure 4.4 presents the cell-size distributions in various part of the retina, with high and low concentration of ganglion cells. The histograms show that the most probable cell size is 20 to 35 μm, though cells as large as 50–60 μm are also present, while there are no cells smaller than 10 μm.

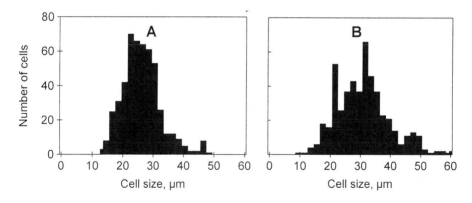

Figure 4.4. Size distributions of ganglion cells in the retina of a bottlenose dolphin *Tursiops truncatus*. **A**. Sample from a region of high cell density in the temporal area of the retina. **B**. Sample from a region of low cell density in the ventral part of the retina. Each sample contained 500 cells.

The large size of ganglion cells, mainly 20 to 30 μm or more, is also characteristic of retinas of many other cetacean species: the harbor porpoise *Phocoena phocoena* (Mass and Supin, 1986), the Amazon river dolphin *Inia geoffrensis* (Mass and Supin, 1989), the minke whale *Balaenoptera acutorostrata* (Murayama et al., 1992a, 1992b), the Dall's porpoise *Phocoe-*

noides dalli (Murayama et al., 1995), the gray whale *Eschrichtius gibbosus* (Mass and Supin, 1997), and the tucuxi dolphin *Sotalia fluviatilis* (Mass and Supin, 1999). The large sizes are not characteristic of neurons at all levels of the visual system in cetaceans (the lateral geniculate body, the visual cortex, and so on). They are typical only of retinal ganglion cells. For example, the largest pyramidal cells in Golgi preparations of the lateral gyrus (visual area) of the bottlenose dolphin cerebral cortex did not exceed 20–30 μm (Garey et al., 1985). Maximal cell sizes found in other parts of the dolphin brain were also no more than 20–45 μm (Jacobs et al., 1971, 1979). Thus, very large neurons are not a feature of the cetacean brain as a whole.

To date, there is no commonly adopted explanation of why the ganglion cells in the cetacean retina are so large. It is possible that giant ganglion cells, with their thick axons conducting nerve spikes at high velocity, are needed for fast signal transmission through long nerve pathways in a large body. However, large terrestrial mammals (for example, bulls or elephants) have retinal ganglion cells of no more than 25–30 μm (Hebel and Hollander, 1979; Stone and Halasz, 1989).

It might be suggested that the system of large ganglion cells in cetaceans is in some way analogous to the Y system in terrestrial mammals (Fukuda and Stone, 1974). However, unlike the Y neurons of terrestrial mammals, which account for no more than 1% of all retinal cells, the proportion of large cells and fibers in cetaceans is incomparably higher.

Apart of large cell sizes, a characteristic feature of the ganglion layer in the cetacean retina is low cell density. The large neurons are separated by wide spaces (see Fig. 4.3). As will be shown below, this feature is also characteristic of the optic nerve. This property has been noticed by many investigators, both in Nissl wholemounts and in Golgi preparations (Pilleri and Wandeler, 1964; Shibkova, 1969; Dral, 1977). However, groups of closely spaced cells were also observed by these authors.

The question of morphological types of ganglion cells in cetaceans has not yet entirely been answered. Cross Nissl-stained sections of the retina in the bottlenose dolphin *Tursiops truncatus* (Dral, 1975a, 1977) and Dall's porpoise *Phocoenoides dalli* (Murayama et al., 1995) have shown two types of ganglion cells. One type was represented by lightly stained, round or oval cells with large nuclei, fine tigroid in the cytoplasm, and only one clearly visible process, the axon. Cells of another type were dark, similar in shape to pyramidal neurons, contain rough and dark tigroid, with clearly visible axon/dendrite bases. No idea of the functional significance of these cell types was proposed.

4.1.1.4. Optic Nerve Structure

The large-size ganglion cells of the cetacean retina send their thick axons into the optic nerve. The diameters of these fibers are significantly greater than those in terrestrial mammals. For example, the maximum optic-fiber diameter in cats and monkeys is no more than 8 μm, while in a variety of dolphin species, more than 6% of optic fibers are larger than 15 μm in diameter (Dawson et al., 1982; Gao and Zhou, 1992). The mean fiber diameter in a few cetacean species was assessed (Gao and Zhou, 1992) as 1.6 to 2.5 times greater than in the cat (Hughes and Wässle, 1976) and 3 to 5 times greater than in humans (Jonas et al., 1990). The only exception is presented by the Chinese river dolphin *Lipotes vexillifer*, which has thin fibers, though ganglion cells as large as 75 μm were described in its retina (Gao and Zhou, 1987)

The low density of ganglion cells in the cetacean retina corresponds to a low density of axons in the optic nerve. In dolphins, cross sections of the optic nerve revealed a fiber density of 48,000 fibers/mm^2 (Dawson et al., 1983). For comparison: in humans and monkeys, the optic-fiber density approaches 220,000 fibers/mm^2. While the cross-section area of the optic nerve in dolphins is larger than in terrestrial mammals, the total number of optic fiber is markedly lower. More than 50% of the cross-section area of the dolphin optic nerve is occupied by interneuronal space (compared with 12 to 20% in terrestrial mammals), rather than glia. It has been supposed that this feature may be associated with peculiarities of tissue metabolism in diving animals.

The number of fibers in the optic nerve differs in various cetacean species. The smallest number of fibers (from 14,000 to 16,000) were found in the optic nerve of the Ganges river dolphin *Platanista gangetica* and the Amazon river dolphin *Inia geoffrensis*, which inhabit turbid and low-transparent waters (Morgane and Jacobs, 1972; Dral and Beumer, 1974). In the bottlenose dolphin *Tursiops truncatus*, the number of optic fibers ranges from 147,000 to 185,000 (Morgane and Jacobs, 1972; Dawson, 1980; Dawson et al., 1982). Other odontocetes feature fiber counts similar to that in the bottlenose dolphin. In mysticetes, the numbers of optic fibers were estimated from 250,000 to 420,000. Data on optic fiber counts in a variety of marine mammals have been collected by Gao and Zhou (1992).

4.1.2. Visual Abilities of Cetaceans: Psychophysical Studies

It was long believed that dolphins – animals with excellent hearing and capability of echolocation – have a poorly developed visual system that plays a minor role in their lives. However, observation of the visual activity of dolphins in air where echolocation is hardly possible has demonstrated the opposite. The ability to catch fish in air, to perform precisely directed jumps through obstacles to reach a target located at significant height above the water, to discriminate their trainers and fine details of objects, all indicate that vision in dolphin is quite well developed. In conditions of long captivity, dolphins are known to decrease their echolocation activity, and, as their interest in events above the water increases, visual orientation takes on a leading role.

The ability to keep cetaceans (dolphins and small whales) in captivity, which has developed over recent decades, has facilitated more detailed observations and a series of precise experimental studies. The resulting data have altered our views on the organization and abilities of their visual system and provided a basis for regarding the visual modality as playing an important role in various aspects of cetacean life: in social interactions, in the search and discrimination of prey, in spatial orientation, in reproductive activity, and in defense. A few reviews (Madsen and Herman, 1980; Herman and Tavolga, 1980; Herman, 1980; Mobley and Helweg, 1990) summarize data on the visually driven behavior of cetaceans. It is self-evident that in spite of excellent hearing and echolocation abilities, hearing cannot substitute vision for discrimination between individuals and other cetacean species based on their colors and individual markers, which is important in mating and social interactions in the school. Only vision provides the ability to discriminate objects in air and to precisely and rapidly assess the distance to objects in air, where echolocation does not operate.

4.1.2.1. Visual Acuity

Apart from the widely known observations mentioned above, good visual abilities of cetaceans were demonstrated by psychophysical experiments for assessing their visual acuity. The first experimental data on visual acuity under water were obtained on the white-sided Pacific dolphin *Lagenorhinchus obliquidens* (Spong and White, 1971) and the killer whale *Orcinus orca* (White et al., 1971). Visual acuity of these two species was assessed as 6 and 5.5 min, respectively. However, it turned out later that these early studies contained a number of methodological errors affecting the equalization of stimulus and background brightness, which led to overestimation of the visual acuity. Nonetheless, as demonstrated in subsequent experiments, visual acuity of dolphins is quite satisfactory both in water and air.

An attempt to measure quantitatively the visual acuity in dolphins was made by Pepper and Simmons (1973). They found aerial acuity of 18 min, which can be assessed as a moderately good visual ability. Noordenbos and Boogh (1974) made an attempt to measure quantitatively the visual acuity in the bottlenose dolphin *Tursiops truncatus*. Their data indicated rather poor resolution of 27 min. This estimate, however, was not confirmed later, so it seems that it was a result of the insufficient performance of the animal.

A precise psychophysical measurement of visual acuity in the bottlenose dolphin *Tursiops truncatus* was performed by Herman et al. (1975). To test visual acuity, gratings of variable strip widths (spatial frequencies) were used. Either a test (grating) or the standard (gray) target was presented to the animal, which was trained to press one of two paddles depending on the presented target – that is, the experimental paradigm included elements of a two-alternative forced choice, though only one stimulus was presented in each trial. Thresholds were estimated by the grating strip width providing 75% correct response performance. Test stimuli were presented both in air and in water, at distances from 1 to 2.5 m. It was found that the best resolution in water was 8.2 min at a distance of 1 m; in air, the best resolution was 12.5 min at a distance of 2.5 m.

The obtained ratio of the aerial to underwater visual acuity of 1.33 is equal to the ratio of the refraction indices of these two media. When an optic system able to function properly both in air and water (it is possible if the front surface of the optics is flat) is transmitted from air to water and back, the size of projected images changes by this ratio: under water, the image size is 1.33 times larger than in air (an effect known to any diver: under water, all objects seem larger than in air). Therefore, it seems that the visual acuity in the bottlenose dolphin is determined by its retinal resolution, and the difference between aerial and underwater visual acuity results from different sizes of retinal images.

4.1.2.2. Visual Field

All cetaceans have more or less laterally positioned eyes. Therefore, their underwater visual field is fairly large, almost panoramic. In many dolphins, the visual field reaches 130 deg along azimuth (Yablokov et al., 1972), in the sperm whale *Physeter catodon* it is around 125 deg (Sliper, 1962). Thus, the visual system underwater serves other spatial sectors than the echolocation system, which functions mainly in a narrow frontal sector. So the visual system of cetaceans provides significant support to the echolocation orientation.

In many dolphins and whales, the eyes are to some extent directed forward and downward (ventronasally). On viewing an object in air, a dolphin can move its eye forward by some 10 to 15 mm; in this eye position, the visual fields of the both eyes overlap in the frontal sector by 20 to 30 deg, giving binocular vision (Dral, 1977; Dawson, 1980).

It should be stressed, however, that the presence of a binocular sector in the visual field is not a direct evidence of the presence of truly binocular (stereoscopic) vision. At least in the bottlenose dolphin, noncrossed visual fibers (which exist in all mammals with binocular vision and provide a basis for binocular integration) have not been found yet (Jacobs et al., 1975; Tarpley et al., 1994). So the question of existence of binocular stereoscopic vision in dolphins still remains open, though some behavioral studies indicated a possibility that dolphins can recognize two- and three-dimensional objects (Hunter, 1988).

4.1.2.3. Visual-Auditory Intermodal Transfer

It has been shown (Herman and Pack, 1992, 1998; Pack and Herman, 1995) that bottlenose dolphins *Tursiops truncatus* can perceive stimulus shape information for fairly complex configuration using both the visual and echolocation systems, with the possibility of intermodal transfer: experience of visual perception could be used for echolocation recognition and vice versa. It was shown in matching-to-sample tests when a dolphin had to select a test object identical to the sample one. To perform the matching-to-sample test, a sample object was exposed to either the dolphin's visual or echolocation inspection. After that, a pair of objects was presented, one to each side of the sample object; one of the test objects was identical to the sample one. Test objects were also presented to either the visual or echolocation sense. To expose objects to echolocation, they were placed in underwater boxes made of sound-transparent but not light-transparent material; to expose objects to vision, they were presented in air. A wide variety of complexly shaped objects were used (construction of PVC pipes). The dolphin had to touch a paddle near the matching-to-sample object; so it was a true two-alternative forced-choice paradigm.

It was found that the dolphin successfully performed intermodal recognition of the test object – that is, when the sample object was presented to vision and test objects were presented to echolocation, or vice versa. The performance was not worse than in within-modal tasks – that is, when both sample and test objects were presented to either vision or echolocation. The success of intermodal transfer did not depend on whether the sample was presented to vision and tests to echolocation, or vice versa. Moreover, the task was solved successfully if an image on a television screen was used instead of the real object for visual inspection. The intermodal transfer was also successful when both the sample and test objects were new objects, never inspected by the dolphin before the test trial (Herman and Pack, 1998).

Less successful intermodal transfer was described by Harley et al. (1995, 1996). The animal performed intermodal recognition when sample and test objects were previously inspected by both echolocation and vision; the echoic-to-visual transfer was more successful than the visual-to-echoic one. The animal failed to perform when new sets of objects were presented the first time. However, in the latter case, the dolphin was not able to within-modal recognition either; in this situation, inability to perform intermodal recognition does not deny the ability to intermodal transfer.

Thus, the dolphin's visual system is not only involved in supporting the echolocation system in underwater orientation but also functions in tight concordance with the latter. On the other hand, the experiments had shown the capability of visual discrimination and recognition of complex forms.

4.1.3. Topographic Distribution of Retinal Ganglion Cells

Ganglion cells are distributed nonuniformly in the mammal retina: the concentration of ganglion cells (number of cells per area unit) in some retinal areas is low, whereas in other areas it is much higher. Regions of high ganglion cell concentration are of special interest since they provide the most detailed analysis of visual images. The shape and position of these regions are very informative for understanding the organization of the visual fields of a certain animal species.

4.1.3.1. Some Principles of Retinal Topography in Mammals

A large body of experimental data obtained mostly in traditional subjects of laboratory studies (monkeys, cats, rats, rabbits, and so on) has demonstrated that the pattern of high cell-density regions is associated with the eye position, eye mobility, illumination level, and lifestyle. The main characteristics of retinal ganglion cell topography have been reviewed by Perry (1982), Stone (1983a), and Hughes (1985).

The major principles of retinal topography in mammals have been laid out by Stone (1983a). The area of the highest concentration of ganglion cells in animals with predominantly frontal vision appears as the fovea of monkeys or the *area centralis* of cats. These areas are located at the vertical meridian of the visual field representing the decussation line between the crossed and uncrossed visual projections. These regions belong to the binocular sector of the visual field and provide the highest visual resolution and binocular vision.

The primate fovea features a particularly high density of ganglion-cell receptive fields (Webb and Kaas, 1976; Wässle et al., 1990). In animals with frontally positioned eyes but not having the fovea, the zone of the highest receptor and ganglion-cell density composes the *area centralis*, as a rule, oval or ellipsoid in shape and located also at the decussation line separating the crossed and uncrossed projections. The most detailed studies of the *area centralis* were performed in cats (see reviews cited above). Regions resembling the *area centralis* have been observed in other terrestrial mammals: mice (Drager and Olsen, 1981), hamsters (Tiao and Blakemore, 1976), and a number of marsupials (Tancred, 1981; Kolb and Wang, 1985; Dunlop et al., 1994; Arrese et al., 1999).

In animals with laterally positioned eyes, the region of increased ganglion-cell density typically looks like a rather narrow horizontal strip, the visual streak. Within the streak, the ganglion-cell density is not uniform; as a rule, there is a narrow area of the highest cell density. It is not homologous to the area centralis, as it is not associated with binocular vision and should be regarded only as a nonuniformity within the visual streak (Stone, 1983a). The visual streak has been described in rabbits (Provis, 1979), chipmunks (Wakakuwa et al., 1985), ground squirrels (Long and Fisher, 1983), several marsupial species (Hughes, 1974; Tancred, 1981; Wong et al., 1986; Arrese et al., 1999) and ungulates (Hebel, 1976; Hebel and Hollander, 1979). It has been suggested that the visual streak allows better scanning of the horizon without eye movements (Tancred, 1981; Stone, 1983a). A hypothesis was proposed (Hughes, 1977) suggesting that the visual streak develops in species living in open space. However, not all species fit this idea.

Other patterns of high cell-density areas were also described. Thus, the sloth has a vertical rather than horizontal streak (Costa et al., 1987; 1989). The elephant retina has horizontal and vertical streaks, the latter being in the field of binocular vision in which the trunk is moved (Stone and Halasz, 1989). Some marsupials have a separate visual streak and an *area centralis* (Tancred, 1981).

Detailed study of the *area centralis* in the cat using both morphological and physiological methods has revealed that it consists of a central region and a kind of horizontal streak, which are different morphological and functional zones of the retina (Stone and Keens, 1980). These two specialized areas of the retina are formed by different types of ganglion cells. The *area centralis* contains ganglion cells of intermediate size (X-type cells), sending their projections mostly to the lateral geniculate body. The visual streak contains small W-type cells sending their projections to the superior colliculus. Further comparative studies in various terrestrial mammals have led to the conclusion that a kind of the horizontal streak is a common property of the retinas of many mammals (Stone, 1983a), including carnivores (Peichl, 1992; Williams et al., 1993) and primates (Fisher and Kirby, 1991). An exception, apparently, is the *Homo sapiens*.

The lowest level of topographic retinal specialization was found in monotremes, which lack regions of increased ganglion-cell density in their retina (Stone, 1983b).

All the data described above indicate the obvious importance of ganglion-cell distribution patterns for understanding the visual functions of a certain species. Therefore, the ganglion-cell distribution was investigated in a number of cetacean species, as described below.

4.1.3.2. Identification of Ganglion Cells in Cetaceans

An important problem arising during studies of ganglion-cell distributions is that of identifying different cell types in the ganglion layer of the retina. To obtain the true pattern of the ganglion-cell distribution, these cells must be distinguished from other types of cells, particularly neuroglial cells and displaced amacrine cells, which can present in the ganglion layer along with ganglion cells. In the case of terrestrial mammals, this problem is fairly complex and was discussed over the years by several investigators (Stone, 1981, 1983a; Hughes, 1975; Provis, 1979; Wong et al., 1986; Wong and Hughes, 1987; Wässle et al., 1987). These studies demonstrated that a certain proportion of cells in the ganglion layer are microneurons (displaced amacrine cells). Different mammals have different proportions and topographic distributions of these cells (Perry, 1982; Wong et al., 1986; Wässle et al., 1990). Being not distinguished from ganglion cells, these cells may markedly affect the assessment of cell density and total cell number in the ganglion layer.

Studies performed on traditional laboratory subjects with the use of a variety of methods (degeneration, electron microscopy, retrograde transport) allowed the elaboration of criteria to identify ganglion cells and distinguish them from neuroglial and amacrine cells. These criteria are based mainly on quantity and properties of Nissl substance in the cell. Ganglion cells are characterized by the presence of a large amount of cytoplasm with a significant amount of Nissl substance, and a clearly visible nucleus and nucleolus. The problem of discrimination of amacrine from small ganglion cells has been solved using immunocytochemical methods (Brecha et al., 1988; Chun et al., 1988; Wässle et al., 1990).

When studies of the marine mammal retina began, there were no reports of neuroglial or displaced amacrine cells or other nonganglion cell types, and there were no criteria for identifying ganglion cells in cetaceans. However, the problem was eased by some properties of the cetacean retina described above – namely, the large size of cells in the ganglion layer. In terrestrial mammals, the problem of how to distinguish ganglion cells from displaced amacrine cells concerns mainly small neurons (less than 10 μm). In cetaceans, neurons in the ganglion layer are not less than 8–10 μm, and small cells are very few in number, while large cells may exceed 75–80 μm. These large cells are easy to discriminate from neuroglial and amacrine cells using the criteria described above.

Undoubtedly, methods using various tracers give more reliable identification criteria than Nissl staining: cells labeled by retrograde axon transport from the optic nerve are definitely the ganglion cells. However, comparison of counts obtained on Nissl preparations with those obtained by labeling ganglion cells by retrograde transport showed both methods to give very similar results (Collin and Pettigrew, 1988). Studies of ganglion-cell density distributions in cats and rabbits based on the Nissl method were also supported by more sophisticated methods. This indicates that cell identification on Nissl preparations was quite adequate.

As applied directly to cetaceans, adequacy of ganglion-cell identification can be additionally verified by comparison of counts of overall number of ganglion cells in the retina with those of optic nerve fibers. It will be shown below that these counts are in good agreement.

4.1.3.3. Topography of Ganglion-Cell Distribution in Cetaceans

For a long time, the existence of regions of ganglion-cell concentration in the cetacean retina remained under question. It was because the cetacean retina does not have an avascular or low vascularized area, which indicates the presence of the fovea or *area centralis* in terrestrial mammals and humans. Therefore, visual examination of the fundus did not reveal anything that could be interpreted as a fovealike region (Dawson et al., 1987a).

Nonuniformity in the distribution of ganglion cells across the retinal surface in the bottlenose dolphin *Tursiops truncatus* was first demonstrated by Peers (1971) using fragments of different parts of a retina. Increased cell density up to 193–271 cells/mm^2 were found in the nasal and temporal sectors of the retina.

Detailed data on the retinal topography in dolphins were obtained when retinal wholemounts (flatmounts) became to be used for investigation of distribution of ganglion cells. The wholemount is a total retina flatmounted on a slide, the ganglion layer upward. Wholemounts allow selective staining of the superficial ganglion layer with no staining of deeper layers, which makes it possible to observe all cells within the ganglion layer and only those cells. In such preparations, the entire retinal surface can be examined, thus yielding data on the ganglion-cell topographic distribution, overall number of cells, cell size, and shape in any area of interest.

Retinal wholemounts were first used to identify *area centralis* in the cat retina (Bishop et al., 1962). Later Jonathan Stone developed a wide field of topographic studies on retinal wholemounts (Stone, 1978, 1981). This approach was used by many authors, and majority of data on retinal organization mentioned in Section 4.1.3.1 were obtained by this method.

A number of cetacean species were investigated using the retinal wholemount method, mostly marine dolphins: the common dolphin *Delphinus delphis* (Dral, 1983), the bottlenose dolphin *Tursiops truncatus* (Dral, 1975a, 1977; Mass, 1993; Mass and Supin, 1995a, 1995b), the harbor porpoise *Phocoena phocoena* (Mass and Supin, 1985, 1986), the Dall's porpoise *Phocoenoides dalli* (Murayama et al., 1995). These studies revealed the presence of distinctive areas of ganglion-cell concentration in the cetacean retinas; in these areas, the ganglion-cell density much exceeded that in other retinal regions (Fig. 4.5).

The most characteristic feature of all these dolphin species was that, un-

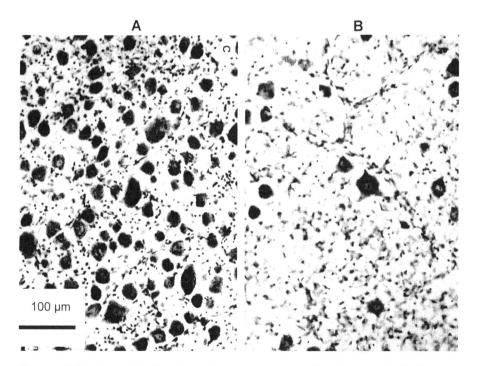

Figure 4.5. Ganglion cell patterns in a wholemounted retina of a bottlenose dolphin *Tursiops truncatus*. **A**. High-density area. **B**. Low-density area.

unlike all the studied terrestrial mammals, marine dolphins have not one area of high ganglion-cell density but two such areas. The high-density areas are located at the horizontal diameter of the retina, one in its nasal sector and the other in the temporal sector.

Figure 4.6 exemplifies a ganglion-cell density distribution map from a bottlenose dolphin *Tursiops truncatus* (Mass and Supin, 1995a, 1995b). To compose the map, numbers of ganglion cells were counted in samples spaced by steps of 1 mm, throughout all the wholemount area. The counts were converted to cell density in cells/mm^2 and plotted on the map. The map demonstrates two clearly defined high-density areas in the nasal and temporal parts of the retina. Both high-density areas are positioned at a distance of 15 to 16 mm from the optic disk, and the maximum cell densities in both these areas are about 700 cells/mm^2. There is also a beltlike area of moderately high cell density which connects the two high-density spots, passing under the optic disk – a structure similar to the visual streak. Cell density in this region reaches 300 cells/mm^2. In remaining parts of the retina, cell density is much

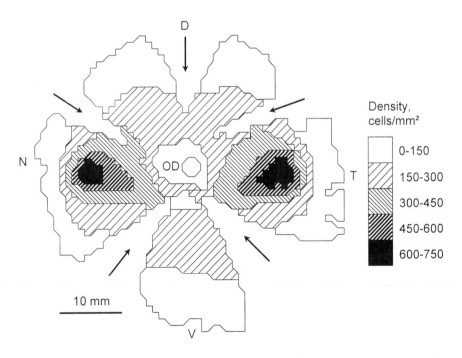

Figure 4.6. Ganglion-cell density distribution in a retinal wholemount from a bottlenose dolphin *Tursiops truncatus*. Cell density (cells/mm^2) is indicated by hatching according to the scale on the right. Arrows indicate radial cuts made to flatten the retina on the slide. D, V, N, and T – dorsal, ventral, nasal, and temporal poles of the retina, respectively; OD – optic disk.

lower, with minimal values (below 150 cells/mm²) in the retinal periphery and around the optic disk.

More precisely, the cell-density distribution can be demonstrated by plots showing the cell density as a function of distance along a horizontal diameter (Fig. 4.7). The plots reveal two peaks of cell density in the nasal and temporal sectors, both at a distance of around 17 mm from the optic disc, as well as a sharp decrease of cell density at the periphery and near the optic disc.

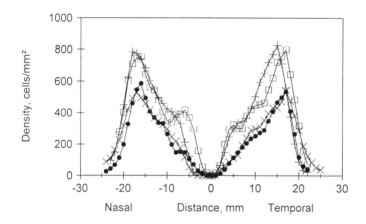

Figure 4.7. Profiles of ganglion cell density across the horizontal diameter in four retinal wholemounts from bottlenosed dolphins *Tursiops truncatus*, marked by four different symbols. Distances are plotted from the center of the optic disc.

For characterization of the visual-field properties, it is more convenient to present the ganglion-cell density distribution not in the retinal coordinates, as in Fig. 4.6, but in angular coordinates of the visual field. For this purpose, maps of the type shown in Fig. 4.6 were transformed in such a way as to eliminate the radial cuts made to flat-mount the retina and to restore the whole retinal hemisphere. Then cell densities presented in cells/mm² can be converted to measures in cells/deg² as

$$D = d(\pi r / 180),$$ (4.1)

where D is the cell density in cells/deg², d is the density in cells/mm², and r is the posteronodal distance (PND) – that is, the distance from the posterior nodal point of the optic system to the retinal surface. As shown above, the optic system of the dolphin's eye underwater is almost centrally symmetric, with the center of the system coinciding with both the lens and retinal hemisphere centers (Fig. 4.1). Such a system has a single nodal point in its center. Thus, PND in the dolphin eye can be assessed as the distance from the lens center to the retina. In the bottlenose dolphin, this distance is 14 to 15 mm. Taking this PND value, transformation of the map presented in Fig. 4.6 results in the map shown in Fig. 4.8. As the map shows, location of the two high-density areas at a distance of 15 to 17 mm from the optic disk corresponds to 50 to 55 deg from the center. The peak cell density of around 700 cells/mm^2 corresponds to 40 to 45 cells/deg^2.

For some time, it was debatable which of the two high-density areas in the dolphin retina has the highest cell density and thus should be considered as the area of the highest resolution. Early studies in the bottlenose dolphin *Tursiops truncatus* (Dral, 1977) and the common dolphin *Delphinus delphis* (Dral, 1983) found higher cell density in the nasal rather than in the temporal

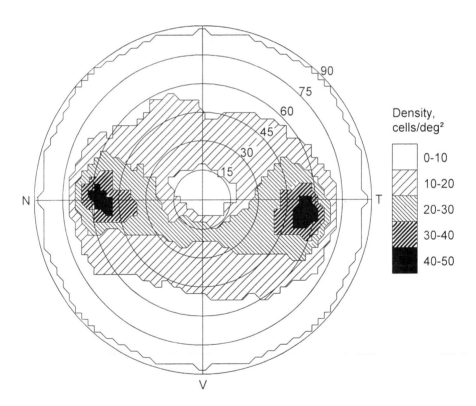

Figure 4.8. Ganglion-cell density distribution as projected onto the visual field in a bottlenose dolphin *Tursiops truncatus*. Cell density (cells/deg^2) is shown by hatching according to the scale on the right. Distances from the center of the retina are shown in spherical coordinates (degrees) in the top right quadrant. D, V, N, and T – dorsal, ventral, nasal, and temporal poles of the retina (not visual field!), respectively.

area. However, studies on more extensive material showed that at least in the bottlenose dolphin, this difference could result from individual variation: in some preparations, a slightly higher density was observed in the nasal area while in others higher density was in the temporal area; the difference was small in both cases (Mass and Supin, 1995a, 1995b). In another dolphin species, the harbor porpoise *Phocoena phocoena*, the cell density in the temporal region of the retina (that is, the region serving the frontal visual field) was consistently higher than in the nasal region: 28 and 20 cells/deg^2, respectively (Mass and Supin, 1986, 1990). In the Dall's porpoise *Phocoenoides dalli*, ganglion cell density in the temporal area was also a little higher than in the nasal one (Murayama et al., 1992a, 1995).

Apart from dolphins (odontocetes), retinal topography has also been studied in two mysticete species: the gray whale *Eschrichtius gibbosus* (Mass, 1996; Mass and Supin, 1990, 1997) (Fig. 4.9) and the Minke whale *Balaen-*

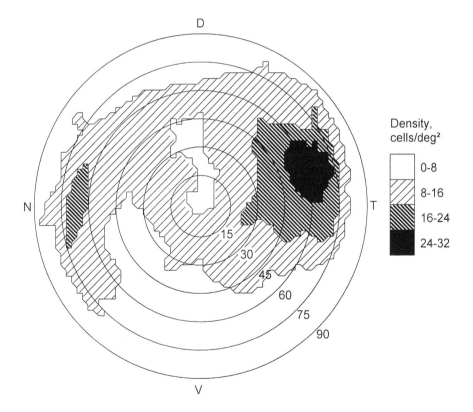

Figure 4.9. Ganglion-cell density distribution as projected onto the visual field in a gray whale *Eschrichtius gibbosus.* Cell density (cells/deg^2) is shown by hatching according to the scale on the right. Distances from the center of the retina are shown in spherical coordinates (degrees) in the bottom right quadrant. D, V, N, and T – dorsal, ventral, nasal, and temporal poles of the retina, respectively.

optera acutorostrata (Murayama et al., 1992a, 1992b). In these species, two areas of high ganglion cell density were also found. The cell density in the temporal region (frontal visual field) was significantly higher than in the nasal region. Thus, the pattern of ganglion-cell distribution with two high-density areas can be considered as a common feature of many cetaceans, both odontocetes and mysticetes.

Among other cetaceans, the Amazon river dolphin *Inia geoffrensis* presents a specific case of topographical organization of the retina. This animal lives in strongly turbid and low-transparent water, a lifestyle that has affected the organization of its visual system. Many authors regard the visual system of this animal as reduced because of small eye size. However, given the fully developed retina (see Section 4.1.1.3), it may be assessed in terms of adaptation of the visual system to short-distance vision, since in turbid water it is impossible to see anything at a large distance.

A characteristic pattern of ganglion cell distribution in the retina of the Amazon river dolphin is only one area of increased cell density (Fig. 4.10). However, contrary to terrestrial mammals, this area is located not in the central but in the ventral part of the retina – that is, in the region responsible for the upper part of the visual field (Mass and Supin, 1988, 1989).

The density of ganglion cells in the high-density area of the Amazon river dolphin reaches 400–500 cells/mm^2. However, because of the small size of

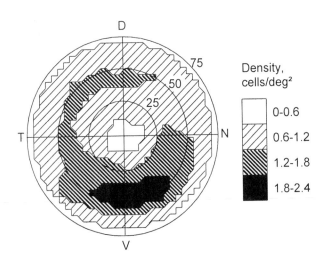

Figure 4.10. Ganglion-cell density distribution as projected onto the visual field in an Amazon river dolphin *Inia geoffrensis*. Cell density (cells/deg^2) is shown by hatching according to the scale on the right. Distances from the center of the retina are shown in spherical coordinates (degrees) in the top right quadrant. D, V, N, and T – dorsal, ventral, nasal, and temporal poles of the retina, respectively.

the eyeball and the short PND (about 4.5 mm), this corresponds to a density of only 2.4–3 cells/deg^2 when projected onto the visual field (Fig. 4.10).

Because of essential difference between marine cetaceans (marine dolphins and whales) and the Amazon river dolphin in their retinal topography, a question arises whether this difference is associated with their different systematic position (the Amazon river dolphin belongs to the *Iniidae* family) or with their visual ecology (the Amazon river dolphin inhabits river water, which is much lower transparent than sea water). To tackle this question, a study of the tucuxi dolphin *Sotalia fluviatilis* may be helpful. This species belongs to the same *Delphinidae* family as many marine dolphins but inhabits turbid river water.

Investigation of the retinal topography of the tucuxi dolphin (Mass, 1998; Mass and Supin, 1999) has shown that it features the same pattern of ganglion-cell distribution as marine dolphins – that is, two areas of ganglion-cell concentration, in the temporal and nasal sectors of the retina. The main difference between retinal organization of marine dolphins and the tucuxi dolphin was markedly lower cell density in the latter. In most wholemounts, it was less than 200 cells/mm^2. With a rather small eyeball and short PND (9.7 mm), this cell density corresponds to 5.7 cells/deg^2 – markedly less that in marine dolphins. Apparently, this is connected with adaptation to short-distance vision in low-transparent water.

Similar patterns were found in other river-inhabiting dolphins: the finless porpoise *Neophocoena phocoenoides* and the Chinese river dolphin *Lipotes vexillifer*. Their retinas contained two areas of increased cell density, in the temporal and nasal quadrants; however, the cell density was rather low, maximum a little more (in the finless porpoise) or a little less (in the Chinese river dolphin) than 200 cells/mm^2 (Gao and Zhou, 1987).

Since the presence of two high cell-density areas in the retina is a feature widely occurring among cetaceans, a question arises of the functional significance of this mode of retinal organization. As shown above (Section 4.1.3.1), only one such area is typical of most terrestrial mammals. One possible answer is that such retinal organization facilitates vision both under and above water. It has often been noted that a dolphin, when it looks at an underwater object, takes a position by the side to the object: it places the object of interest into a posterolateral part of the visual field, which projects onto the nasal area of high ganglion-cell density (Fig. 4.11). On the contrary, when a dolphin raises its head above the water surface to look at an above-water object, it places the object in the ventronasal part of the visual field, which projects onto the temporal area of high cell density (Dral, 1972, 1977; Dawson, 1980). Our experience confirms these observations. Thus, there are indications that underwater and above-water vision exploit different areas of the retina.

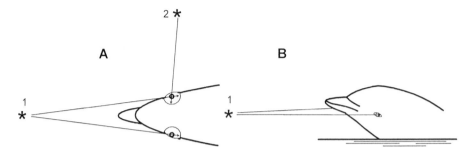

Figure 4.11. Typical positions of a bottlenose dolphin *Tursiops truncatus* when looking at an above-water (1) and underwater (2) object. **A**. Dorsal view (both underwater and aerial vision). **B**. Lateral view (aerial vision only).

The temporal location of the high-density area responsible for above-water vision in dolphins is important for understanding the mechanisms of aerial vision, in particular, of preventing aerial myopia. The optics of the dolphin's eye (see Fig. 4.1) is such that this area is illuminated through one (nasal) of two pinholes formed when the pupil is strongly constricted. The mutual position of the pupil, lens, and high-density retinal area is such that light passes through the hole of the constricted pupil just through the center of the lens to the retinal high-density area. Thus, only the central part of the lens participates in light refraction, which markedly reduces refractive errors (Kröger and Kirschfeld, 1993). Just opposite the hole of the constricted pupil, the corneal surface is of lower curvature. This significantly reduces aerial myopia for projections on this part of the retina. This hypothesis was supported by ophthalmoscopic observations (Dral, 1972).

Under water, both areas of high cell density are in conditions of emmetropic vision. Although behavioral observations demonstrated mainly the use of the nasal high-density area (the lateral part of the visual field), it may be a result of the observation conditions. There is no reason to exclude a role for the temporal high-density area in underwater vision. This area serves the frontal part of the visual field, which is the most important for a forward-moving animal, and the eye optics provides underwater emmetropia for the entire retina. It is possible that the existence of two retinal areas of high cell density compensates for the limited head mobility in marine dolphins: even with the high eye mobility, a single retinal area in an animal with low head mobility allows it to inspect by its gaze only a limited part of the visual field, while two such areas allow it to scan a much wider part of the visual field.

The same comments regarding two retinal areas of high cell density can be made about mysticetes. Above-water aerial vision may be an important mechanism of their orientation.

As to the Amazon river dolphin, which has the high cell-density area in the ventral part of the retina, this location can be explained by properties of the inhabited water. Low transparent turbid water not only prevents clear vision at large distances but also strongly absorbs light, so illumination is very low at a depth. In such conditions, significant illumination is possible only in upper water layers, in the upper part of the visual field of a normally oriented animal. Just this part of the visual field is served by the high-density area in the retina of the Amazon river dolphin.

4.1.3.4. Estimation of Visual Acuity by The Ganglion-Cell Density

Visual acuity is determined mainly by two factors: the quality of the eye optics and the retinal resolution. These two factors are expected to be in correspondence in a normal eye; it is unlikely that a retina with much better resolution than that of the eye optics or vice versa would develop during evolution. Thus, the retinal resolution can be used as a convenient measure of visual acuity when other methods of measuring the animal's visual acuity are not available.

Retinal topography makes it possible to estimate the retinal resolution. The density of ganglion cells (but not other cell types) is currently regarded as a measure of the visual resolution in a particular site of the retina. The density of other cell types, particularly, photoreceptors, may be much higher than the ganglion-cell density. Given that information is transmitted to the brain through ganglion cells, each integrating signals from a large number of photoreceptors, it is the quantity of ganglion cells that determines how detailed the description of a visual image is.

Since early studies, the monotonic reduction in visual acuity from the fovea to the periphery of the visual field has been noted to correlate with the density of ganglion cells (van Buren, 1963). These data confirm that the retinal resolution is determined by the density of ganglion cells, not photoreceptors. This suggestion was further supported by data showing that the magnification factor for retinal projection to the visual cortex decreased with distance from the fovea, this decrease being proportional to the reduced ganglion cell density, not photoreceptor density (Rolls and Cowey, 1970; Wässle et al., 1990).

Good correlation was also found between retinal resolution estimates obtained from the ganglion cell density and visual-acuity estimates obtained by psychophysical methods. In particular, very close values of retinal resolution were obtained in cats as determined by ganglion-cell density (Hughes, 1985) and visual acuity tested both psychophysically by grids of various spatial frequencies (Mitchell et al., 1977) and by evoked-potential recording produced by grid visual stimuli (Harris, 1978).

An important refinement is necessary. The concept of the ganglion-cell density generally relates to the distribution of cell bodies, the cell part that is the easiest to count and measure in morphological preparations. However, the retinal resolution is, strictly speaking, determined by distribution of not cell bodies but cell receptive fields. In many cases, this difference can be ignored since the receptive fields of ganglion cells are mostly located around the cell bodies, within the dendritic fields. There are, however, some exceptions to this rule, when positions of ganglion cell bodies and their receptive fields are markedly different. This occurs in the fovea in higher primates, including humans: there are no ganglion-cell bodies within the fovea itself. This construction reduces the light scatter at the path to photoreceptors, giving higher visual acuity. All ganglion cells serving the foveal region are located outside it and receive signals from foveal photoreceptors via long processes of the receptor cells (the Henle fibers), so the density of ganglion-cell receptive fields is the highest just within the fovea in spite of lack of cell bodies therein (Wässle et al., 1990). Thus, the main principle underlying the evaluation of retinal resolution from the ganglion cell density is true, as long as this principle is correctly applied. However, the foveal structure is easy identifiable, which prevents the possibility of wrong assessments. In all other cases, there is no reason to expect that the receptive fields of ganglion cells are systematically shifted from their cell bodies by to a significant distance. Thus, measurement of ganglion-cell density based on counts of cell bodies is an adequate approach.

Using this approach, the retinal resolution can be defined as the mean angular distance between neighboring ganglion cells. The latter can be calculated as

$$s = 1 / \sqrt{D} \tag{4.2}$$

or, basing on equation (4.1),

$$s = 180 / \pi r \sqrt{d} , \tag{4.3}$$

where s (deg) is the angular distance, D is the cell density in cells/deg^2, d is the cell density in cells/mm^2, and r (mm) is the PND. Strictly speaking, these equations are precisely applicable only to square arrays because the mean distance between array units depends not only on the unit density but also, to some extent, on the array pattern that determines the "packing" density. However, the latter factor very little affects the intercell distance in most of real patterns of ganglion-cell distribution, so the average inter-cell distance is always very close to the value predicted by equations (4.2) and (4.3) (Mass and Supin, 2000b).

Apart from intercellular distance, another metric of the retinal resolution is the resolvable spatial frequency – that is, the number of resolvable cycles of a grid pattern per unit angle. Taking that at least two cells are necessary to encode each cycle of a grid,

$$f = 1/2s,$$
(4.4)

where f (cycles/deg) is the resolvable ripple density and s (deg) is the mean intercellular distance.

Based on these equations, retinal resolution can be calculated for any part of the retina. Of course, resolution in the high-density areas is of the most interest since these areas determine the visual acuity of the animal. Such calculations were made for a number of cetacean species, using the ganglion-cell density data presented above.

In the bottlenose dolphin *Tursiops truncatus*, the highest ganglion-cell density was estimated as 40 to 45 cells/deg^2; the mean value can be taken as 43 cells/deg^2. This results in a mean intercellular distance of 0.15 deg = 9 min and resolvable spatial frequency of 3.3 cycle/deg. This value can be taken as a measure of visual acuity in this species, though only under water, because the PND used for the calculations (from the lens center to the retina) was taken based on the assumption that the lens is the only refractive structure; this assumption is valid for underwater vision only.

In air, the size of a retinal image is less than that in water because of refraction at the corneal surface. At a flat surface, the ratio of underwater and in-air retinal images should be equal to the ratio of refractive indices of water and air, which is 1.33. Assuming that the corneal surface opposite to the high-resolution areas is low convex (see Section 4.1.1.1), it can be taken that a retinal image in air is $1/1.33 = 0.75$ of that in water. Thus, the aerial retinal resolution in the bottlenose dolphin is expected to be $0.15/0.75 = 0.2$ deg = 12 min, and the resolvable spatial frequency is 2.5 cycle/deg.

These evaluations of underwater and aerial retinal resolution are in very close agreement with psychophysical data on visual acuity of the bottlenose dolphin: 8.2 to 9.8 min in water and 12.5 to 12.7 min in air (Herman et al., 1975). This agreement supports the reliability of the use of ganglion cell density data to assess the visual acuity.

Similar estimates for retinal resolution (thus, for visual acuity) were obtained for a number of other cetaceans. In particular:

• In the harbor porpoise *Phocoena phocoena*, the ganglion-cell densities of 20 and 28 cells/deg^2 in the nasal and temporal high-density areas correspond to underwater resolution of 13–14 min and 11–12 min (around 2.2 and 2.6

cycle/deg, respectively); aerial resolution is expected to be around 18 and 15 min (1.7 and 2 cycle/deg) in the nasal and temporal areas, respectively;

• In the Dall's porpoise *Phocoenoides dalli*, the peak cell density in the temporal area of around 540 cells/mm^2 corresponds (at a posteronodal distance of about 12.5 mm) to the underwater resolution of around 0.2 deg or 2.6 cycle/deg;

• In the gray whale *Eschrichtius gibbosus*, underwater retinal resolution is 12 to 13 min in the nasal high-density area (around 2.4 cycle/deg) and 10 to 11 min in the temporal high-density area (around 2.9 cycle/deg);

• A little better is underwater resolution in the Minke whale *Balaenoptera acutorostrata*: around 7 min, or 4 cycle/deg.

These values for retinal resolution (a few minutes or a few cycle/deg) are in general within the range of visual acuity characteristic of many terrestrial mammals, except primate foveal vision, which features a resolution of about 1 min (30 cycle/deg).

Much worse is the retinal resolution in river dolphins:

• In the Amazon river dolphin *Inia geoffrensis*, the highest cell density of 2.4 to 3 cells/deg^2 results in an underwater retinal resolution of around 0.6 deg or 0.75 cycle/deg;

• In the tucuxi dolphin *Sotalia fluviatilis*, the highest cell density of 5.7 cells/deg^2 results in an underwater retinal resolution of about 0.4 deg or 1.2 cycle/deg.

Since the river dolphins live in conditions (strongly turbid water) where objects are visible at a distance not more than a few tens of cm, visual acuity of an order of 1 deg seems adequate for them.

4.1.4. Visual Projections to the Cerebral Cortex

None of the morphological studies succeeded in identifying the visual projection area in the cerebral cortex. As mentioned above (Section 2.11), the cerebral cortex of cetaceans is agranular; it does not contain the layer IV, which is a characteristic feature of sensory projection areas.

Sensory projections to the cerebral cortex in dolphins were identified in the bottlenose dolphin *Tursiops truncatus* and the harbor porpoise *Phocoena phocoena* by the evoked-potential method (Sokolov et al., 1972; Supin et al., 1978). In those studies, EP to light flashes were recorded through electrodes implanted into the cerebral cortex. It was found that EP to light stimuli could be recorded only in a limited cortical area occupying a large portion of the lateral gyrus of the cerebral cortex (Fig. 4.12). No light-evoked responses were observed outside this area. Thus, this part of the cerebral cortex can be regarded as a cortical projection area of the visual system (the visual cortex).

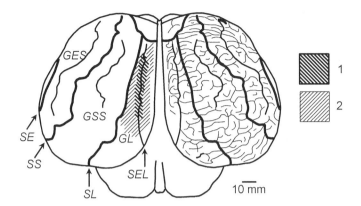

Figure 4.12. Localization of the visual projection area in the cerebral cortex of the bottlenose dolphin *Tursiops truncatus*. Dorsal view of the dolphin's brain. On the right hemisphere, pattern of cortical sulci and gyri is shown. On the left hemisphere, only first-order sulci and position of the areas of light-evoked responses are shown. *SE – sulcus ectosylvius, SS – s. suprasylvius, SL -- s. lateralis, SEL – s. endolateralis, GES – gyrus ectosylvius, GSS – g. suprasylvius, GL – g. lateralis.* 1 and 2 – areas, producing responses of types I and II (primary and secondary visual areas), respectively.

The size of the visual projection is markedly smaller than that of the auditory system, though it occupies a significant area of the cortex.

Within this area, a variety of EP waveforms were recorded which could be subdivided into two types. The type I response contained an early wave complex with an onset latency of 22 to 25 ms and a peak latency of about 30 ms when the stimulus was a bright light flash (Fig. 13A). The first wave of the early complex could be either positive or negative, depending on the recording electrode position (records 1 and 2 in Fig. 4.13A). Taking into consideration the retinal delay, which in mammals is usually 15 to 20 ms, this latency is rather short and indicates a direct conduction of the afferent volley from the lateral geniculate body (the visual thalamic relay station) to the cerebral cortex. Thus, the cortical subarea generating this response type should be considered as a primary projection visual area. In the dolphin's cerebral cortex, this subarea occupies a part of the cortex located mainly in the depth of the endolateral sulcus (area 1 in Fig. 4.12).

Apart from the early short-latency complex, EP always contained later waves lasting up to 100 ms after the stimulus. Type II responses contained only long-latency waves; their onset latency was at least 30 to 35 ms (Fig. 4.13B). The waveform of this response type varied depending on the recording electrode position, but in any case its duration was at least 50 to 70 ms after the onset latency – that is, about 100 ms after the stimulus. This was the same late wave complex that accompanied the early response in the pri-

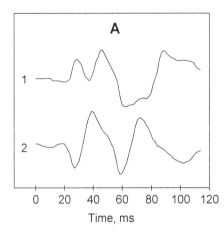 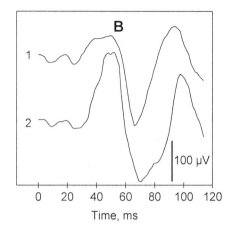

Figure 4.13. EP recorded from the visual cortical area in a bottlenose dolphin *Tursiops truncatus.* **A.** Responses containing both short- and long-latency components (type I). **B.** Responses containing only long-latency components (type II). 1 and 2 – various waveforms of each response type.

mary projection area. This late complex appeared in a much wider area rather than the primary sub-area. The area of type-II responses occupies a major part of the lateral gyrus (area 2 in Fig. 4.12). It can be considered as a secondary visual projection area in the cerebral cortex.

A characteristic feature of late waves of visual EPs, both the type-II response and the late components of the type-I response, was their capability to form a rhythmic after-discharge. After a single light flash, sometimes only one slow wave cycle followed (Fig. 4.14 (1)); however, in most cases, a rhythmic sequence of a few wave cycles appeared as a rhythmic after-discharge (Fig. 4.14 (2) and (3)). It should be taken into consideration that records presented in Fig. 4.14 were obtained using the averaging procedure; therefore, the presence of rhythmic waves in the averaged records indicated their coherency with the stimulus during a long time, up to 0.5 s and longer. The after-discharge frequency was around 20 Hz.

Localization of the visual projection area in the lateral gyrus was confirmed by experiments with neuron labeling by retrograde axonal transport (Garey and Revishchin, 1990). Tracer injections into the lateral gyrus of the cortex resulted in appearance of labeled neurons in the lateral geniculate body (LGB), the thalamic relay station of the visual system.

In the found visual cortical area, synaptic organization markedly differs from that in most other mammals because of low development of the "incipient" layer IV (Glezer and Morgane, 1990; Glezer et al., 1990, 1992). Synaptic contacts are much more numerous in superficial (I and II) rather

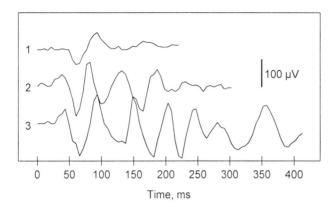

Figure 4.14. Rhythmic after-discharges evoked by a single light stimulus (diffuse light flash) in the visual area of the cerebral cortex in a bottlenose dolphin *Tursiops truncatus*. 1–3 – various durations of the rhythmic responses.

than in deep (V and VI) layers, and some important neurotransmitters are distributed in a corresponding manner. Among other features of this cortical area, there may be mentioned slight differentiation of cortical layers (weakly expressed stratification) and cells (poor neuronal differentiation), predominance of simple neuronal forms, atypical (transitional) pyramidal neurons, and some other features of rather primitive cytoarchitecture (Morgane and Glezer, 1990). Distribution of neurons with various mediators in the dolphin's visual cortex much more resembles that of archetypal terrestrial mammals (bats, hedgehogs) than progressive terrestrial mammal (Glezer et al., 1992).

All the data on visual representation in the dolphin's cerebral cortex were obtained using diffuse light stimuli (flashes), which bring no information concerning the retinotopic organization within the visual projection area. No attempts were made to study retinotopic organization using punctiform or patterned visual stimuli. Such detailed investigation would require the implanting of a very large number of electrodes, which is contradicted by ethical restrictions.

The found position of the visual projection area in the dolphin cerebral cortex somewhat differs from that in other mammals. Apart from neurological data in humans indicating localization of the visual cortical area in the occipital lobe (the calcarine fissure), localization of the visual projection area of the cerebral cortex was experimentally studied in a number of mammal orders: monotremats (Lende, 1964; Allison and Goff, 1972), marsupials (Lende, 1963), insectivores (Lende and Sadler, 1967; Hall and Diamond, 1968), rodents (Krieg, 1946; LeMessurier, 1948; Lende and Woolsey, 1956; Hall et al., 1971), lagomorphs (Woolsey, 1947), carnivores (e.g., Rose and

Woolsey, 1949), and primates (e.g., Diamond et al., 1970). In most of mammals, the visual projection area, including all its subdivisions, is located in the occipital or medio-occipital part of the cerebral cortex. In carnivores, it occupies a caudal part of the lateral gurus. In dolphins, the visual projection area is also located in the lateral gyrus, however, not in the caudal but in its middle (parietal) part. Thus, the visual projection area in dolphins (maybe in other cetaceans also) is somewhat shifted to the frontal direction as compared to other mammals.

4.2. VISION IN PINNIPEDS

Unlike cetaceans, which spend their entire lives in water, pinnipeds spend a significant part of time on land. Their vision is an important sensory modality in both water and air. On land, vision plays an important role in reproductive behavior, in maintaining intrapopulation relationships, as well as for orientation. In water, vision is important for detection and recognition prey and for spatial orientation during migrations.

Pinnipeds exhibit a great diversity of species differing in their systematic positions, ecologies, and lifestyles. Thus, walruses are bentofags and rely mainly on their well-developed vibrissal apparatus for finding and recognition prey; using their vibrissal sensitivity, they can identify objects in terms of their shape and size, with little participation of vision (Kastelein and van Gaalen, 1988; Kastelein et al., 1990). Vision is little used by walruses in water, and in air it is used mainly for social interactions and orientation (Oliver et al., 1983; Kastelein and Weipkema, 1989).

Seals and sea lions, like walruses, have a well-developed vibrissal apparatus (Ladygina et al., 1992) that is capable of discriminating objects by shape, texture, and size (Dehnhardt, 1990, 1994). Nevertheless, it is probable that they to a large extent rely on their vision when searching for food and actively use their vision for orientation.

4.2.1. Eye Morphology

4.2.1.1. Eye Anatomy and Optics

Eye anatomy in pinnipeds, despite significant difference from cetaceans, has some common features reflecting the adaptation to underwater vision (Fig. 4.15). In particular, the main refractive element is the strongly convex, almost spherical, or slightly elliptical lens. Due to a rather large anterior

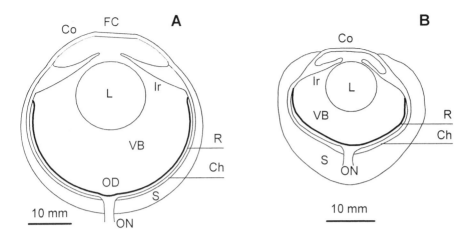

Figure 4.15. Diagrammatic drawing of the eye anatomy in pinnipeds. **A**. The northern fur seal *Callorhinus ursinus*. **B**. The walrus *Odobenus rosmarus*. C – cornea, FC – flattened region of the cornea, L – lens, Ir – iris, VB – vitreous body, R – retina, Ch – choroid, OD – optic disc, ON – optic nerve.

chamber, the eyeball in general does not look as shortened in the axial direction. Nevertheless, a major part of the eyecup has a shape close to a sphere segment centered at the lens, so a significant proportion of the retina is equally distant from the lens center. Thus, similarly to cetaceans, the eye optics in pinnipeds is in general centrally symmetric.

The eye size is large in most pinnipeds, except walruses (Jamieson and Fisher, 1972). Most pinnipeds have a pear-shaped pupil. Pupil size can change over a very wide range; at bright illumination, it constricts to a pinhole. The ciliary muscle in pinnipeds is well developed, much better than in cetaceans (Jamieson and Fisher, 1972; Sivak et al., 1989; West et al., 1991), though accommodation is weak or absent (Sivak et al., 1989).

Unlike cetaceans, keratoscopic studies in pinnipeds demonstrated a region of the cornea with very slight curvature, almost flat (see Fig. 4.15). This region is located near the center of the cornea. The cornea is very thick (Pütter, 1903; Wilson, 1970; our observations); perhaps the large thickness helps to maintain the flattened shape of the cornea. Detailed keratoscopic studies were performed in the California sea lion *Zalophus californianus* (Dawson et al., 1987b). While peripheral convex part of the corneal surface had refraction as large as 21.7 D, the central region of about 6.5 mm in diameter had almost zero refraction. This flat part of the corneal surface serves as an emmetropic window in which refraction remains almost constant both in air and under water. In some pinnipeds (the hooded seal *Cystophora cristata*), there is no clearly delimited flat area at the corneal surface; nevertheless, a thick cornea in conjunction with very large eyeball size results in a very low curvature of the central part of the cornea, thus providing an emmetropic window for both aerial and underwater vision (Sivak et al., 1989).

The presence of a flattened region of the cornea in pinnipeds is a very intriguing feature considered from the point of view of general principles of eye construction. The convex shape of the cornea in most mammals is a direct and inevitable sequence of excessive intraocular pressure, which serves to maintain the shape and size of the eyeball. Unfortunately, direct data on intraocular pressure in pinnipeds are still absent. However, there is a good reason to suppose that the excessive intraocular pressure in pinnipeds is minimal, at least in the anterior chamber; otherwise, it would be hardly possible to maintain a flat surface of the elastic cornea. If so, it remains unknown how the eyeball shape is kept constant in pinnipeds. In northern fur seals we observed that, unlike other mammals, the vitreous body is of a rigid rather than a gelatinous consistency. This unusual mode of maintaining the eye shape are not known in terrestrial mammals.

The tapetum in pinnipeds is one of the best developed among both terrestrial and aquatic mammals (Walls, 1942). It consists of a number (20 to 34, according to different authors) of layers and covers a large proportion of the fundus (Nagy and Ronald, 1970, 1975; Jamieson and Fisher, 1971).

4.2.1.2. Retinal Structure

Studies of the retina in pinnipeds are limited to just a few reports (Landau and Dawson, 1970; Jamieson and Fisher, 1971; Nagy and Ronald, 1970, 1975). In general, the retinal laminar structure in pinnipeds is the same as in terrestrial mammals. All layers are present in the pinniped retina, though there are several specific features, mainly in the inner nuclear and ganglion layers.

The inner nuclear layer does not have clear boundaries and is rather chaotically organized. This contrasts with terrestrial mammals, in which this layer is strictly ordered. All reports noticed very large horizontal cells within this layer. Giant processes of these cells spread to a large distance. These giant horizontal cells do not form ordered rows; they are distributed irregularly in the close vicinity of bipolar as well as amacrine cells. Bipolar cells typically are of a round shape with oval nuclei and diffuse chromatin filaments. They are located mostly in the outer part of the inner nuclear layer. Amacrine cells are large, irregular in shape; they are located close to the inner plexiform layer.

The ganglion layer consists of a single row of rather large ganglion cells separated by wide intercellular spaces. In the harp seal _Pagophilus gro-_

enlandicus, most of ganglion cells were found to be of intermediate or large size (Nagy and Ronald, 1970). In the walrus *Odobenus rosmarus* and the northern fur seal *Callorhinus ursinus*, the retina is also dominated by rather large ganglion cells, from 10 to 30 µm (Mass, 1992; Mass and Supin, 1992; Fig. 4.16). These sizes are somewhat smaller than in cetaceans; nevertheless, they are significantly larger than in most terrestrial mammals.

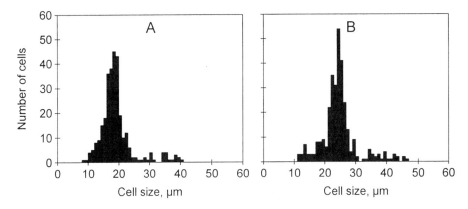

Figure 4.16. Size distributions of ganglion cells in the retina of northern fur seal *Callorhinus ursinus*. **A**. Sample from a region in the temporal area of high cell density. **B**. Sample from a region of low cell density in the periphery of the retina. Each sample contained 300 cells.

In the northern fur seal, a group of giant cells, larger than 30 µm, is distinguishable; Fig. 4.16 demonstrates that along with the major distribution peak at 20 to 25 µm, a small additional peak is visible at about 35 µm. In another pinniped, the walrus, different-size classes of ganglion cells were not distinguished.

All investigated pinnipeds had predominantly rod retina. The question of existence of cones has been a matter of discussion for a long time. In early studies of the pinniped retina, cones were not found (Landau and Dawson, 1970; Nagy and Ronald, 1970). However, later investigations using light and electron microscopy demonstrated the presence of two types of photoreceptors, presumably of rods and cones, in the harbor seal *Phoca vitulina* (Jamieson and Fisher, 1971) and the harp seal *Pagophilus groenlandicus* (Nagy and Ronald, 1975). However, immunochemical studies in seals (Peichl and Moutairou, 1998) revealed only one opsin type in the cone receptors, the L-opsin (middlle-to-long-wave sensitive) and did not revealed the S-opsin (short-wave sensitive). This feature distinguish pinnipeds from majority of other mammals that have at least two spectrally sensitive cone types (middle-

types (middle- and short-wavelength sensitive) or three cone types in primates (Jacobs, 1993). On the other hand, it is noteworthy that this feature of the pinniped retina is similar to that of cetaceans, in spite of their quite different phylogeny.

4.2.2. Visual Abilities of Pinnipeds

The visual abilities of pinnipeds have been studied by a variety of methods. Although no evoked-potential studies of vision were ever performed in pinnipeds, there is a significant body of psychophysical studies. Apart from that, some amount of data was obtained using pupillometric and electroretinographic investigations.

4.2.2.1. Light Sensitivity and Pupillary Responses

Being underwater foraging animals, phocid and otariid pinnipeds are capable of very high light sensitivity, when dark-adapted. This was shown by pupillometric and psychophysical studies in the northern elephant seal *Mirounga angustirostris*, the harbor seal *Phoca vitulina*, and the California sea lion *Zalophus californianus* by Levenson and Schusterman (1997, 1999).

Substantial constriction of pupils in pinnipeds to regulate retinal illumination was noticed by Lavigne and Ronald (1972). Levenson and Schusterman (1997) performed precise pupillometric studies (filming by an infra-red sensitive videocamera) which have shown that the range of illuminance resulting in pupillary responses markedly differs for shallow-diving and deep-diving pinnipeds. In shallow-diving pinnipeds (the harbor seal and California sea lion), the pupil reached its maximal size at illuminance level of around 3×10^{-4} foot-candle and constricted when illuminance increased up to 85 foot-candle. This indicates that their visual system is adapted to function at least in this moderately wide (about 5.5 log units) range. Fairly different were data in a deep-diver, the northern elephant seal. Its pupil responded to an illuminance level within a range from 85 to 3×10^{-9} foot-candle (about 10.5 log units). Pupillary responses within this range are indicative of the capability of the visual system to function within this extremely wide range. This capability is obviously associated with the lifestyle of the northern elephant seal, which during foraging dives from the surface where illumination may be very high to a depth of a few hundred meters, where illumination is extremely low.

The range of pupillary area variation in different pinniped species is also deserving of attention. In shallow-diving species, this range was rather small:

from 162 to 2.3 mm^2 in the harbor seal *Phoca vitulina* (70.5-times variation) and from 220 to 8.4 mm^2 in the California sea lion *Zalophus californianus* (26-times variation). In a deep diver, the northern elephant seal *Mirounga angustirostris*, the pupil area varied within an extremely wide range, from a giant area of 422 mm^2 in dark-adapted conditions (about 23 mm in diameter!) to a pinhole opening of 0.9 mm^2 in light-adapted conditions: the range of variation was almost 470 times. The authors suggested that the wide variation of the pupil area is important for deep-diving pinnipeds since there is little time for retinal adaptation when the animal dives during a few minutes to a little-illuminated depth and then returns to brightly illuminated surface.

Direct measurement of light sensitivity in a few pinniped species was performed in psychophysical experiments by Levenson and Schusterman (1999). Using a go/no-go experimental paradigm, they presented in air light stimuli of various intensity and measured light-sensitivity thresholds at various stages of light-dark adaptation. For this purpose, the animal was light-adapted in daylight and then was placed in a dark enclosure; thresholds were measured at various times after beginning the dark adaptation. For comparison, thresholds in humans were measured in the same conditions. In shallow-diving pinnipeds, the harbor seal and California sea lion, dark adaptation was markedly more rapid than in the human, and thresholds reached lower levels: −10.39 and −10.70 log units re 1 W/m^{2}sr, as compared to −10.10 log units in the human. Extremely rapid adaptation and low thresholds were obtained in a deep diver, the northern elephant seal: during a few minutes, thresholds dropped to a level of −11.09 log units, a log unit lower than in the human. Both the pupil response and retinal adaptation probably contributed to this sensitivity change.

4.2.2.2. Pattern Discrimination and Visual Acuity

Psychophysical investigations have demonstrated that seals are capable not only of visual discriminating the shape of planar figures but also of more complicated analysis of visual images (Hanggi and Schusterman, 1995).

Psychophysical measurements of visual acuity were performed in a few pinniped species with the use of test grids of varying spatial frequency. Schusterman and Balliet (1970a, 1971) measured visual acuity in California sea lions *Zalophus californianus* both in air and water using a version of the go/no-go paradigm differing from the standard version by the use of animal vocalization and silence as yes- and no-responses instead of a movement and staying. Animals were trained to respond by vocalization when a grid consisting of light and dark strips of equal width was presented and to remain silent when a gray field was presented. The width of strips yielding 75% correct responses was taken as an estimate of visual acuity. Measurements both in water and air at distances from 1.9 to 5.5 m gave almost identical results: visual acuity was within a range from 4.7 to 7 min, in average around 5.5 min. Similar measurements with the use of the same technique were carried out in the Steller sea lion *Eumetopias jubata* and the harbor seal *Phoca vitulina*. Visual acuity in water was estimated as 7.1 min in the Steller sea lion and 8.3 min in the harbor seal (Schusterman and Balliet, 1970b).

It should be noted that the threshold estimates presented above were obtained using an arbitrarily chosen threshold criterion of 75% in the go/no-go paradigm, with noticeable false-alarm probability: in the study of Schusterman and Balliet (1970a), a false-alarm percentage of 10 to 30% was mentioned. Given rather shallow psychometric curves in that study, thresholds could be estimated from 4–5 to 10–11 min, depending on threshold criterion. Similar estimates of visual acuity (several minutes of arc) of the harbor seal *Phoca vitulina* were obtained by Jamieson and Fisher (1970).

Aerial visual acuity of southern fur seals *Arctocephalus pusillus* and *A. australis* was measured in the two-alternative paradigm using a "minimum visible" technique (the animal had to discriminate a black circle on a white background from a white plate) and "minimum-separable" technique (the animal had to discriminate a vertical grating from a horizontal one). The latter technique seems more adequate to measure visual acuity. It resulted in estimates of visual acuity as 6.6 min in *A. pusillus* and 7.2 min in *A. australis* (Bush and Dücker, 1987). Thus, most of seals and sea lions have a rather good visual acuity of an order of a few minutes of arc.

Bush and Dücker (1987) also studied the ability of the southern fur seals to discriminate objects by size. The animals were able to discriminate two circles with a difference ratio as small as 1:1.24 (*A. australis*) or 1:1.19 (*A. pusillus*).

4.2.2.3. Color Discrimination

As mentioned above (Section 4.2.1.2), there are evidences of two types of photoreceptors, presumably rods and L-cones, in pinnipeds (Jamieson and Fisher, 1971; Nagy and Ronald, 1975; Peichl and Moutairou, 1998). A number of studies confirmed the existence of a double set of photoreceptors in pinnipeds, both rods and cones. The double set of photoreceptors may be a basis for color vision. Spectral sensitivity of pinniped retinas was studied by a number of authors to find whether the receptor sets are able to provide color vision. In the harp seal *Pagophilus groenlandicus*, the Purkinje shift was found, which is indicative of two kinds of photoreceptors with different sensitivity (Lavigne and Ronald, 1972). However, even if the pinniped retina contains cones, it is obviously dominated by rod photoreceptors. Spectro-photometric measurements indicated that the rod photopigment was sensitive to short wavelengths; its absorption peak was at about 496 nm (Lavigne and Ronald, 1975), and in the elephant seal the rod pigment is sensitive to even shorter wavelengths, 486 nm (Lythgoe and Dartnall, 1970).

An attempt of psychophysical measurement of spectral sensitivity in photopic conditions (that is, determined by cones) was performed by Lavigne and Ronald (1972) in the harp seal *Pagophilus groenlandicus*. They measured light-sensitivity threshold at various light wavelengths and found the best photopic sensitivity at 525 nm. However, the spectral curve obtained in those experiments did not have a characteristic shape of photopigment absorption spectra, which made it difficult to determine precisely the maximum sensitivity.

Crognale et al. (1998) applied the electroretinographic (ERG) technique to spectral-sensitivity measurements in the harbor seal *Phoca vitulina*. In photopic conditions, they stimulated the retina by alternatively flickered gray (reference) and monochromatic color (test) light and adjusted the test-light intensity until the response to the test light was equivalent to the response to the reference one. Thus, relative sensitivity to the monochromatic light of a certain wavelength was determined, and repeating this procedure at various wavelengths yielded the spectral-sensitivity curve. The obtained curve fitted the standard photopigment-absorption curve very well, and it peaked at 510 nm. This value can be taken as a peak sensitivity of the cone photopigment.

Crognale et al. (1998) also tested for the presence of more than one cone photopigment in the harbor seal retina. For this purpose, they measured by the same method sensitivity to a short-wavelength (450 nm) test light as compared to long-wavelength (550 nm) reference light in both the presence and absence of long-wavelength (585 nm) adaptation light. They supposed that if there were two (or more) photopigments contributing to the obtained spectral-sensitivity function, the chromatic adaptation should differently depress the sensitivities of these pigments, thus changing the shape of the overall spectral sensitivity function. The result was negative: the chromatic adaptation did not change the relative sensitivity to short-wavelength and long-wavelength lights. This result indicates that the harbor seal has only a single spectral mechanism operating under photopic conditions, in agreement with data indicating existence of only one cone type (L-cones) in the pinniped retina.

Another evidence for the double set of photoreceptors is a shift of the critical flicker frequency depending on illumination (Bernholz and Mattews, 1975).

The absence of more than one cone type means that the harbor seal cannot have color vision under purely photopic conditions – that is, color vision based solely on cone action. However, it does not exclude a kind of color vision under mesopic conditions due to the ability to compare the signals from rods and cones, which have markedly different spectral sensitivity.

Direct evidence of elements of color vision was also obtained in psychophysical studies. Color discrimination was found in a spotted seal *Phoca larga* (Wartzok and McCormick, 1978). They reported successful discrimination between blue and orange lighted targets in a two-alternative forced-choice task.

More precise quantitative investigation was performed by Busch and Dücker (1987) with the southern fur seals *Arctocephalus pusillus* and *A. australis*. In a two-alternative task, the animals had to discriminate between simultaneously presented color and gray targets; in order to exclude discrimination by brightness, the density of gray targets was varied within a wide range. It was shown that the seals were capable of distinguishing blue and green objects from gray ones: discrimination was successful at any density of gray targets. However, the animals could not distinguish red objects from gray ones at a certain density of gray objects, when brightness of the gray and color objects were equalized.

Similar studies were performed in the California sea lion *Zalophus californianus* by Griebel and Schmid (1992). In a two-alternative task, the animals had to choose the color plate when a color plate and and agray plate were presented simultaneously. The density of gray plates was varied within a wide range. The animals successfully distinguished blue and green plates from the gray ones at any density of the gray plate, which is an indication of the use of the plate color as a discrimination cue. The animals failed to discriminate a red plate from the gray one when the gray density was properly adjusted.

Thus, the results of all psychophysical studies indicate the presence of dichromatic (blue-green) color vision in pinnipeds. It remains unclear whether this kind of color vision is provided by two cone types, as in most dichromatic mammals, or one cone type and rods.

4.2.3. Topographic Distribution of Retinal Ganglion Cells and Retinal Resolution

Long ago it was suggested that the retina of the harbor seal *Phoca vitulina* has an *area centralis* (Cheivitz, 1889). However, later ophthalmoscopic examinations do not supported this idea (Johnson, 1901), and over many years, further attempts to identify a kind of *area centralis* or other regularities in the retinal topography in pinnipeds were not successful. Even much later attempts to reconstruct the ganglion-cell distribution in the retina of the California sea lion *Zalophus californianus* using serial transverse sections of the retina (Landau and Dawson, 1970) did not give an unequivocal result: numerous regions of slightly increased ganglion cell density were found, no regularity in their positions was determined, and the existence of an *area centralis* in the pinniped retina was denied.

Only the use of the retinal wholemount technique has clarified the question. Different pinniped groups were found to have different clearly defined types of retinal topography. A typical retinal structure with a well developed *area centralis* was found in an otariid species, the northern fur seal *Callorhinus ursinus* (Mass and Supin, 1992) (Fig. 4.17). This area features many times higher ganglion cell density than in the surrounding retina (Fig. 4.18). The area of increased ganglion cell density was strictly defined and rather small, as compared with a very large total retinal surface. The high-density area is shifted to the temporal pole of the retina, about 12 mm from the optic

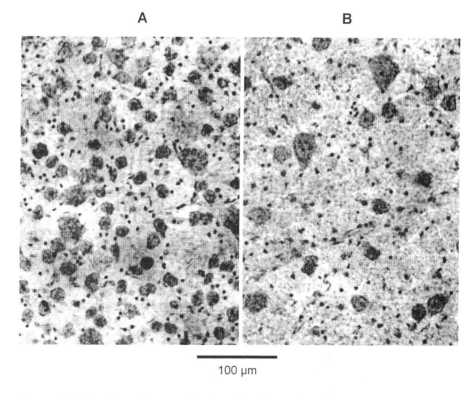

100 µm

Figure 4.17. Ganglion cell patterns in the retina of a northern fur seal *Callorhinus ursinus* (retinal wholemount). **A**. High-density area. **B**. Low-density area.

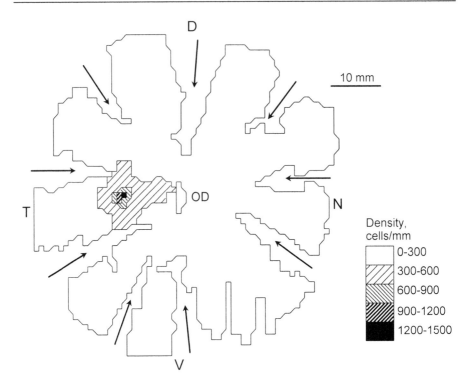

Figure 4.18. Ganglion-cell density distribution in a retinal wholemount from a northern fur seal *Callorhinus ursinus*. Cell density (cells/mm^2) is indicated by hatching according to the scale on the right. Arrows indicate radial cuts made to flatten the retina on the slide. D, V, N, and T – dorsal, ventral, nasal, and temporal poles of the retina, respectively; OD – optic disk.

disk. In this area, the cell density exceeded 1000 cells/mm^2, and the density dropped down sharply outside this region, to a level less than 100–200 cells/mm^2 in the major part of the retina. Given frontally positioned eyes, the location of the *area centralis* corresponds exactly to the projection of the frontal visual field. Thus, unlike dolphins, the retinal topography in the northern fur seal does not differ fundamentally from that in terrestrial carnivores and several other terrestrial mammals.

More precisely, the cell-density distribution can be demonstrated by plots showing the cell density as a function of distance along a horizontal and vertical diameters (Fig. 4.19). The plots reveal a very sharp peak of cell density at a distance of 11 to 12 mm from the optic disc and a steep decrease of cell density outside the high-density area.

In order to convert the cell-distribution map in the retinal wholemount coordinates into a map in the visual-field coordinates, PND in the northern fur seal, as in dolphins, was taken as a distance from the lens center to the

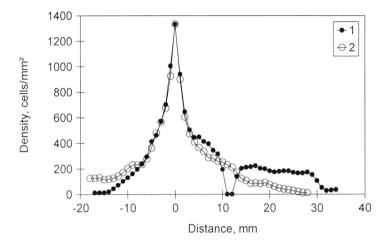

Figure 4.19. Horizontal (1) and vertical (2) profiles of ganglion-cell density across a retinal wholemount from a northern fur seal *Callorhinus ursinus*. Distances are plotted from the center of the area centralis (negative – temporal and dorsal, positive – nasal and ventral directions). Zero cell density at a distance of 11 to 12 mm in the nasal direction from the area centralis is the optic disk.

retinal surface. In the large fur seal's eye, it is 22 to 23 mm (mean 22.5 mm). Taking this PND, the highest ganglion-cell density of more than 1000 cells/mm^2 in the *area centralis* corresponds to more than 160 cells/deg^2, and the *area centralis* is located at a distance of 30 to 35 deg from the optic disk (Fig. 4.20).

The found cell density of 160 cells/deg^2 corresponds to a mean intercellular distance of about 0.08 deg = 4.8 min. This calculation is made using the lens center as the posterior nodal point, which is a feature of underwater vision when the lens is the only refractive element. Thus, the value of 4.8 min or 6.3 cycle/deg can be adopted as underwater retinal resolution (and supposedly, the visual acuity) of the northern fur seal. In air, the flat corneal surface produces decrease of retinal images by a factor of 0.75 (this is a ratio of the refraction index of air to that of water). Therefore, aerial retinal resolution can be estimated as 4.8/0.75 = 6.4 min, or 4.7 cycle/deg. The latter estimate is close to visual acuity estimates obtained in psychophysical measurements in other seal species, mainly 5 to 8 min (see Section 4.2.2.2). This resolution is markedly better than in many dolphins: as shown above (Section 4.1.2.1), resolution in marine dolphins is mostly 8 to 10 min in water and 12 to 15 min in air; in river dolphins, it is even worse.

Substantially different is the retinal topography in a representative of another pinniped family, *Odobenidae* – the walrus *Odobenus rosmarus*. The walrus retina also features a nonuniform distribution of ganglion cells and

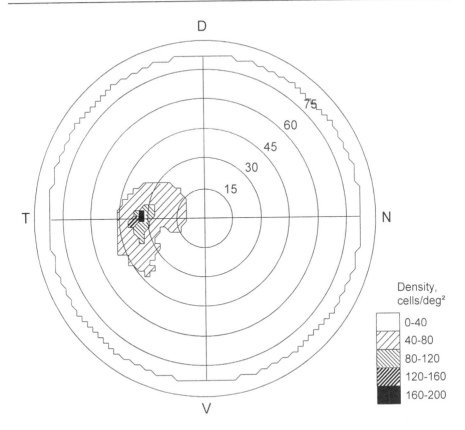

Figure 4.20. Ganglion-cell density distribution as projected onto the visual field in a northern fur seal *Callofhinus ursinus*. Cell density (cells/deg²) is shown by hatching according to the scale on the right. Distances from the center of the retina are shown in spherical coordinates (degrees) in the top right quadrant. D, V, N, and T – dorsal, ventral, nasal, and temporal poles of the retina, respectively.

the presence of an area of increased ganglion cell density. However, this area is less strictly defined than in the northern fur seal (Mass, 1992). The area of increased cell density is a horizontally elongated oval, resembling the visual streak in terrestrial mammals (Fig. 4.21). Ganglion-cell density is the highest in the temporal part of this streak; here it reaches 1000 cells/mm² – almost the same value as in the northern fur seal. However, the walrus eye is markedly smaller (see Fig. 4.15); its PND is about 12.5 mm. With this posteronodal distance, the highest cell density in the walrus retina corresponds to not more than 50 cells/deg².

The cell density of 50 cells/deg² corresponds to a mean intercellular distance of 0.14 deg = 8.5 min, or 3.6 cycle/deg. This value can be adopted as a retinal resolution (and supposedly, the visual acuity) in water. In air, the

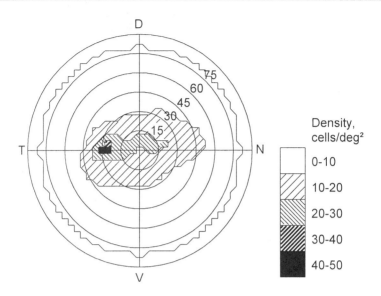

Figure 4.21. Ganglion-cell density distribution as projected onto the visual field in a walrus *Odobenus rosmarus*. Cell density (cells/deg²) is shown by hatching according to the scale on the right. Distances from the center of the retina are shown in spherical coordinates (degrees) in the top right quadrant. D, V, N, and T – dorsal, ventral, nasal, and temporal poles of the retina, respectively.

resolution is expected to be worse by a factor of 0.75 – about 11 min or 2.7 cycle/deg – it is markedly weaker than in the northern fur seal, but of the same order as in many marine dolphins.

Thus, in general, retinal topographies in pinnipeds strictly resembles those in some groups of terrestrial mammals. In the northern fur seal, it is similar to carnivores in the sense that the *area centralis* is the most prominent feature of the ganglion-cell distribution pattern; in the walrus, it resembles that of many terrestrial mammals possessing a visual streak in the retina.

In this context, comparison with data on retinal organization in the sea otter may be of interest (Mass and Supin, 2000a). The sea otter, as all other otters, belongs to the *Carnivora* order, not to one of the orders of true aquatic mammals. Nevertheless, it is deeply adapted to the marine mode of life and may be considered as an intermediate form of such adaptation, to some extent similar to ancestors of recent pinnipeds. The sea otter's retina contains both a kind of visual streak and a well-defined *area centralis* overlapped on the temporal part of the streak; it saved the main properties of the retinal organization of carnivores. Many of these features have been saved in pinnipeds as well.

4.2.4. Visual Projections to the Cerebral Cortex

Localization of the visual projection area in the cerebral cortex was studied by Ladygina and Popov (1986) in two pinniped species: the northern fur seal *Callorhinus ursinus* (*Otariidae*) and the Caspian seal *Pusa caspica* (*Phocidae*). In that study, EP to diffuse light flashes were recorded from the cerebral cortex of deeply anaesthetized animals through small halls drilled in the skull. In both species, high-amplitude and short-latency EPs to light stimuli were found in the lateral gyrus of the cerebral cortex (Fig. 4.22). In the northern fur seal, the area of EP recording occupied mostly the caudal part of this gyrus and extended to a small part of the adjacent suprasylvian gurus (Fig. 4.22A). In the Caspian seal, this area occupied the caudal and parietal part of the lateral gyrus, also involving a small part of the adjacent suprasylvian gurus (Fig. 4.22B). This area can be considered as the visual projection area in the cerebral cortex.

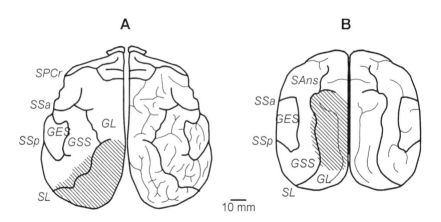

Figure 4.22. Localization of the visual projection area in the cerebral cortex of the northern fur sear *Callorhinus ursinus* (**A**) and the Caspian seal *Pusa caspica* (**B**). Dorsal view of the brains. On the right hemisphere, pattern of cortical sulci and gyri is shown. On the left hemisphere, only first-order sulci and position of the area of light-evoked responses (hatched area) are shown. *SSa – sulcus suprasylvius anterior, SSP – s. suprasylvius posterior, SL – s. lateralis, SPCr – s. precruciatus SAns – s. ansatus, GES – gyrus ectosylvius, GSS – g. suprasylvius, GL – g. lateralis.*

Taking into consideration the difference in gyrification patterns between the two species (compare Fig. 4.22A and B), position of the visual cortical area is rather similar in both the northern fur seal representing the *Otariidae* family and in the Caspian seal representing the *Phocidae* family. This position is also similar to that in terrestrial carnivores (e.g., Rose and Woolsey, 1949).

It is noteworthy that contrary to cetaceans, the visual cortical area in pinnipeds is larger than the auditory one. It can be an indication of different importance of auditory and visual orientation in cetaceans and pinnipeds: greater importance of visual orientation in pinnipeds.

4.3. VISION IN SIRENIANS

Sirenians (the order *Sirenia*) are a relict group of unique, completely aquatic mammals adapted to a herbivorous lifestyle. Among only a few species of this order (the West Indian manatee *Trichechus manatus*, the Amazonian manatee *T. inunguis*, the West African manatee *T. senegalensis* and the dugong *Dugong dugong*), only two former species were studied to some extent in respect of their vision.

Knowledge about visual capabilities of sirenians, including the manatee, is very limited. Piggins et al. (1983) summarized available findings. Researchers' estimates of the manatee's visual abilities range from "wretched" (Walls, 1942) to a far more optimistic suggestion, based on visual behavior, that the West Indian manatee *Trichechus manatus* possesses binocular stereoscopic vision (Hartman, 1979).

A more correct estimation based on the experimental data was provided by Piggins et al. (1983) on the Amazonian manatee *Trichechus inunguis*. Their data have shown that the near emmetropic eye of *T. inunguis* is capable of focusing on and tracking objects situated at a distance of a meter and further under water, but no attempt was made to track a stimulus above water. The authors suggested that the eye of *T. inunguis* is adopted for underwater vision, and it is more adapted for sensitivity than acuity.

4.3.1. Eye Anatomy and Retinal Structure

The eye anatomy (Fig. 4.23) was described in the Amazonian manatee *Trichechus inunguis* by Piggins at al. (1983) and that in the Florida manatee (a subspecies of the West Indian manatee) *Trichechus manatus latirostris* by Mass et al. (1997). The eye is almost spherical and rather small: Piggins et al. (1983) estimated the axial length of the Amazonian manatee eye as 13.4 mm and noticed a shallow anterior chamber with the lens set relatively forward. In the Florida manatee, Mass et al. (1997) estimated the axial length

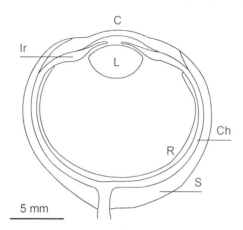

Figure 4.23. Drawing of a longitudinal section of the eye of a Florida manatee *Trichechus manatus latirostris*. C – cornea, L – lens, Ir – iris, R – retina, Ch – choroid, S – sclera

between the external surfaces of the cornea and sclera as 16 mm, equatorial (nasotemporal) width as 17 mm, and internal eyecup diameter as 15 to 15.5 mm. The distance from the center of the lens to the retina was estimated as 10 to 10.5 mm. The cornea was also of rather small diameter: not more than 6.5 to 7 mm. The iris was well developed, the pupil was slightly oval, the vertical axis longer. The lens was set forward in the manatee eye, and the anterior chamber was shallow. Contrary to spherical lenses of cetaceans and pinnipeds, the lens in the manatee's eye is lenticular, about 3 mm in the axial dimension and 5 mm in the transverse diameter. However, because of the small lens diameter, the curvature of its front and rear surfaces is almost the same as in other aquatic mammals with a comparable eye size. The optic nerve is rather thin, 1.5 to 2 mm in diameter.

Data on retinal organization of sirenians are rare. The first common description of the manatee's retina was published by Pütter (1903). Rochon-Duvigneaud (1943) and Walls (1942) described the retina of the manatee and dugong as pure rod. Later, Piggins et al. (1983) also indicated that the manatee's retina has a structure typical of a nocturnal animals and that cones are rare or absent. Ganglion cells were noticed to be few in number, although their number were not be estimated quantitatively.

The detailed description of the *T. manatus* retina was made by Cohen et al. (1982), who established that the laminar structure of the retina is full developed. Using light and electron microscopy, they have found both rodlike and conelike photoreceptors. Moreover, two cone subclasses were found which indicated a possibility of color vision. The large number of cone cells suggests the visual acuity superior to that described previously as "wretched" by Walls (1942).

4.3.2. Psychophysical Studies

Psychophysical studies of vision in manatees have been very limited but have provided important data on their visual capabilities. The ability of West Indian manatees to color discrimination was shown by Griebel and Schmid (1996). The subjects were trained to discriminate between a colored stimulus and a gray one in a two-alternative forced-choice paradigm: a colored and a gray target were presented simultaneously, and the animal had to make its choice by touching the colored target by its snout. To exclude discrimination by target brightness, gray targets varied from low to high brightness. The animals distinguished successfully both blue and green targets from gray ones at any gray-target brightness, thus indicative of color discrimination. The animal could not, however, discriminate red targets from gray ones when the brightness of the latter was adjusted properly. They also failed to discriminate an intermediate blue-green color from certain degrees of gray. These results indicate that manatees possess dichromatic blue-green color vision.

Further studies have shown the West Indian manatees are capable of brightness discrimination (Griebel and Schmid, 1997). Measurements were made in the same two-alternative forced-choice paradigm as described above, the animals being trained to touch the brighter of two simultaneously presented gray targets. Reflectivity of targets varied within a wide range, from 3.5 to 89%. Within all this range, the ability to discern brightness difference corresponded to a Weber's fraction of about 0.35.

4.3.3. Topographic Distribution of Ganglion Cells and Retinal Resolution

The topographic organization of retinal ganglion cells was studied by Mass et al. (1997) in the Florida manatee *T. manatus latirostris* using the Nissl-stained retinal wholemount technique. It appeared that ganglion-cell distribution was not uniform but varied smoothly across the retina. The pattern of ganglion-cell distribution is presented in Fig. 4.24. This pattern of ganglion-cell distribution can be described as bell-shaped – that is, cell density was higher in a large part around the center of the retina (except far periphery), and the highest cell density was located below the optic disk. The cell density in this region exceeded 250 cells/mm^2, up to 270 cells/mm^2.

The ganglion cell distribution profile featured a weak gradient in the cell density within a major part of the retina, and no sharply restricted spot of cell concentration could be identified. Only at the retina periphery and near

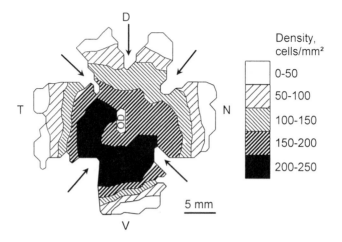

Figure 4.24. Ganglion-cell density distribution in a retinal wholemount from a Florida manatee *Trichechus manatus latirostris*. Cell density (cells/mm²) is indicated by hatching according
to the scale on the right. Arrows indicate radial cuts made to flatten the retina on the slide. D,
V, N, and T – dorsal, ventral, nasal, and temporal poles of the retina, respectively; OD - optic
disk.

the optic disk, ganglion-cell density fell markedly. There is no clearly restricted area of high cell density similar to the *area centralis* or visual streak
of terrestrial mammals.

To calculate the retinal resolution, the posterior nodal distance was
adopted as the distance from the lens center to the central part of the retina.
This distance was about 10.5 mm. With this PND, the cell density of 270
cells/mm² corresponds to 9 cell/deg² and a mean intercellular distance of
0.33 deg = 20 min. Thus, the retinal resolution on the Florida manatee can be
taken as 20 min or 1.5 cycle/deg. This is rather low resolution, markedly
lower than in marine dolphins and pinnipeds, but a little better than in river
dolphins.

4.4. SUMMARY

In cetaceans, the eye anatomy features the central symmetry of the eyecup
relative to the center of the spherical lens. The eye optics provides underwater emmetropia, whereas some aerial myopia is possible due to additional
refraction at the cornea surface. Accommodation occurs not by lens defor-
mation but probably by its axial displacement. The pupil constriction occurs
in such a way as to form an U-shaped slit and then two pin-holes at its ends.

Against these pupil holes, the cornea has lower curvature, thus minimizing the aerial myopia. Eye movements are well developed, two eyes being able to move independently. The eye retina contains mainly rod photoreceptors and a minor number of cones. A characteristic feature of the cetacean retina is large sizes of ganglion cells separated by wide intercellular spaces. Psychophysically measured visual acuity of the bottlenose dolphin was found as 8.2 min in water and 12.5 min in air; the difference between aerial and underwater acuity is equal to the difference between aerial and underwater retinal images due to different refractive indices of the two media. Successful intermodal transfer is possible between vision and echolocation. Studies of topographic distribution of ganglion cells in the retina revealed two areas of ganglion-cell concentration (the best-vision areas) located in the temporal and nasal quadrants, whereas other mammals have only one such area. The position of these areas opposite to the pinholes of the constricted pupil and low-curvature regions of the cornea helps to reduce the aerial myopia. The only exception was the Amazon river dolphin, which has only one best-vision area in the ventral part of the retina. The two best-vision areas are used differently for aerial and underwater vision. Retinal resolution of the bottlenose dolphin derived from the ganglion cell density in the best-vision areas was estimated as around 9 min in water and 12 min in air which is very close to psychophysical data. In a number of other marine cetaceans, retinal resolution of the same order was found; in river dolphins, the resolution is much worse. The visual projection to the cerebral cortex of dolphins occupies a middle part of the lateral gyrus and contains two subareas, one producing short-latency EP (the primary projection area) and another producing longer-latency EP (the secondary projection area).

In pinnipeds, similarly to cetaceans, the main element of the eye optics is the spherical lens. The central part of the cornea has a flat region that provides equal refraction both in air and water. The retinal structure is similar to that of terrestrial mammals in many respects. It contains predominantly rods and a small number of cones. Similarly to cetaceans, very large ganglion cells are typical of the pinniped retina. Psychophysical, pupillometric, and electroretinographic studies revealed a very wide range of light sensitivity and rapid dark adaptation, particularly in deep-diving pinnipeds. Psychophysical measurements of visual acuity of a few otariid and phocid species gave estimates of approximately 5 to 7 min both in air and in water. Color discrimination is possible within a green-blue range although only one cone type was found in the pinniped retina. Topographic distribution of ganglion cells was studied in the northern fur seal and walrus. In the northern fur seal, it featured only one high-density area in the temporal part of the retina (*area centralis*), similarly to terrestrial carnivores. It provides retinal resolution of 4.8 min in water and 6.4 min in air. In the walrus, the best-vision area looked as a visual streak with the best resolution of 8.5 min in water and 11 min in air. Visual projections to the cerebral cortex of otariids and phocids occupies a posterior part of the lateral gyrus, similarly to terrestrial carnivores.

In sirenians, the eye optics is based on a small lenticular lens. The retina contains predominantly rods and a small number of cones. Psychophysical studies revealed green-blue color discrimination and brightness discrimination. Retinal ganglion cell topography features a smooth pattern of cell distribution without a clearly defined best-vision area. Underwater retinal resolution was assessed as 20 min.

Chapter 5
SOMATIC SENSE IN AQUATIC MAMMALS

5.1. SOMATIC SENSE IN CETACEANS

Somatic sense in cetaceans did not attract as much attention of investigators as the auditory and visual ones. So available data in this field are restricted. Morphological studies have shown the presence of both incapsulated and free nerve endings in the dolphin's skin, especially numerous at the head and snout, around the blowhole, and around the anus and genital slit (Palmer and Weddel, 1964; Harrison and Thurley, 1974; Ling, 1974; Herman and Tavolga, 1980). This is an indication of developed tactile sensitivity of these regions. In addition, some river dolphins have vibrissae along the upper lip area (Layne and Caldwell, 1964; Ling, 1977). The trigeminal nerve and the trigeminal nucleus are very well developed; the trigeminal nerve is exceeded in size only by the auditory nerve (Breathnach, 1960), which is also an indication of a predominating role of tactile sensitivity in the head region as compared to other body regions.

The first physiological investigation of somatosensoty sensitivity in dolphins (the bottlenose dolphin *Tursiops truncatus*) was undertaken by Lende and Welker (1972), who recorded evoked potentials of exposed cerebral cortex to a variety of stimuli, such as vibrating, tapping, stroking, or dripping water on the skin. EP were recorded only in the contralateral cerebral cortex in a frontally located area, which was regarded as the somatosensory cortex. Qualitative estimations indicated the highest skin sensitivity on the head, particularly the upper and lower lips, regions around the eyes, and the blowhole; the trunk and tail were less effective in producing EP.

A similar investigation was performed by Ridgway and Carder (1990), but with the use of thin extracranial electrodes instead of those at the exposed cortex. Using this technique, they succeeded at recording somatosensory EP to various kinds of contralateral tactile stimulation. When the lip, snout, or melon was stimulated, EP amplitudes were much higher than that to stimulation of the trunk flank.

Similar results have been obtained in a study on a common dolphin *Delphinus delphis* with the use of the electrodermal response arising to tactile stimulation (Kolchin and Bel'kovich, 1973). They used skin stimulation by a thin (0.3 mm) wire pressed to the skin with a certain force. The best sensitivity (thresholds of about 10 mg/mm^2 – around 10^2 Pa) was found in small (a few cm in diameter) areas around the blowhole and eyes; the lower jaw and melon were a little less sensitive (thresholds of 10–20 mg/mm^2); the skin along the back was even less sensitive (20–40 mg/mm^2).

In studies of Supin et al. (1978), EP to electric skin stimulation were recorded in the cerebral cortex of bottlenose dolphins *Tursiops truncatus* and harbor porpoises *Phocoena phocoena* using implanted intracortical recording electrodes. The task of that study was to delimit the somatosensory cortical projection area, so no attention was paid to the estimation of the sensitivity of various skin regions. EP to contralateral electric skin stimulation (Fig. 5.1) were recorded in an area posterior to the crucial sulcus – that is,in the

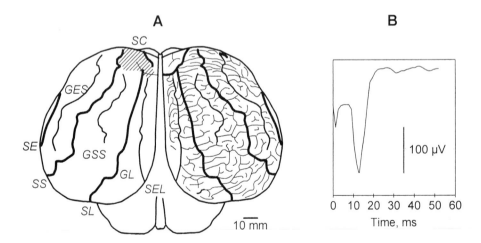

Figure 5.1. **A.** Localization of the somatosensory projection area in the cerebral cortex of the bottlenose dolphin *Tursiops truncatus*. Dorsal view of the dolphin brain. On the right hemisphere, the pattern of sulci and gyri is replicated. On the left hemisphere, only first-order sulci and gyri are shown: *SE – sulcus ectosylvius, SS – s. suprasylvius, SL – s. lateralis, SEL – s. endolateralis, SC – s. cruciatus,* GES – *gyrus ectocylvius,* GSS – *g. suprasylvius,* GL – *g. lateralis.* The hatched area is that producing EP to electric skin stimulation. **B.** Typical EP waveform in the somatosensory projection area to electric skin stimulation (a short zero-latency peak is the stimulus artifact).

postcrucial gyrus, spreading also onto anterior parts of lateral and suprasylvian gyri. In this area, short-latency primary EP appeared; their onset latency was 8 to 9 ms, and the first-wave peak latency was 12 to 13 ms. This area was considered as the primary projection area in the cerebral cortex. No attempt was made to investigate somatotopic organization of somatosensory projections in the cerebral cortex in cetaceans.

5.2. SOMATIC SENSE IN PINNIPEDS

5.2.1. Morphological and Psychophysical Data

Tactile sensitivity in pinnipeds is well developed. This is true for all three pinniped families, *Otariidae*, *Phocidae*, and *Odobenidae*. Particularly, the vibrissae serve as an important sensitive tactile organ (Schusterman, 1968). The structure of the vibrissae in pinnipeds is well developed (Ling, 1966, 1977; Hyvärinen and Katajisto, 1984; Hyvärinen, 1989), as well as well developed their innervation (Stephens et al., 1973; Hyvärinen, 1995). In a number of psychophysical studies, the abilities of vibrissae apparatus were studied quantitatively.

In experiments with a walrus *Odobenus rosmarus* wearing eyecups, it was shown that it was able to distinguish between 3-D circles and triangles using the vibrissae (Kastelein and van Gaalen, 1988). The animal was trained to inspect the objects by its vibrissae and to perform different movements (head nodding or shaking) depending on the presented figure, a circle or a triangle, respectively. When the figures were 20-mm thick, the animal distinguished the figures with almost 100% probability at all tested surface areas, from 50.2 to 0.8 cm^2. At figure thickness of 3 mm, the task was solved until the surface area of the objects decreased down to 0.4 cm^2 and was not solved at the area as small as 0.2 cm^2.

In experiment of Denhaardt (1990, 1994) it was shown that California sea lions *Zalophus californianus* are able to discriminate a variety of forms (a semicircle, hexagon, "sandglass", rectangle, and square, each less than 2 cm in size) being visually deprived by eyecups and using the vibrissae to inspect the objects. Different stimuli were differentially rewarded; when two different objects were presented to the animal, it selected the better rewarded one with almost 100% probability.

Dehnhardt et al. (1998) measured the sensitivity of the wiskers of a harbor seal *Phoca vitulina* to water movements. In a go/no-go paradigm, the sea lion was trained to respond to hydrodynamic stimuli generated by an oscilating sphere (in the fange of 10 to 100 Hz). Optical and acoustical cues were excluded by eyecups and noise-producing headphones. The animal could detect water velocity at speeds as low as 245 μm/s – a very high sensitivity. Thus, the seal's wiskers can function as a hydrodynamic receptor system.

5.2.2. Somatosensory Projections to the Cerebral Cortex

Localization of the somatosensory projection area in the cerebral cortex was determined in two pinniped species, the Caspian seal *Pusa caspica* (*Phocidae*) and the northern fur seal *Callorhinus ursinus* (*Otariidae*) by Ladygina and Popov (1984, 1986). They recorded EP from exposed cerebral cortex in deeply anaesthetized animals. Tactile stimuli were touching the skin by a magnetoelectric vibrator, which was fed by short electric pulses synchronized with EP collection.

In both the northern fur seal and Caspian seal, the area of EP recording was located anteriorly of the anterior suprasylvian sulcus (Fig. 5.2). Dorsally, the area was bounded by the ansate sulcus, anteriorly by the coronar and postcruciate sulci. Within this area, short-latency EPs arose in response to contralateral tactile stimuli.

Although EP obtained by stimulation of different points of the body surface overlapped markedly, it was possible to localize cortical points producing EP of the shortest latency and the highest amplitude – projections of dif-

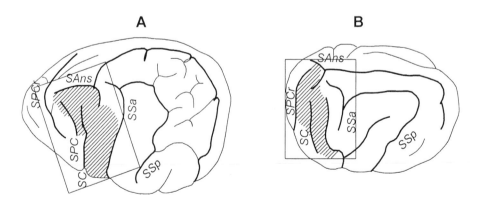

Figure 5.2. Localization of the somatosensory projection area in the cerebral cortex of the northern fur seal *Callorhinus ursinus* (**A**) and the Caspian seal *Pusa caspica* (**B**). Lateral view of the left hemisphere; *SPCr – sulcus postcruciatus, SC – s. coronarius, SPC – s. precoronarius, SAns – s. ansatus, SSA – s. suprasylvius anterior, SSP – s. suprasylvius posterior*. The hatched area is the somatosensory projection area; the rectangle delimits a region presented in more detail in Fig. 5.3.

ferent parts of the body. Thus, a preliminary somatosensory projection map was obtained (Fig. 5.3). It showed a distinct ordering of somatosensory projections to the cerebral cortex: the projection of the nasal area of the head was followed by the projection of the vibrissae, then the orbital region, near-ear region, neck, chest, trunk, and pelvis; beside this axis are projections of the fore- and hindlimb.

Nevertheless, the EP investigation is not the best method for detailed

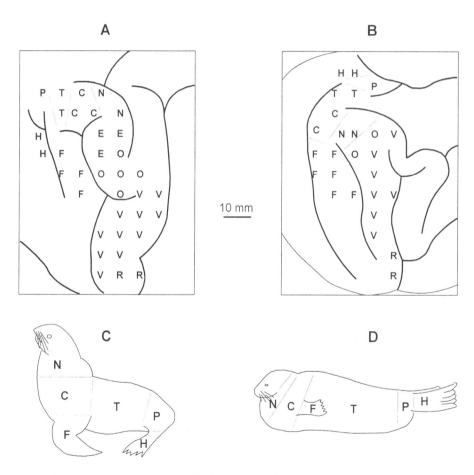

Figure 5.3. Projections of parts of the body surface to the somatosensory cerebral cortex of the northern fur seal *Callorhinus ursinus* (**A**, **C**) and the Caspian seal *Pusa caspica* (**B**, **D**). **A** and **B** A region of the cortical surface as delimited by rectangle in Fig. 5.2. Letters indicate points of EP recording to stimulation of different parts of the body surface as shown in **C** and **D**: N – neck, C – chest, F – forelimb; T – trunk, P – pelvis, H – hindlimb. The head region is subdivided to subareas, which are not shown in **C** and **D**: R – rhinal area (nose), V – vibrissae, O – orbital area, E – ear. Thin straight lines are arbitrarily driven boundaries between the parts of the body surface (in **C** and **D**) and their cortical projections (in **A** and **B**).

study of somatotopic projections since far-field EP can be recorded at some distance from the site of generation. Therefore, a more detailed study of the somatotopic projection to the cerebral cortex (micromapping) was performed in the northern fur seal *Callorhinus ursinus* using a more precise method – namely, microelectrode recording of multiunit neuronal activity in the cortex (Ladygina *et al.*, 1986, 1992). In that study, thin tungsten microelectrodes with tip sizes of 10 to 15 μm were used to record noiselike multiunit activity of cortical neurons in deeply anaesthetized animals. This activity enhanced greatly when a skin site corresponding to the recording point was stimulated by its touching by a thin needle. Appearance of such a response indicated the somatotopic correspondence between the stimulated and recorded points. Punctiform tactile stimulation allowed to encounter receptive fields of the recorded points with a good precision.

Figure 5.4 demonstrates a representative example of correspondence between cortical recording points and receptive fields at the body surface. All the receptive fields were found only at the contralateral side of the body surface. When the recording point was shifted regularly along the cortical surface, the receptive field position was also shifted regularly along the body surface. All the cortical surface was investigated continuously with 1-mm steps; however, only a few demonstrative rows of points are exemplified in Fig. 5.4A. Their receptive fields shown in Fig. 5.4B–D, also formed regular rows, thus illustrating the point-to-point correspondence of the body and cortical surfaces.

Figure 5.4 illustrates also some regularities of the body-surface projection onto the cerebral cortex. The trunk is represented into a dorsal part of the projection area, near the ansate sulcus (points 1–6 and 7–12); the back is represented dorsally (points 2–6). Below, there are representations of the neck (points 13–16) and the head (points 17–46). Within the head projection, the dorsally located points (17–22) represent the ear-orbital region, more ventrally located ones (23–42) represent the vibrissae region, and far more ventrally located points (43–46) represent the nasal region.

Within the area of the vibrissae projection, there also is a correspondence between positions of cortical points and their receptive fields. Rostro-caudal rows of points (23–30, 31–38, 39–42) correspond to vertical vibrissae rows, and dorso-ventral rows of cortical points correspond to horizontal vibrissae rows.

At different parts of the body surface, receptive field sizes differ to a very large extent. At the trunk, receptive field are more than 100 cm^2 (fields 1–16 in Fig. 5.4), the largest fields are located at the back and abdomen. At the head, receptive fields mainly are less than 10 cm^2 (fields 17–22); at the nose and lips they are less than 1 cm^2. The vibrissae receptive fields are very local. Each cortical points responds only to stimulation of a single vibrissa, and several cortical points within an area of a few mm^2 respond to stimulation of a particular vibrissa.

For composing the whole projection map of the somatosensory cortex, all its surface was tested with 1-mm steps. Since the testing involved not only convex but also walls of cortical sulci, a special topographic basis was used to map both the superficial cortex and sulcus walls. For this purpose, the major cortical sulci (Fig. 5.5A) were represented in such a way as if they were "opened"; their walls were in the figure plane (Fig. 5.5B). Such "flattening" of the cortex resulted in gaps of superficial cortical regions.

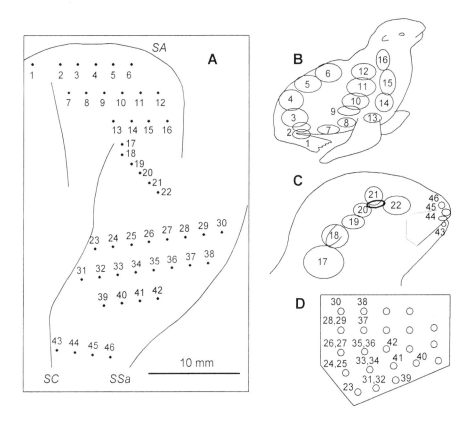

Figure 5.4. Correspondence of receptive fields on the body surface to cortical points in the cerebral cortex of a northern fur seal *Callorhinus ursinus*. **A**. The cortical area of the left hemisphere, as delimited by the rectangle in Fig. 5.2A. *SA – sulcus ansatus, SC – s. coronarius, SSa – s. suprasylvius anterior.* Tested points are indicated by dots. **B**. Positions of receptive fields at the right side of the body surface; fields are enumerated according to the numbers of the recorded cortical points. **C**. The same as (**B**), enlarged drawing of the head; the polygon delimits the region of maxillar vibrissae. **D**. Enlarged drawing of the vibrissae region, vibrissae are shown by circles, numbers indicate the responding cortical points.

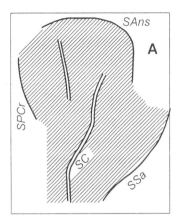
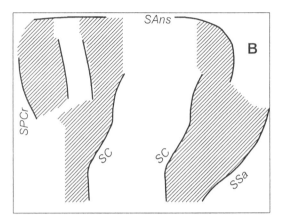

Figure 5.5. Topographic basis for the somatotopic projection map in the northern fur seal *Callorhinus ursinus* (the left hemisphere). **A**. The region of the cortical surface containing the somatosensory projection area. **B**. drawing of the same region transformed in such a way as to move apart the sulci borders and to flatten the sulci walls at the figure plane. The same part of the convexital surface is hatched in both **A** and **B**. *SAns – sulcus ansatus, SPCr – s. postcruciatus, SC – s. coronarius, SSa – s. suprasylvius anterior.*

With the use of this topographic basis, the somatotopic map of the somatosensory cortex in the fur seal looks like that presented in Fig. 5.6. The map shows that the somatosensory projections occupy both the convexital and intrasulcus regions (compare with Fig. 5.5B). In particular, a major part of the forelimb projection is located inside the precoronar sulcus. Somatosensory projections occupy also a large part of the coronar sulcus and the rostral wall of the anterior suprasylvian sulcus.

An important feature of the somatosensory cortical projection in the northern fur seal is the duplicate representation of almost all the body surface within the projection area. These two projections are disposed as two arcs, one inside the other, with the outside boundary limited by the postcruciate, ansate, and anterior suprasylvian sulci. Within this area, each point of the body surface is represented twice, in the outer and inner arc-shaped projection belts.

The sequence of projections of main body parts is as follows. The head projection is located ventrally. Within the head projection, the mouth and nose are presented the most ventrally, then the lips, cheek, and vibrissae area (the skin under vibrissae, not vibrissae properly), the orbital area, head dorsum, and near-ear area, all the parts presented twice as two belts. The vibrissae projection occupies a separate area. The pattern of the trunk projection is similar; when moving from the ansate sulcus downward, the sequence is the back, flank, abdomen, and then flank again. The outer projection belt is more

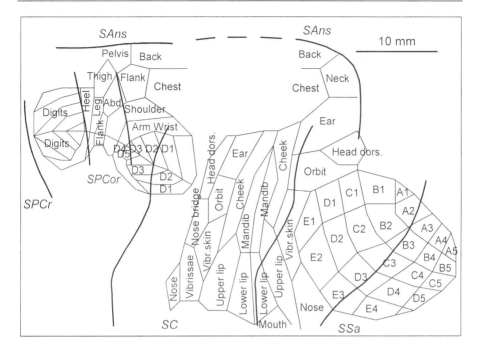

Figure 5.6. Somatotopic projection map of the somatosensory cerebral cortex of the fur seal *Callorhinus ursinus* (the left hemisphere). In the forelimb projection, D1–D5 are projections of digits from 1 to 5. In the vibrissae projection area, the projected vibrissae designated as A1, A2,... B1,... etc., where A to D are horizontal vibrissae rows and 1 to 5 are vertical vibrissae columns. *SC – sulcus coronarius, SPCr – s. postcruciatus, SPCor – s. precoronarius, SSa – s. suprasylvius anterior, SAns – s. ansatus.*

extensive, it occupies a major part of the convexital cortex and, partially, the walls of the coronar and anterior suprasylvian sulci. The inner projection belt occupies mainly the walls of the precoronar and coronar sulci.

A remarkable feature of the projection map is that different parts of the body surface are represented in the cortex with different magnification. A major part of the somatosensory area is occupied by the head representation. Fore- and hindlimb also are represented more extensively than the trunk although less extensively than the head. The limbs have separate representations of their dorsal and ventral surface, which are linked by representation of the distal border of the limb.

Within the head projection, an extremely large part is occupied by the representation of maxillar vibrissae. Each vibrissa has its own projection region in the cortex, a few mm in size. The vibrissa projections in the cortex are arranged in an ordered manner.

A widely adopted characteristic of somatotopic projections is the magnification factor. This factor is the ratio of a distance between two neighboring cortical points to the distance between the corresponding points of the body surface; the higher the factor, the more extensive the projection. In the presented somatotopic map, the magnification factor varies from 0.01 in the trunk projection (that is, a 1 mm shift of a recording point in the cortex results in 100 mm shift of the receptive field) to 0.3 to 0.5 in the vibrissae projection (1 mm shift of a recording point results in a 2 to 3 mm shift of the receptive field). The limb and neck projection have magnification factors of 0.02 to 0.05; the head (except vibrissae) projection have magnification factors of 0.05 to 0.2.

The magnification factor is a good indicator of relative significance of the tactile sense from different parts of the body. It shows that, in the northern fur seal, similarly to many other mammals, the head and the limbs provide much more tactile information than the trunk. In particularl, the extremely wide representation of maxillar vibrissae reflects the great importance of this source of tactile information.

A widely used way to illustrate peculiarities of somatosensory projections is to draw the so-called cortical cartoon which is a schematic picture of the animal's body positioned on the cortical map, each part of the body positioned on its cortical projection. Although such "cartoons" do not present precise information of the somatotopic projection, they show the general projection pattern and magnification of different body parts. Such "cartoons" in the somatosensory cortex of the northern fur seal are presented in Fig. 5.7. The two "cartoons" demonstrate the duplicate representation of the body surface in the somatosensory cortex. They show the general disposition of projections: the trunk upward, the limbs and head downward, the caudal part of the body rostrally. The "cartoons" also illustrate different magnification factors for different body parts, particularly the large representation of the head and extremely large representation of the vibrissae.

The duplicate somatotopic representation in the fur seal merits special attention. It should be noted that many mammals have multiple sensory projections in the cerebral cortex, named as SI, SII, SIII. However, some features of the cortical somatosensory projection area indicate that all this area should be considered as the primary projection, SI. All parts of this area feature characteristic properties of the primary projection area such as (1) vigorous neuronal responses to tactile stimuli and clearly defined receptive fields, (2) strictly contralateral location of all the receptive fields, and (3) position of the projection area relative to cortical sulci and gyri similar to the position of SI area in carnivores.

The presence of two somatotopic projections is not contradictory to this conclusion. Examples of duplicate representation of some important parts of the body within the SI area are known in mammals: vibrissae representation in the opossum (Pubols et al., 1976), the forepaw in the gray squirrel (Sur et

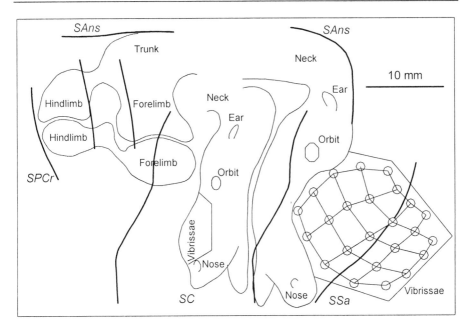

Figure 5.7. Cortical "cartoons" in the somatosensory area of the northern fur seal *Callorhinus ursinus* (the left hemisphere).

al., 1978; Nelson et al., 1979; Krubitzer et al., 1986), the hand in prosimian primates (Sur et al., 1980; Carlson and Welt, 1980; Carlson and FitzPatric, 1982; FitzPatric et al., 1982; Carlson et al., 1986) and jerboa (Sokolov et al., 1986). Some monkeys have duplicate projection of all the body within the SI area (Merzenich et al., 1978; Nelson et al., 1980; Sur *et al.,* 1982). The duplicate projection in the SI area was considered as an indication of the very important role of the tactile sense. The same may be valid for the northern fur seal.

5.2.2. Tactile Sensitivity of Vibrissae

Extensive representation of maxillar vibrissae in the cerebral cortex indicates an important role of this organ in the tactile orientation of the northern fur seal. Further evidence were provided by quantitative data on sensitivity of the vibrissae (Supin et al., 1986). In that study, focal EP to vibrissae stimulation were recorded in the somatosensory cortex of deeply anaesthetized northern fur seals *Callorhinus ursinus*. The recording microelectrode was always positioned in the cortical projection of the stimulated vibrissa.

For quantitative stimulation of vibrissae, a device based on an electromechanical transducer (a motor for ink-writing recorder) was used that allowed an electric signal to be transform to the proportional rotation of a lever. Depending on the position of the device relative to the stimulating vibrissa, the rotation of the lever resulted either in a rotational shift of the vibrissa around its root with a minimal longitudinal shift (Fig. 5.8A) or in a longitudinal (push-pool) movement of the vibrissa with a minimal rotation (Fig. 5.8B). The device allowed quantitatively controlled rotational shifts from 5×10^{-4} to 0.5 rad and longitudinal shifts from 5×10^{-3} to 5 mm.

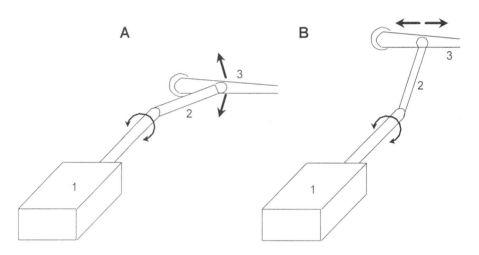

Figure 5.8. A device for the quantitative mechanical stimulation of vibrissae. 1 – electromechanic transducer, 2 – lever, 3 – vibrissa. **A**. Rotational shift of the vibrissa. **B**. Longitudinal shift of the vibrissa.

EP recorded in the cortical representation of the stimulated vibrissa (Fig. 5.9) were stimulus dependent: EP amplitude increased with increasing the shift of the vibrissa, either rotational or longitudinal, until it reached a certain limit. As Fig. 5.9 exemplifies, a just-detectable EP appeared at a vibrissa shift of 0.001 rad and reached its maximal amplitude at a shift of 0.1 rad. Using such records, it was possible to draw the stimulus-response curves and to find response thresholds.

Figure 5.10 presents EP amplitude dependence on tactile stimulus intensity (the value of vibrissa shift) for several vibrissae in the northern fur seal. Maxillar vibrissae differ in size to a large extent, from 12–15 mm (A4, B5, D5) to more than 100 mm (D1). Nevertheless, all of them have more or less similar thresholds, better than 0.001 rad of rotational shift and 0.01 to 0.02 mm of longitudinal shift. It may be an indication of equal importance of all

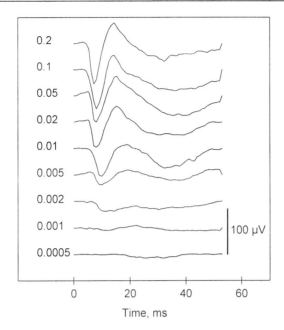

Figure 5.9. EP to vibrissa rotational movement in the somatosensory cortex of the northern fur seal *Callorhinus ursinus*. Vibrissa shift (rad) is indicated near records.

of them for tactile orientation, in agreement with their similar representation in the cerebral cortex.

5.3. SUMMARY

Limited morphological, psychophysical, and EP data indicate developed tactile sensitivity in dolphins, particularly at the head (the lips, orbital region, blowhole, snout, melon). The somatosensory projection area of the cerebral cortex was found in the postcruciate gyrus.

In pinnipeds, tactile sensitivity is well developed, particularly the vibrissae sensitivity, which allows fine discrimination of 3-D objects. The somatosensory projection area occupies a part of the cerebral cortex around the coronar sulcus and features an ordered somatotopic projection. Micromapping studies revealed extended projections of the head and limbs as compared to the trunk; within the head projection area, projections of the lips and nose and particularly maxillar vibrissae are the most extended. Projections of almost all parts of the body surface are duplicated in the SI projec-

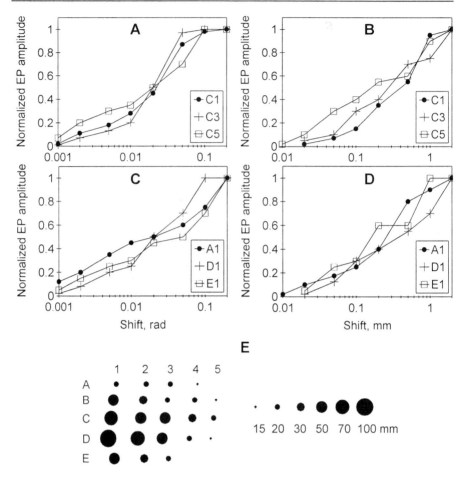

Figure 5.10. Cortical EP amplitude dependence on vibrissa shift in a northern fur seal *Callorhinus ursinus.* **A** and **B**. Data for different vibrissae of a row C (C1 to C5, as indicated). **C** and **D**. Data for different vibrissae of a column 1 (A1 to E1, as indicated). **A** and **C**. Responses to rotational shifts of the vibrissae. **B** and **D**. Responses to longitudinal shifts. **E**. Positions of maxillar vibrissae roots (right side) and designation of rows (A to E) and columns (1 to 5). The spot diameter approximately indicates the vibrissa length, according to the scale on the right.

tion area. Magnification factors of the somatosensory projections vary from 0.01 in the trunk projection to 0.3–0.5 in the vibrissae projection. All the maxillar vibrissae, both the longest and the shortest ones, are equally sensitive displaying thresholds of around 0.001 rad of rotational displacement and 0.01 to 0.02 mm of longitudinal shift.

REFERENCES

Abbas PJ, Gorga MP. AP responses in forward-masking paradigms and their relationship to responses of auditory-nerve fibers. J Acoust Soc Am 1981; 69: 492–499

Abdala C, Folsom RC. The development of frequency resolution in humans as revealed by auditory brain-stem response with notched-noise masking. J Acoust Soc Am 1995; 98: 921–930

Abel S. Duration discrimination of noise and tone bursts. J Acoust Soc Am 1972; 51: 1219–1223

Achor J, Starr A. Auditory brainstem responses in the cat. Electroenceph Clin Neuriphysiol 1980; 48: 174–190

Alderson AM, Diamantopoulos E, Downman CBB. Auditory cortex of the seal (*Phoca vitulina*). J Anat 1960; 94

Allison T, Goff WR. Electrophysiological studies of the echidna, *Tachiglossus aculcatus*. III. Sensory and interhemispheric evoked responses. Arch Ital Biol 1972; 110: 195–216

Andersen S. Auditory sensitivity of the harbour porpoise, *Phocoena phocoena*. Invest Cetacea 1970; 2: 255–258

Arrese C, Dunlop SA, Harman AM, Braekevelt CR, Ross WM, Shand J, Beazley LD. Retinal structure and visual acuity in a polyprotodont marsupial, the fat-tailed dunnart (*Sminthopsis crassicaudata*). Brain Behav Evol 1999; 53: 111–126

Au WWL. Echolocation signals of the Atlantic bottlenose dolphin (*Tursiops truncatus*) in open waters. In: *Animal Sonar Systems*, RG Busnel, JF Fish, eds. New York: Plenum, 1980, pp. 251–282

Au WWL. Target detection in noise by echolocating dolphins. In: *Sensory Abilities of Cetaceans. Laboratory and Field Evidence*, JA Thomas, RA Kastelein, eds. New York: Plenum, 1990, pp. 203–216

Au WWL. *The Sonar of Dolphins*. Springer-Verlag, New York, 1993

Au WWL, Moore PWB. Receiving beam patterns and directivity indices of the Atlantic bottlenose dolphin *Tursiops truncatus*. J Acoust Soc Am 1984; 75: 255–262

Au WWL, Moore PWB. Critical ratio and critical band width for the Atlantic bottlenose dolphin. J Acoust Soc Am 1990; 88: 1635–1638

Au WWL, Moore PWB, Pawloski DA. Detection of complex echoes in noise by an echolocating dolphin. J Acoust Soc Am 1988; 83: 662–668

Au WWL, Nachtigall PE, Pawloski JL. Acoustic effects of the ATOC signal (75 Hz, 195 dB) on dolphins and whales. J Acoust Soc Am 1997; 101: 2973–2977

Au WWL, Pawloski JL. Detection of rippled noise by an Atlantic bottlenose dolphin. J Acoust Soc Am 1989; 86: 591–596

Awbrey FT, Thomas JA, Kastelein RA. Low-frequency underwater hearing sensitivity in belugas, *Delphinapterus leucas*. J Acoust Soc Am 1988; 84: 2273–2275

Babushina YeS, Zaslavskiy GL, Yurkevich LI. Air and underwater hearing characteristics of the northern fur seal: Audiograms, frequency and differential thresholds. Biofizika (Biophysics) 1991; 36: 909–913

Bain DE, Dalhiem ME. Effects of masking noise on detection thresholds of killer whales. In *Consequences of the Exxon Valdes Oil Spill*, T Loughlin, ed. New York: Academic Press, 1994, pp. 243–256

Bateson G. Observation of a cetacean community. In: *Mind in Waters*. New York: Scribner's, 1974, pp. 146–169

Batra R, Kuwada S, Maher VL. The frequency following response to continuous tones in humans. Hearing Res 1986; 21: 167–177

Batteau DW. The role of the pinna in human sound localization. Proc R Soc Lond 1967; 168: 158–180

Belendiuk D, Butler RA. Spectral cues which influence monaural localization in the horizontal plane. Percept Psychophys 1977; 22: 353–358

Bernholz CD, Mattews ML. Critical flicker frequency in a harp seal: evidence for duplex retinal organization. Vision Res 1975; 15: 733–736

Bibikov NG. Auditory brainstem responses in the harbor porpoise *Phocoena phocoena*. In: *Marine Mammal Sensory Systems*, JA Thomas, RA Kastelein, AYa Supin, eds. New York: Plenum, 1992, pp. 197–211

Bilsen FA, ten Kate JH, Buunen TJF, Raatgever J. Responses of single units in the cochlear nucleus of the cat to cosine noise. J Acoust Soc Am 1975; 58: 858–866

Bilsen FA, Wieman JL. Atonal periodicity sensation for comb filtered noise signals. In: *Psychophysical and Behavioral Studies in Hearing*, G van der Brink, FA Bilsen, eds. Delft: Delft Univ. Press, 1980, pp. 379–382

Bishop PO, Kozak W, Vakkur J. Some quantitative aspects of the cat's eye: Axis and plane of reference, visual field coordinates and optics. J Physiol 1962; 163: 466–502

Blauert J. Sound localization in the medial plane. Acoustica 1969/1970; 22: 205–213

Breathnach AS. The cetacean central nervous system. Biol Rev 1960; 35: 187–230

Brecha N, Johnson D, Peichl L, Wässle H. Cholinergic amacrine cells of the rabbit retina contain glutamat decarboxilase and γ-aminobutirate immunoreactivity. Proc Nat Acad Sci USA 1988; 85: 6187–6191

Brill RL. The jaw-hearing dolphin: preliminary behavioral and acoustical evidence. In: *Animal Sonar: Processes and Performance*, PE Nachtigall, PWB Moore, eds. New York: Plenum Press, 1988, pp. 281–287

Brill RL. The effect of attenuating returning echolocation signals at the lower jaw of a dolphin (*Tursiops truncatus*). J Acoust Soc Am 1991; 89: 2851–2857

Brill RL, Sevenich ML, Sullivan TJ, Sustman JD, Witt RE. Behavioral evidence for hearing through the lower jaw by an echolocating dolphin, *Tursiops truncatus*. Marine Mammal Sci 1988; 4: 223–230

Brown CJ, Abbas PJ. A comparison of AP and ABR tuning curves in guinea pig. Hearing Res 1987; 25: 193–204

Bullock TH, Domning DP, Best RC. Evoked brain potentials demonstrate hearing in a manatee (Sirenia: *Trichechus inunguis*). J Mammol 1980; 61: 130–133

Bullock TH, Grinnell AD, Ikezono F, Kameda K, Katsuki Y, Nomoto M, Sato O, Suga N, Yanagisava K. Electrophysiological studies of the central auditory mechanisms in cetaceans. Z Vergl Physiol 1968; 59: 117–156

Bullock TH, O'Shea TJ, McClune MC. Auditory evoked potentials in the West Indian manatee (Sirenia: *Trichechus manatus*). J Comp Physiol 1982; 148: 547–554

Bullock TH, Ridgway SH. Evoked potentials in the central auditory system of alert porpoises to their own and artificial sounds. J Neurobiol 1972; 3: 79–99

Burdin VI, Markov VI, Reznik AM, Skornyakov VM, Chupakov AG. Determination of the just noticeable intensity difference for white noise in the Black sea bottlenose dolphin

(*Tursiops truncatus ponticus* Barabash). In: *Morphology and Ecology of Marine Mammals*, KK Chapskii and VE Sokolov, eds. New York: Wiley, 1973, pp. 169–173

Burkard R, Hecox K. The effect of broadband noise on human brainstem auditory evoked response. II. frequency specificity. J Acoust Soc Am 1983; 74: 1214–1223

Busch H, Dücker G. Das visuelle Leistungsvermogen der Seebären (*Arctocephalus pusillus* und *Arctocephalus australs*). Zool Anz 1987; 219: 197–224

Busnel R-G, Fish JF, eds. *Animal Sonar Systems*. New York: Plenum, 1980

Butler RA. Monaural and binaural localization of noise bursts vertically in the median sagittal plane. J Aud Res 1969; 3: 230–235

Butler RA, Flannery R. The spatial attributes of stimulus frequency and their role in monaural localization of sound along the horizontal plane. Percept Psychophys 1980; 28: 449–457

Carlson M, FitzPatric KA. Organization of the hand area in the primary somatic sensory cortex (SmI) of the prosimian primate, *Nycticebus coucang*. J Comp Neurol 1982; 204: 280–295

Carlson M, Huerta MF, Cusik CG, Kaas JH. Studies on the evolution of multiple somatosensory representations in primates: The organization of anterior parietal cortex in the new world callitrichid, *Saguinus*. J Comp Neurol 1986; 246: 409–426

Carlson M, Welt C. Somatic sensory cortex (SmI) of the prosimian primate *Galago crassicaudatus*: Organization of mechanoreceptive input from the hand in relation to cytoarhitecture. J Comp Neurol 1980; 189: 249–271

Cheivitz JH. Untersuchungen über die Area centralis retinae. Arch Anat Physiol Lpz Anat Abteil Suppl 1889; 139–196

Chen T-J, Chen S-S. Brain stem auditory-evoked potentials in different strains of rodents. Acta Physiol Scand 1990; 138: 529–538

Chun MH, Wässle H, Tulunay-Keesey U. Colocalization of [^{3}h] muscimol uptake and choline acetyltransferase immunoreactivity in amacrine cells of the cat retina. Neurosci Lett 1988; 94: 259–263

Clark CW. The acoustic repertoire of the southern right whale: A quantitative analysis. Anim Behav 1982; 30: 1060–1071

Clark CW. Acoustic behavior of mysticete whales. In: *Sensory Abilities of Cetaceans. Laboratory and Field Evidence*, JA Thomas, RA Kastelein, eds. New York: Plenum, 1990, pp. 571–583

Clopton BM. Detection of increments in noise intensity by monkeys. J Exp Anal Behav 1972; 17: 437–481

Cohen JL, Tucker GS, Odell DK. The photoreceptors of the West Indian manatee. J Morphol 1982; 173: 197–202

Collin SP, Pettigrew JD. Retinal ganglion cell topography in teleosts: A comparison between Nissl-stained material and retrograde labeling from the optic nerve. J Comp Neurol 1988; 276: 412–422

Corwin JT, Bullock TH, Schweitzer J. The auditory brainstem responses in five vertebrate classes. Electroenceph Clin Neurophysiol 1982; 54: 629–641

Costa BLSA, Pessoa VF, Bousfield JD, Clarke RJ. Unusual distribution of ganglion cells in the retina of the three-toed sloth (*Bradypus variegatus*). Brazilian J Med Biol Res 1987; 20: 741–748

Costa BLSA, Pessoa VF, Bousfield JD, Clarke RJ. Unusual distribution of ganglion cells in the retina of the two-toed sloth (*Choloepus didactilus*). Brazilian J Med Biol Res 1989; 22: 233–236

Crognale MA, Levenson DH, Ponganis PJ, Deegan II JF, Jacobs GH. Cone spectral sensitivity in the harbor seal (*Phoca vitulina*) and implications for color vision. Can J Zool 1998; 76: 2114–2118

Dahlheim ME, Ljungblad DK. Preliminary hearing study on gray whales (*Eschrichtius robustus*) in the field. In: *Sensory Abilities of Cetaceans. Laboratory and Field Evidence*, JA Thomas, RA Kastelein, eds. New York: Plenum, 1990, pp. 335–346

Dallos P, Cheatham MA. Compound action potential (AP) tuning curves. J Acoust Soc Am 1976; 59: 591–597

Davis H, Hirsh SK. The audiometric utility of brainstem responses to low-frequency sounds. Audiology 1976; 15: 181–195

Dawson WW. The cetacean eye. In: *Cetacean Behavior: Mechanisms and Functions*, LM Herman, ed. New York: Wiley, 1980, pp. 53–100

Dawson WW, Adams CK, Barris MC, Litzkow CA. Static and kinetic properties of the dolphin pupil. Am J Physiol 1979; 237: R301–R305

Dawson WW, Brindford LA, Perez JM. Gross anatomy and optics of the dolphin eye (*Tursiops truncatus*). Cetology 1972; 10: 1–12

Dawson WW, Carder DA, Ridgway SH, Schmeisser ET. Synchrony of dolphin eye movement and their power density spectra. Comp Biochem Physiol 1981; 68A: 443–449

Dawson WW, Hawthorne MN, Jenkins RL, Goldston RT. Giant neural system in the inner retina and optic nerve of small whales. J Comp Neurol 1982; 205: 1–7

Dawson WW, Hope GM, Ulshafer RJ, Hawthorne MN, Jenkins RL. Contents of the optic nerve of a small cetacean. Aquatic Mammals 1983; 10(2): 45–56

Dawson WW, Perez JM. Unusual retinal cells in the dolphin eye. Science 1973; 181: 747–749

Dawson WW, Schroeder JP, Dawson JF. The ocular fundus of two cetaceans. Marine Mammal Sci 1987a; 3: 1–13

Dawson WW, Schroeder JP, Dawson JC, Nachtigall PE. Cyclic ocular hypertension in cetaceans. Marine mammal Sci 1992; 8: 135–142

Dawson WW, Schroeder JP, Sharpe SN. Corneal surface properties of two marine mammal species. Marine Mammal Sci 1987b; 3: 186–197

de Boer E. Auditory time constants: a paradox? In: *Time Resolution in Auditory System*, A Michelsen, ed. 1984, pp 141–158

Dehnhardt G. Preliminary results from psychophysical studies of the tactile sensitivity in marine mammals. In: *Sensory Abilities of Cetaceans. Laboratory and Field Evidence*, JA Thomas, RA Kastelein, eds. New York: Plenum, 1990, pp. 435–446

Dehnhardt G. Tactile size discrimination by a California sea lion (*Zalophus californianus*) using its mystacial vibrissae. J Comp Physiol 1994; 175: 791–800

Dehnhardt G, Mauck B, Bleckmann H. Seal wiskers detect water movements. Nature 1998; 394: 235–236

Diamond IT, Snyder M, Killackey H, Jane J, Hall WC. Thalamocortical projections in the tree shrew (*Tupaia glis*). J Comp Neurol 1970; 139: 273–306

Dolphin WF. Steady-state auditory-evoked potentials in three cetacean species elicited using amplitude-modulated stimuli. In: *Sensory Systems of Aquatic Mammals*, RA Kastelein, JA Thomas, PE Nachtigall, eds. Woerden, The Netherlands: De Spil, 1995, pp 25–47

Dolphin WF, Au WWL, Nachtigall P. Modulation transfer function to low-frequency carriers in three species of cetaceans. J Comp Physiol A 1995; 177: 235–245

Dolphin WF, Mountain DC. The envelope following response: Scalp potential elicited in the Mongolian gerbil using sinusoidally AM acoustic signals. Hearing Res 1992; 58: 70–78

Don M, Eggermont JJ. Analysis of the click-evoked brainstem potentials in man using high-pass noise masking. J Acoust Soc Am 1978; 63: 1084–1092

Donaldson GS, Ruth RA. Derived band auditory brain-stem response estimates of traveling wave velocity in humans. I. Normal-hearing subjects. J Acoust Soc Am 1993; 93: 940–951

Drager UC, Olsen JF. Ganglion cell distribution in the retina of the mouse. Invest Ophthalmol 1981; 20: 285–293

Dral ADG. Aquatic and aerial vision in the bottle-nosed dolphin. Neth J Sea Res 1972; 5: 510–513

Dral ADG Some quantitative aspects of the retina of *Tursiops truncatus*. Aquatic Mammals 1975a; 2: 28–31

Dral ADG. Vision in Cetacea. J Zool Animal Med 1975b; 5: 510–573

Dral ADG. On the retinal anatomy of cetacea (mainly *Tursiops truncatus*). In: *Functional Anatomy of Marine Mammals*, RJ Harrison, ed. Vol. 3. London: Academic, 1977, pp. 81–134

Dral ADG The retinal ganglion cells of *Delphinus delphis* and their distribution. Aquatic Mammals 1983; 10(2): 57–68

Dral ADG, Beumer L. The anatomy of the eye of the Ganges river dolphin *Platanista gangetica*. Z Saugetierkunde 1974; 39: 143–167

Dubrovskiy NA. On the two auditory systems in dolphins. In: *Sensory Abilities of Cetaceans. Laboratory and Field Evidence*, JA Thomas, RA Kastelein, eds. New York: Plenum, 1990, pp. 233–254

Dubrovsky NA, Zorikov TV, Kvighinadze OSh, Kuratishvili MM. Mechanisms of signal discrimination and identification in the auditory system of *Tursiops truncatus*. In: *Marine Mammal Sensory Systems*, JA Thomas, RA Kastelein, AYa Supin, eds. New York: Plenum, 1992, pp. 235–240

Dunlop SA, Ross WM, Beazley LD. The retinal ganglion cell layer and optic nerve in a marsupial, the honey possum (*Tarsipes rostratus*). Brain Behav Evol 1994; 44: 307–323

Eddins D. Amplitude modulation detection of narrow-band noise: Effect of absolute bandwidth and frequency region. J Acoust Soc Am 1993; 93: 470–479

Eddins DA, Hall JW, Grose JH. Detection of temporal gaps as a function of frequency region and absolute noise bandwidth. J Acoust Soc Am 1992; 91: 1069–1077

Edds PL. Vocalizations of the blue whale, *Balaenoptera musculus*, in the St. Lawrence River. J Mammol 1982; 63: 345–347

Edds PL. Characteristics of finback *Balaenoptera physalus* vocalization in the St. Lawrence Estuary. Bioacoustics 1988; 1: 131–149

Eggermont JJ., 1976. Electrocochleography. In: *Handbook of Sensory Physiology*, Vol. V/III, WD Keidel, WD Neff, eds., Berlin: Springer-Verlag, pp. 625–705

Eggermont JJ. Compound action potential tuning curves in normal and pathological human ears. J Acoust Soc Am 1977; 62: 1247–1251

Eggermont JJ. Narrow-band AP latencies in normal and recruiting human ears. J Acoust Soc Am 1979; 65: 463–470

Ehret G, Merzenich MM. Complex sound analysis (frequency resolution, filtering and spectral integration) by units of the inferior colliculus of the cat. Brain Res Rev 1988; 13: 139–163.

Elberling C. Action potentials along the cochlear partition recorded from the ear canal in man. Scand Audiol 1974; 3: 3–19

Elliot DN, McGee TM. Effect of cochlear lesions upon audiogram and intensity discrimination in cats. Ann Otol Rhinol Laryngol 1965; 74: 386–408

Evans EF. Frequency selectivity at high signal levels of single units in cochlear nerve and cochlear nucleus. In: *Psychophysics and Physiology of Hearing*, EF Evans, JP Wilson, eds. London: Academic, 1977, pp. 185–192

Evans EF. Auditory processing of complex sounds: an overview. Phil Trans R Soc Lond 1992; B 336: 295–306

Evans E.F, Nelson PG. The responses of single neurones in the cochlear nucleus of the cat as a function of their location and the anesthetic state. Exp. Brain Res 1973; 17: 402–427

Evans EF, Wilson JP. Frequency selectivity of the cochlea. In: *Basic Mechanisms of Hearing*, AR. Møller, ed. New York: Academic, 1973, pp. 519–551

Evans WE, Haugen RM. An experimental study of the echolocation ability of a California sea lion, *Zalophus californianus* (Lesson). Bull Southern California Acad Sci 1963; 62: 165–175

Evans WE, Herald ES. Underwater calls of a captive Amazon manatee, *Trichechus inunguis*. J Mammol 1970; 51: 820–823

Fasick JI, Cronin TW, Hunt DM, Robinson PR. The visual pigments of the bottlenose dolphin (*Tursiops truncatus*). Vis Neurosci 1998; 15: 643-651

Fay RR. *Hearing in Vertebrates: A Psychophysics Databook.* Hill-Fay, Vinnetka, IL, 1988

Fay RR. Structure and function in sound discrimination among vertebrates. In *The Evolutionary Biology of Hearing*, New York: Springer-Verlag, 1992, pp. 229–263

Fischer QS, Kirby MA. Number and distribution of retinal ganglion cells in anubis baboons (*Papio anubis*). Brain Behav Evol 1991; 37: 189–203

Fitzgibbons PJ. Temporal gap detection in noise as a function of frequency, bandwidth and level. J Acoust Soc Am 1983; 74: 67–72

Fitzgibbons PJ, Wightman, FL. Gap detection in normal and hearing-impaired listeners. J Acoust Soc Am 1982; 72: 761–765

FitzPatric KA, Carlson M, Charlton J. Topography, cytoarchitecture, and sulcal patterns in primary somatic sensory cortex (SmI) of the prosimiam primate, *Perodicticus potto*. J Comp Neurol 1982; 204: 296–310

Flannery R, Butler RA. Spectral cues provided by pinna for monaural localization in the horizontal plane. Percept Psychophys 1981; 29: 438–444

Fleischer G. Evolutionary principles of the mammalian middle ear. Adv Anat Embryol Cell Biol 1978; 55: 1–70

Fletcher H. Auditory patterns. Rev Mod Phys 1940; 12: 47–65

Fobes JL, Smock CC. Sensory capacities of marine mammals. Psychol Bull 1981; 89: 288–307

Formby C, and Muir K. Modulation and gap detection for broadband and filtered noise signals. J Acoust Soc Am 1988; 84, 545–550

Fraser FC, Purves PE. Hearing in cetaceans. Bull British Mus (Nat Hist) 1954; 2: 103–116

Fraser FC, Purves PE. Hearing in whales. Endeavour 1959; 18: 93–98

Fraser FC, Purves PE. Hearing in cetaceans: Evolution of the accessory air sacs and the structure and function of the outer and middle ear in recent cetaceans. Bull British Mus (Nat Hist) 1960; 7: 1–140

Frisina RD, Smith RL, Chamberlan SC. Encoding of amplitude modulation in the gerbil cochlear nucleus: I. A hierarchy of enhancement. Hearing Res 1990; 44: 99–122

Fukuda Y, Stone J. Retinal distribution and central projection of W, X and Y cells of the cat's retina. J Neurophysiol 1974; 37: 749–772

Fuzessery ZM, Pollak GD. Determinants of sound locations selectivity in bat inferior colliculus: A combined dihotic and free-field stimulation study. J Neurophysiol 1985; 54: 757–781

Galambos R, Makeig S, Talmachoff PJ. A 40-Hz auditory potentials recorded from the human scalp. Proc Nat Acad Sci USA 1981; 78: 2643–2647

Gao A, Zhou K. On the retinal ganglion cells of *Neophocoena* and *Lipotes*. Acta Zool Sin 1987; 33: 316–332

Gao G, Zhou K. The number of fibers and range of fiber diameters in the cochlear nerve of three odontocete species. Can J Zool 1991; 69: 2360–2364

Gao G, Zhou K. Fiber analysis of the optic and cochlear nerves of small cetaceans. In: *Marine Mammal Sensory Systems*, JA Thomas, RA Kastelein, AYa Supin, eds. New York: Plenum, 1992, pp. 39–52

Garey LJ, Revishchin AV. Structure and thalamocortical relations of the cetacean sensory cortex: Histological, tracer and immunochemical studies. In: *Sensory Abilities of Cetaceans. Laboratory and Field Evidence*, JA Thomas, RA Kastelein, eds. New York: Plenum, 1990, pp. 19–30

Garey LJ, Winkelman E, Brauer K. Golgi and Nissl studies of the visual cortex of the bottlenosed dolphin. J Comp Neurol 1985; 240: 305–321

Gentry RL. Underwater auditory localization in the California sea lion (*Zalophus californianus*). J Aud Res 1967; 7: 187–193.

Gerstein ER, Gerstein L, Forsythe SE, Blue JE. The underwater audiogram of the West Indian manatee (*Trichechus manatus*). J Acoust Soc Am 1999; 105: 3575–3583

Glasberg BR, Moore BCJ. Comparison of auditory filter shapes derived with three different maskers. J Acoust Soc Am 1984; 75: 536–544

Glasberg BR, Moore BCJ. Derivation of auditory filter shapes from notch-noise data. Hearing Res 1990; 47: 103–138

Glasberg BR, Moore BCJ. Effect of envelope fluctuations on gap detection. Hearing Res 1992; 64, 81–92

Glezer II, Hof PR, Leranth C, Morgane PJ. Morphological and histochemical features of odontocete visual neocortex: Immunocytochemical analysis of pyramidal and non-pyramidal populations of neurons. In: *Marine Mammal Sensory Systam*, JA Thomas, RA Kastelein, AYa Supin, eds. New York: Plenum, 1992, pp. 1–38

Glezer II, Morgane PJ, Leranth C. Immunocytochemistry of neurotransmitters in visual neocortex of several toothed whales: Light and electron microscopic study. In: *Sensory Abilities of Cetaceans. Laboratory and Field Evidence*, JA Thomas, RA Kastelein, eds. New York: Plenum, 1990, pp. 39–66

Gorga MP, Abbas PJ. Forward-masked AP tuning curves in normal and acoustically traumatized ears. J Acoust Soc Am 1981; 70: 1322–1330

Gorga MP, McGee J, Walsh EJ, Javel E, Farley GR. ABR measurement in the cat using a forward-masking paradigm. J Acoust Soc Am 1983; 73: 255–261

Gould HJ, Sobhy OA. Using the derived auditory brain stem response to estimate traveling wave velocity. Ear Hearing 1992; 13: 96–101

Green DM. Masking with two tones. J Acoust Soc Am 1965; 37: 802–813

Green DM. Temporal factors in psychoacoustics. In *Time Resolution in Auditory System*, A Michelsen, ed. 1984, pp. 123–140

Green DM, Swets JA. *Signal Detection theory and Psychophysics*. Huntington, NY: Krieger, 1966

Griebel U, Schmid A. Colour vision in the California sea lion (*Zalophus californianus*). Vision Res 1992; 32: 477–482

Griebel U, Schmid A. Color vision in the manatee (*Trichechus manatus*). Vision Res 1996; 36: 2747–2757

Griebel U, Schmid A. Brightness discrimination ability in the West Indian manatee (*Trichechus manatus*). J Exp Biol 1997; 200: 1587–1592

Grinnell AD, Grinnell VS. Neural correlates of vertical localization by echolocating bats. J Physiol 1965; 181: 830–851

Grose JG, Eddins DA, Hall JW. Gap detection as a function of stimulus bandwidth with fixed high-frequency cutoff in normal-hearing and hearing-impaired listeners. J Acoust Soc Am 1989; 86: 1747–1755

Grünthal E. Über den Primatencharakter des Gehirns von *Delphinus delphis*. Monatschr Psychiatr Neurol 1942; 105: 249–274

Hall JD. Johnson CS. Auditory thresholds of a killer whale, *Orcinus orca* Linnaeus. J Acoust Soc Am 1971; 51: 515–517

Hall JW. Auditory brainstem frequency following responses to waveform envelope periodicities. Science 1979; 205: 1297–1299

Hall WC, Diamond IT. Organization and function of the visual cortex in hedgehog. Brain Behav Evol 1968; 7: 215

Hall WC, Kaas JH, Killackey H, Diamond IT. Cortical visual areas in the grey squirrel (*Sciurus carolinensis*): A correlation between cortical evoked potential maps and architectonic subdivisions. J Neurophysiol 1971; 34: 437–452

Hanggi EB, Schusterman R. Conditional discrimination learning in a male harbor seal (*Phoca vitulina*). In: *Sensory Systems of Aquatic Mammals*, RA Kastelein, JA Thomas, PE Nachtigall, eds. Woerden, the Netherlands: De Spil, 1995, pp. 543–559

Harley HE, Xitco MJ Jr, Roitblat HL. Echolocation, cognition, and the dolphin's world. In: *Sensory Systems of Aquatic Mammals*, RA Kastelein, JA Thomas, PE Nachtigall, eds. Woerden, the Netherlands: De Spil, 1995, pp. 529–542

Harley HE, Roitblat HL, Nachtigall PE. Object representation in the bottlenose dolphin (*Tursiops truncatus*): Integration of visual and echoic information. J Exp Psychol Animal Behavior Processes 1996; 22: 164–174

Harris DM. Action potential suppression, tuning curves and thresholds: Comparison with single fiber data. Hearing Res 1978; 1: 133–154

Harris LR. Contrast sensitivity and acuity of a conscious cat measured by the occipital evoked potential. Vision Res 1978; 18: 175–178

Harrison RJ, King JE. *Marine Mammals*. London: Hutchinson Univ Libr, 1965

Harrison RJ, Thurley KW. Structure of the epidermis in *Tursiops*, *Delphinus*, and *Phocoena*. In: *Functional Anatomy of Marine mammals*, RJ Harrison, ed. Vol. 2. London: Academic, 1974, pp. 45–71

Harrison RV, Aran J-M, Erre JP. AP tuning curves from normal and pathological human and guinea pig cochleas. J Acoust Soc Am 1981; 69: 1374–1385

Hartman DS. Ecology and behavior of the Manatee (*Trichechus manatus*) in Florida. American Society of Mammologists. Special Publication 1979; 5: 153

Hebel R. Distribution of retinal ganglion cells in five mammalian species (pig, sheep, ox, horse, dog). Anat Embryol 1976; 150: 45–51

Hebel R, Hollander H. Size and distribution of ganglion cells in the bovine retina. Vision Res 1979; 19: 667–674

Herbank J, Wright D. Spectral cues used in the localization of sound in the medial plane. J Acoust Soc Am 1974; 56: 1829–1834

Herman LM. Cognitive characteristics of dolphins. In: *Cetacean Behavior: Mechanisms and Functions*, LM Herman, ed. New York: Wiley, 1980, pp. 363–430

Herman LM, Arbeit WR. Frequency difference limens in the bottlenose dolphin: 1–70 kHz. J Aud Res 1972; 2: 109–120

Herman LM, Pack AA. Echoic-visual cross-modal recognition by a dolphin. In: *Marine Mammal Sensory Systems*, JA Thomas, RA Kastelein, AYa Supin, eds. New York: Plenum, 1992, pp. 709–726

Herman LM, Pack AA. Seeing through sound: Dolphins (*Tursiops truncatus*) perceive the spatial structure of objects through echolocation. J Comp Psychol 1998; 112: 292–305

Herman LM, Peacock MF, Yunker MP, Madsen CJ. Bottlenosed dolphin: Double-split pupil yields equivalent aerial and underwater diurnal acuity. Science 1975; 189: 650–652

Herman LM, Tavolga WN. The communication systems of cetaceans. In: *Cetacean Behavior: Mechanisms and Functions*, LM Herman, ed. New York: Wiley, 1980, pp. 149–210

Hollien H, Brandt JF. Effect of air bubbles in the external auditory meatus on underwater hearing thresholds. J Acoust Soc Am 1969; 46: 384–387

Hosokawa H. On the extrinsic eye muscles of the whale with special remarks upon the innervation and function of the musculus retractor bulbi. Sci Rep Whales Res Inst 1951; 6: 1–31

Houtgast T. Masking patterns and lateral inhibition. In: *Facts and Models in Hearing*, E Zwicker, E Terhardt, eds. Berlin: Springer, 1974, pp. 258–265

Houtgast T. Auditory-filter characteristics derived from direct-masking and pulsation-threshold data with a rippled-noise masker. J Acoust Soc Am 1977; 62: 409–415

Hughes A. A comparison of retinal ganglion cell topography in the plains and tree kangaroo. J Physiol 1974; 244: 61–63

Hughes A. A quantitative analysis of the cat retinal ganglion cell topography. J Comp Neurol 1975; 63: 107–128

Hughes A. The topography of vision in mammals of contrasting life style: Comparative optics and retinal organization. In: *Handbook of Sensory Physiology: The Visual System in Vertebrates*, F Crescitelli, ed. Vol VII/5. Berlin: Springer, 1977, pp. 613–756

Hughes A. Population magnitudes and distribution of the major modal classes of cat retinal ganglion cells as estimated from HRP filling and systematic survey of the soma diameter spectra for classical neurons. J Comp Neurol 1981; 197: 303–339

Hughes A. New perspectives in retinal organization. In: *Progress in Retinal Research*, N Osborne, G Chader, eds. Vol 4. Oxford: Pergamon, 1985, pp. 243–313

Hughes A, Wässle H. The cat optic nerve: Fiber total count and diameter spectrum. J Comp Neurol 1976; 169: 171–184

Hunter G. Visual delayed matching of two-dimensional forms by a bottlenosed dolphin. Thesis, University of Hawaii, Manoa, 1988.

Hyvärinen H. Living in darkness: whiskers as sense organs of the ringed seal (*Phoca hispida Saimensis*). J Zool (Lond) 1989; 218: 663–678

Hyvärinen H. Structure and finction of the vibrissae of the ringed seal (*Phoca hispida* L.). In: *Sensory Systems of Aquatic Mammals*, RA Kastelein, JA Thomas, PE Nachtigall, eds. Woerden, The Netherlands: De Spil, 1995, pp. 429–445

Hyvärinen H, Katajisto H. Functional structure of the vibrissae of the ringed seal (*Phoca hispida Schr*). Acta Zool Fenn 1984; 171: 27–30

Jacobs DW. Auditory frequency discrimination in Atlantic bottlenose dolphin, *Tursiops truncatus* Montague: A preliminary report. J Acoust Soc Am 1972; 52: 696–698

Jacobs DW, Hall JD. Auditory thresholds of a fresh water dolphin, *Inia geoffrensis* Blainvsille. J Acoust Soc Am 1972; 51: 530–533

Jacobs GH. The distribution and nature of color vision among the mammals. Biol Rev Camb Philos Soc 1993; 68: 413–471

Jacobs MS, McFarland WL, Morgane PJ. The anatomy of the brain of the bottlenosed dolphin (*Tursiops truncatus*). Rhinic lobe (rhinencephalon): The archicortex. Brain Res Bull 1979; 4: 1–108

Jacobs MS, Morgane PJ, McFarland WL. The anatomy of the brain of the bottlenosed dolphin (*Tursiops truncatus*). Rhinic lobe (rhinencephalon): I. The paleocortex. J Comp Neurol 1971; 141: 205–272

Jacobs MS, Morgane PJ, McFarland WL. Degeneration of visual pathways in the bottlenosed dolphin. Brain Res 1975; 88: 346–352

Jamieson GS, Fisher HD. Visual discriminations in the harbor seal *Phoca vitulina*, above and below water. Vision Res 1970; 10: 1175–1180

Jamieson GS, Fisher HD. The retina of the harbor seal *Phoca vitulina*. Can J Zool 1971; 49: 19–23

Jamieson GS, Fisher HD. The pinniped eye: A review. In: *Functional Anatomy of Marine Mammals*, V. 1, RJ Harrison, ed. New York: Acad Press, 1972, pp.245–261

Jesteadt W, Wier CC, Green DM. Intensity discrimination as a function of frequency and level. J Acoust Soc Am 1977; 61: 169–177

Johnson CS. Sound detection thresholds in marine mammals. In: *Marine Bio-Acoustics*, V. 2, WN Tavolga, ed. New York: Pergamon, 1967, pp. 247–260

Johnson CS. Masked tonal thresholds in the bottlenosed porpoise. J Acoust Soc Am 1968a; 44: 965–967

Johnson CS. Relation between absolute threshold and duration of tone pulse in the bottlenosed porpoise. J Acoust Soc Am 1968b; 43: 757–763

Johnson CS. Auditory masking of one pure tone by another in the bottlenosed porpoise. J Acoust Soc Am 1971; 49: 1317–1318

Johnson CS. Hearing thresholds for periodic 60-kHz tone pulses in the beluga whale. J Acoust Soc Am 1991; 89: 2996–3001

Johnson CS. Detection of tone glides by the beluga whale. In: *Marine Mammal Sensory Systems*, JA Thomas, RA Kastelein, AYa Supin, eds. New York: Plenum, 1992, pp. 241–247

Johnson CS, McManus MW, Skaar D. Masked tonal hearing thresholds in the beluga whale. J Acoust Soc Am 1989; 85: 2651–2654

Johnson GL. Contribution to the comparative anatomy of mammalian eye chiefly based on ophthalmoscopic examination. Phil Trans R Soc B 1901; 194: 1–82

Johnson RA. Energy spectrum analysis in echolocation. In: *Animal Sonar Systems*, RG Busnel, JA Fish, eds. New York: Plenum, 1980, pp. 673–693

Johnson RA, Titlebaum EL. Energy spectrum analysis: A processing model of echolocation. J Acoust Soc Am 1976; 60: 484–491

Johnson-Davis D, Patterson RD. Psychophysical tuning curves: Restricting the listening band to the signal region. J Acoust Soc Am 1979; 65: 765–770

Jonas JB, Muller-Bergh JA, Schloetzer-Schrehardt UM, Naumann GOH. Histomorphometry of the human optic nerve. Invest Ophthalmol Visual Sci 1990; 31: 736–744

Kamminga C. Echolocation signal types of odontocetes. In: *Animal Sonar: Processes and Performance*. New York: Plenum, 1988, pp. 9–22

Kamminga C, Wiersma H. Investigations on cetacean sonar. II. Acoustical similarities and differences in odontocete sonar signals. Aquatic Mammals 1981; 8: 41–62

Kastelein RA, Mosterd P, van Ligtenberg CL, Verboom WC. Aerial hearing sensitivity tests with a male Pacific walrus (*Odobenus rosmarus divergens*), in the free field and with headphones. Aquatic Mammals 1996; 22(2): 81–93

Kastelein RA, Nieuwstraten SH, Staal C, van Ligthenberg CL, Versteegh D. Low-frequency aerial hearing of a harbour porpoise (*Phocoena phocoena*). In: *The Biology of the Harbor Porpoise*, AJ Read, PR Wiepkema, PE Nachtigall, eds. Woerden, The Netherlands: De Spil, 1997, pp. 295–312

Kastelein RA, Stevens S, Mosterd P. The tactile sensitivity of the mystacial vibrissae of Pacific walrus (*Odobenus rosmarus divergens*) Part 2: Masking. Aquatic Mammals 1990; 16: 78–87

Kastelein RA, Thomas JA, Nachtigall PE, eds. *Sensory Systems of Aquatic Mammals.* Woerden, The Netherland: De Spil, 1995

Kastelein RA, van Gaalen MA. The sensitivity of the vibrissae of a Pacific walrus (*Odobenus rosmarus divergens*) Part 1. Aquatic Mammals 1988; 14: 123–133

Kastelein RA, van Ligtenberg CL, Gjertz I, Verboom WC. Free field hearing tests on wild Atlantic walruses (*Odobenus rosmarus rosmarus*) in air. Aquatic Mammals 1993; 19(3): 143–148

Kastelein RA, Weipkema PR. A digging trough as occupational therapy for Pacific walruses (*Odobenus rosmarus divergens*) in human care. Aquatic Mammals 1989; 15: 9–17

Katsak D, Schusterman RJ. Aerial and underwater hearing thresholds for 100 Hz pure tones in two pinniped species. In: *Sensory Systems of Aquatic mammals*, RA Kastelein, JA Thomas, PE Nachtigall, eds. Woerden, The Netherlands: De Spil, 1995, pp. 71–79

Katsak D, Schusterman RJ. Low-frequency amphibious hearing in pinnipeds: Methods, measurements, noise, and ecology. J Acoust Soc Am 1998; 103: 2216–2228

Kay RH. Hearing of modulation in sounds. Physiol Rev 1982; 62: 894–975

Kellogg WN. Auditory perception of submerged objects by porpoises. J Acoust Soc Am 1959; 31: 1–6

Kellogg WN, Kohler R, Morris HN. Porpoise sounds as sonar signals. Science 1953; 117: 239–243

Kesarev VS. Structural organization of the dolphin limbic cortex (in Russ.). Arkhiv Anat Histol Embriol 1969; 56(6): 28–35

Kesarev VS. Some data on neuronal organization of the dolphin neocortex (in Russ.). Arkhiv Anat Histol Embriol 1970; 59(8): 71–77

Kesarev VS, Malofeyeva LI, Trykova OV. Ecological specificity of cetacean neocortex. J Hirnforsch 1977; 18: 447–460

Ketten DR. Three-dimensional reconstructions of the dolphin ear. In: *Sensory Abilities of Cetaceans. Laboratory and Field Evidence*, JA Thomas, RA Kastelein, eds. New York, Plenum, 1990, pp. 81–105

Ketten DR.The cetacean ear: Form, frequency, and evolution. In: *Marine Mammal Sensory Systems*, JA thomas, RA Kastelein, AYa Supin, eds. New York: Plenum, 1992a, pp. 53–75

Ketten DR. The marine mammal ear: Specialization for aquatic audition and echolocation. In: *The Evolutionary Biology of Hearing*, D Webster, RR Fay, AN Popper, eds. New York: Springer, 1992b, pp 717–754

Ketten DR. Structure, function, and adaptation of the manatee ear. In: *Marine Mammal Sensory Systems*, JA thomas, RA Kastelein, AYa Supin, eds. New York: Plenum, 1992c, pp. 77–95

Ketten DR. Structure and function in whale ears. Bioacoustics 1997; 8: 103–135

Klishin VO, Pezo Dias R, Popov VV, Supin AYa. Some characteristics of hearing of the Brazilian manatee, *Trichechus inunguis.*

Klishin VO, Popov VV. Two-tone tuning curves in the bottlenosed dolphin *Tursiops truncatus*. Sensornyie Systemy (Sensory Systems) 1996; 10 (2): 30–37

Klishin VO, Popov VV. Hearing characteristics of a harbor porpoise *Phocoena phocoena*. Dokl Biol Sci 2000; 370: 413–415

Klishin VO, Popov VV, Supin AYa. Recovery of responsiveness in the dolphin *Tursiops truncatus* auditory system at paired acoustic stimuli with different spectra. Z Evoluts Biokhim Fisiol (J Evol Biochem Physiol) 1991; 27: 314–319

Klishin VO, Popov VV, Supin AYa. Hearing capabilities of a beluga whale, *Delphinapterus leucas*. Aquatic Mammals 2000; 26: 212–228

Kojima T. On the brain of the sperm whale (*Physeter catodon*). Sci rep Whales Res Inst Tokio 1951; 6: 49–72

Kolb H, Wang HH. Distribution of photoreceptors, dopaminergic amacrine cells and ganglion cells in the retina of the North American opossum. Vision Res 1985; 25: 1207–1221

Kolchin S, Bel'kovich V. Tactile sensitivity in *Delphinus delphis*. Zoologicheskiy Z (Zool. J) 1973; 52: 620–622

Krieg WJS. Connections of the cerebral cortex. I. The albino rat: A topography of the cortical areas. J Comp Neurol 1946; 84: 221–275

Kröger RHH, Kirschfeld K. Accommodation in the bottlenosed dolphin (*Tursiops truncatus*). Abstr Fifth Internat Theriol Congr. Rome: 1989, pp. 367–368

Kröger RHH, Kirschfeld K. The cornea as an optical element in the cetacean eye. In: *Marine Mammal Sensory Systems*, JA Thomas, RA Kastelein, AYa Supin, eds. New York: Plenum, 1992, pp. 97–106

Kröger RHH, Kirschfeld K. Optics of the harbor porpoise eye in water. J Opt Soc Am 1993; 10: 1481–1489

Kröger RHH, Kirschfeld K. Refractive index in the cornea of a harbor porpoise (*Phocoena phocoena*) measured by two-wavelengths laser interferometry. Aquatic Mammals 1994; 20(2): 99–107

Krubitzer LA, Sesma MA, Kaas JH. Microelectrode maps, myeloarchitecture, and cortical connections of three somatotopically organized representations of the body surface in the parietal cortex of squirrels. J Comp Neurol 1986; 250: 403–430

Kruger L. The thalamus of the dolphin (*Tursiops truncatus*) and comparison with other mammals. J Comp Neurol 1959; 111: 133–194

Kuwada S, Batra R., Maher V. Scalp potentials of normal and hearing impaired subjects in response to sinusoidally amplitude modulated tones. Hearing Res 1986; 21: 179–192

Ladygina TF, Mass AM, Supin AYa. Multiple sensory projections in the dolphin cerebral cortex. Z Vyss Nervn Deyat (J Higher nervous Activity) 1978; 28: 1047–1053

Ladygina TF, Popov VV. Organization of projection of the body surface in the cerebral cortex of the Caspian seal and the northern fur seal. Dikl Akad Nauk SSSR (Proc Acad Sci USSR) 1984; 278: 758–761

Ladygina TF, Popov VV. Sensory projections to the cerebral cortex of the Caspian seal and northern fur seal. In: *Electrofiziologiya Sensornykh System Morskikh Mlekopitayushikh (Electrophysiologi of Sensory Systems of marine Mammals)*, VE Sokolov, ed. Moscow: Nauka, 1986, pp. 130–137

Ladygina TF, Popov VV, Supin AYa. Organization of somatic and motor projections to the cerebral cortex of the northern fur seal. In: *Electrofiziologiya Sensornykh System Morskikh Mlekopitayushikh (Electrophysiologi of Sensory Systems of marine Mammals)*, VE Sokolov, ed. Moscow: Nauka, 1986, pp. 137–158

Ladygina TF, Popov VV, Supin AYa. Micromapping of the fur seal's somatosensory cerebral cortex. In: *Marine Mammal Sensory Systems*, JA Thomas, RA Kastelein, AYa Supin, eds. New York: Plenum, 1992, pp. 107–118

Ladygina TF, Supin AYa. Acoustic projection in the dolphin's cerebral cortex. Fisiol. Z. SSSR (Physiol. J. USSR) 1970; 56: 1554–1560

Ladygina TF, Supin AYa. Evolution of cortical areas in terrestrial and aquatic mammals (in Russ.). In: *Morfologiya, Fiziologiya y Acustika Morskikh Mlekopitayushikh (Morphology, Physiology, and Acoustics of Marine mammals)*, VE Sokolov, ed. Moscow: Nauka, 1974, pp. 6–15

Landau D, Dawson WW. The histology of retinas from the *Pinnipedia*. Vision Res 1970; 10: 691–702

Langner G. Periodicity coding in the auditory system. Hearing Res 1992; 60: 115–142

Langworthy OR. Factors determining the differentiation of the cerebral cortex in sealiving mammals (the Cetacea). A study of the brain of the porpoise, *Tursiops truncatus*. Brain 1931; 54: 225–236

Lavigne DM, Ronald K. The harp seal, *Pagophilus groenlandicus* (Erxleben, 1777). XXIII. Spectral sensitivity. Can J Zool 1972; 50: 1197–1206

Lavigne DM, Ronald K. Pinniped visual pigments. Comp Biochem Physiol 1975; 52: 325–329

Layne JN, Caldwell DK. Behavior of the Amazon dolphin, *Inia geoffrensis* B in captivity. Zoologia 1964; 49: 81–108

Legatt AD, Aresso JC, Vaughan HG. Short-latency auditory evoked potentials in the monkey. I. Wave shape and surface topography. Electroenceph Clin Neurophysiol 1986; 64: 41–52

LeMessurier DH. Auditory and visual areas of the cerebral cortex of the rat. Federat Proc 1948; 7: 70–71

Lende RA. Sensory representation in the cerebral cortex of the opossum (*Didelphis virginiana*). J Comp Neurol 1963; 121: 395–403

Lende RA. Representation in the cerebral cortex of a primitive mammal: Sensorimotor, visual and auditory fields in the echidna (*Tachiglossus aculcatus*). J Neurophysiol 1964; 27: 37–48

Lende RA, Sadler KM. Sensory and motor areas in neocortex of hedgehog (*Erinaceus*). Brain Res 1967; 5: 390–405

Lende RA, Welker WI. An unusual sensory area in the cerebral neocortex of the bottlenose dolphin, *Tursiops truncatus*. Brain Res 1972; 45: 555–560

Lende RA, Woolsey CN. Sensory and motor localization in cerebral cortex of porcupine (*Erethison dorsatum*). J Neurophysiol 1956; 19: 544–563

Levenson DH, Schusterman RJ. Pupillometry in seals and sea lions: ecological implications. Can J Zool 1997; 75: 2050–2057

Levenson DH, Schusterman RJ. Dark adaptation and visual sensitivity in shallow and deep-diving pinnipeds. Marine Mammal Sci 1999; 15: 1303–1313

Levitt H. Transformed up-down methods in psychoacoustics. J Acoust Soc Am 1971; 49: 467–477.

Lilly J. Animals in aquatic environment: Adaptation of mammals to the ocean. In: *Handbook of Physiology: Adaptation to the Environment*, Vol. 1, Section 4. Washington: Am Physiol Soc, 1964, pp. 741–747

Ling JK. The skin and hair of the southern elephant seal, *Mirounga leonina* (Linn). 1. The facial vibrissae. Aust J Zool 1966; 14: 855–866

Ling JK. The integument of marine mammals. In: *Functional Anatomy of Marine Mammals*, RJ Harrison, ed. Vol. 2. London: Academic, 1974, pp. 1–44

Ling JK. Vibrissae of marine mammals. In: *Functional Anatomy of Marine Mammals*, RJ Harrison, ed. Vol. 3. London: Academic, 1977, pp. 387–415

Ljungblad DK, Scoggins PD, Gilmartin WG. Auditory thresholds of a captive eastern Pacific bottle-nosed dolphin, *Tursiops* spp. J Acoust Soc Am 1982; 72: 1726–1729

Lyamin OI, Manger PR, Mukhametov LM, Siegel JM, Shpak OV. Rest and activity states in a gray whale. J Sleep Res 2000; 9: 261–267

Lythgoe JN, Dartnall HJA. A deep sea "rhodopsin" in a mammal. Nature 1970; 227: 955–956

Long KO, Fisher SK. The distribution of photoreceptors and ganglion cells in the California ground squirrel, *Spermophilus beecheyi*. J Comp Neurol 1983; 221: 329–340

Madsen CM, Herman LM. Social and ecological correlates of cetacean vision and visual appearance. In: *Cetacean Behavior: Mechanisms and Functions*, LM Herman, ed. New York, Wiley, 1980, pp. 101–147

Mann G. Ojo y vision de las balenas. Biologica 1946; 4: 23–71

Mass AM. Retinal topography in the walrus (*Odobenus rosmarus divergrns*) and fur seal (*Callorhinus ursinus*). In: *Marine Mammal Sensory Systems*, JA Thomas, RA Kastelein, AYa Supin, eds. New York: Plenum, 1992, pp. 119–135

Mass AM. The best vision areas and ganglion cell distribution in the retina of the dolphin *Tursiops truncatus*. Dokl Akad Nauk (Proc Acad Sci) 1993; 330: 396–398

Mass AM. Zones of higher ganglionic cell density and resolving power of the retina in the gray whale *Eschrichtius gibbosus*. Dokl Biol Sci 1996; 350: 472–475

Mass AM Retinal resolution and topography in the Amazonian white dolphin *Sotalia fluviatilis*. Dokl Biol Sci 1998; 359: 141–143

Mass AM, Odell DK, Ketten DR, Supin AYa. Ganglion layer topography and retinal resolution of the Caribbean manatee *Trichechus manatus latirostris*. Dokl Biol Sci 1997; 355: 392–394

Mass AM, Supin AYa. Distribution of ganglion cells in the dolphin retina. Dokl Biol Sci 1985; 284: 612–615

Mass AM, Supin AYa. Topographic distribution of sizes and density of ganglion cells in the retina of a porpoise, *Phocoena phocoena*. Aquatic Mammals 1986; 12(3): 95–102

Mass AM, Supin AYa. Topographic organization of the ganglion layer of the retina of the Amazon dolphin *Inia geoffrensis*. Dokl Biol Sci 1988; 303: 726–729

Mass AM, Supin AYa. Distribution of ganglion cells in the retina of an Amazon river dolphin *Inia geoffrensis*. Aquatic Mammals 1989; 15(2): 49–56

Mass AM, Supin AYa. Best vision zones in the retinae of some cetaceans. In: *Sensory Abilities of Cetaceans. Laboratory and Field Evidence*, JA Thomas, RA Kastelein, eds. New York: Plenum, 1990, pp. 505–517

Mass AM, Supin AYa. Peak density, size and regional distribution of ganglion cells in the retina of the fur seal *Callorhinus ursinus*. Brain Behav Evol 1992; 39: 69–76

Mass AM, Supin AYa. Ganglion cell topography of the retina in the bottlenose dolphin, *Tursiops truncatus*. Brain Behav Evol 1995a; 45: 257–265

Mass AM, Supin AYa. Retinal resolution in the bottlenose dolphin (*Tursiops truncatus*). In: *Sensory systems of Aquatic Mammals*, RA Kastelein, JA Thomas, PA Nachtigall, eds. Woerden, The Netherlands: De Spil, 1995b, pp. 419–428

Mass AM, Supin AYa. Ocular anatomy, retinal ganglion cell distribution, and visual resolution in the gray whale, *Eschrichtius gibbosus*. Aquatic Mammals 1997; 23(1): 17–28

Mass AM, Supin AYa. Retinal topography and visual acuity in the riverine tucuxi (*Sotalia fluviatilis*). Marine mammal Sci 1999; 15: 351–365

Mass AM, Supin AYa. Ganglion cell density and retinal resolution in the sea otter, *Enhydra lutris*. Brain Behav Evol 2000a; 55: 111–119

Mass AM, Supin AYa. Quantitative estimation of retinal resolution depending on ganglion cell density and distribution pattern. Sensory Systems 2000b; 14:

McCormick JG. Relationship of sleep, respiration and anesthesia in the porpoise: a preliminary report. Proc Nat Acad Sci USA 1969; 62: 697–703

McCormick JG, Wever EG, Palin G, Ridgway SH. Sound conduction in the dolphin ear. J Acoust Soc Am 1970; 48: 1418–1428

McCormick JG, Wever EG, Ridgway SH, Palin G. Sound reception in the porpoise as it relates to echolocation. In: *Animal Sonar Systems*, RG Busnel, JF Fish, eds. New York: Plenum, 1980, pp. 449–467

Merzenuch MM, Kaas JH, Sur M, Lin CS. Double representation of the body surface within cytoarchitectonic areas 3b and 1 in "SI" in the owl monkey (*Aotus trivigratus*). J Comp Neurol 1978; 181: 41–74

Mills AM. On the minimum audible angle. J Acoust Soc Am 1958; 30: 237–246

Mitchell C, Fowler C. Tuning curves of cochlear and brain-stem responses in the guinea pig. J Acoust Soc Am 1980; 68: 896–900

Mitchell DE, Giffin F, Timney BN. A behavioral technique for the rapid assessment of the visual capabilities in kittens. Perception 1977; 6: 181–193

Mobley JR, Helweg DA. Visual ecology and cognition in cetaceans. In: *Sensory Abilities of Cetaceans. Laboratory and Field Evidence*, JA Thomas, RA Kastelein, eds. New York: Plenum, 1990, pp. 519–536

Møhl B. Frequency discrimination in the common seal. In: *Underwater Acoustics*, VA Albers, ed. New York: Plenum, 1967a, pp. 43–54

Møhl B. Seal ears. Science 1967b; 157: 99

Møhl B. Auditory sensitivity of the common seal in air and water. J Aud Res 1968a; 8: 27–38

Møhl B. Hearing in seals. In: *The Behavior and Physiology of Pinnipeds*, RJ Harrison, RS Peterson, CE Rice, RJ Schusterman, eds. New York: Appleton-Century-Crofts, 1968b, pp. 172–195

Møhl B, Au WWL, Pawloski J, Nachtigall PE. Dolphin hearing: Relative sensitivity as a function of point of application of a contact sound source in the jaw and head region. J Acoust Soc Am 1999; 105: 3421–3424

Møller AR. Coding of amplitude and frequency modulated sounds in the cochlear nucleus of the rat. Acta Physiol Scand 1972; 86: 223–238

Moore BCJ, Glasberg BR. Auditory filter shapes derived in simultaneous and forward masking. J Acoust Soc Am 1981; 70: 1003–1014

Moore BCJ, Glasberg BR. Suggested formulae for calculating auditory filter bandwidths and excitation patterns. J Acoust Soc Am 1983; 74: 750–753

Moore BCJ, Glasberg BR. Gap detection with sinusoids and noise in normal, impaired and electrically stimulated ears. J Acoust Soc Am 1988; 83: 1093–1101

Moore BCJ, Glasberg BR. Comparison of auditory filter shapes obtained with notched-noise and noise-tone maskers. J Acoust Soc Am 1995; 97: 1175–1182

Moore BCJ, Glasberg BR, Plack CJ, Biswas AK. The shape of the ear's temporal window. J Acoust Soc Am 1988; 83: 1102–1116

Moore BCJ, Peters RW, Glasberg BR. Detection of temporal gaps in sinusoids: Effects of frequency and level. J Acoust Soc Am 1993; 93: 1563–1570

Moore PWB, Au WWL. Underwater localization of pulsed pure tones by the California sea lion (*Zalophus californianus*). J Acoust Soc Am 1975; 58: 721–727

Moore PWB, Hall RW, Friedl WA, Nachtigall PE. The critical interval in dolphin echolocation: What is it? J Acoust Soc Am 1984; 76: 314–317

Moore PWB, Pawloski DA, Dankewicz L. Interaural time and intensity difference thresholds in the Bottlenose dolphin (*Tursiops truncatus*). In *Sensory Systems of Aquatic Mammals*, RA Kastelein, JA Thomas, PE Nachtigall, eds. Woerden, The Netherlands: De Spil, 1995, 11–23

Moore PWB, Schusterman RJ. Discrimination of pure tone intensities by the California sea lion. J Acoust Soc Am 1976; 60: 1405–1407

Moore PWB, Schusterman RJ. Audiometric assessment of northern fur seals *Callorhinus ursinus*. Marine Mammal Sci 1987; 3: 31–53

More EJ, ed. *Bases of auditory brain stem evoked responses*. New York: Grune and Stratton,. 1983

Morgane PJ, Glezer II. Sensory neocortex in dolphin brain. In: *Sensory Abilities of Cetaceans. Laboratory and Field Evidence*, JA Thomas, RA Kastelein, eds. New York: Plenum, 1990, pp. 107–136

Morgane PJ, Jacobs MS. The comparative anatomy of the cetacean nervous system. In: *Functional Anatomy of Marine Mammals*, RJ Harrison, ed. Vol. 1. New York: Academic, 1972, pp 109–239

Moushegian G, Rupert AL, Stillman RD. Scalp recorded early responses in man to frequencies in the speech range. Electroenceph Clin Neurophysiol 1973; 35: 665–667

Murayama T, Fujise Y, Aoki I, Ishii T. Histological characteristics and distribution of ganglion cells in the retina of the Dall's porpoise and minke whale. In: *Marine Mammal Sensory Systems*, JA Thomas, RA Kastelein, AYa Supin, eds. New York: Plenum, 1992a, pp. 137–145

Murayama T, Somiya H, Aoki I, Ishii T. The distribution of ganglion cells in the retina and visual acuity of Minke whale. Nippon Suissan Gakkaishi 1992b; 58: 1057–1061

Murayama T, Somiya H, Aoki I, Ishii T. Retinal ganglion cell size and distribution predict visual capabilities of Dall's porpoise. Marine Mammal Sci 1995; 11: 136–149

Musicant AD, Buttler RA. The influence of pinnae-based spectral cues on sound localization. J Acoust Soc Am 1984; 75: 1195–1200

Nachtigall PE, Au WWL, Pawloski JL, Moore PWB. Risso's dolphin (*Grampus griseus*) hearing thresholds in Kaneohe Bay, Hawaii. In: *Sensory Systems of Aquatic Mammals*, RA Kastelein, JA Thomas, PE Nachtigall, eds. Woerden, The Netherlands: De Spil, 1995, pp. 49–54

Nachtigall PE, Moore PWB, eds. *Animal Sonar. Processes and Performance*. New York: Plenum, 1988

Nagy AR, Ronald K. The harp seal, *Pagophilus groenlandicus* (Erxleben 1777). Can J Zool 1970; 48: 367–370

Nagy AR, Ronald K. A light and electron microscopic study of the structure of the retina of the harp seal *Pagohpilus groenlandicus* (Erxleben 1777). Rapp P Cons Inst Explor Mer 1975; 169: 92–96

Narins PM, Evans EF, Pick GF, Wilson JP. A combfiltered noise generator for use in auditory neurophysiological and psychophisical experiments. IEEE Trans Biomed Eng 1979; BME-26: 43-47

Nelson DA. Two-tone masking and auditory critical bandwidths. Audiology 1979; 18: 279–306

Nelson RJ, Sur M, Felleman DJ, Kaas JH. Representation of the body surface in the postcentral parietal cortex of *Macaca fascicularis*. J Comp Neurol 1980; 192: 611–644

Nelson RJ, Sur M, Kaas JH. The organization of the second somatosensory area (SmII) in the grey squirrel. J Comp Neurol 1979; 184: 473–490

Noordenbos JW, Boogh CJ. Underwater visual acuity in the bottlenosed dolphin *Tursiops truncatus* (Mont). Aquatic Mammals 1974; 2: 15–24

Normark J. Perception of distance in animal echolocation. Nature 1961; 190: 363–364

Norris KS. The evolution of acoustic mechanisms in odontocete cetaceans. In: *Evolution and Environment*, ET Drake, ed. New Haven: Yale Univ, 1968, pp. 297–324

Norris KS. The echolocation of marine mammals. In: *The Biology of Marine Mammals*, HJ Andersen, ed. New York: Academic, 1969, pp. 391–424

Norris KS. Peripheral sound processing in odontocetes. In: *Animal Sonar System*, R-G Busnel, JF Fish, eds. New York: Plenum, 1980, pp. 495–509

Norris KS, Prescott JH, Asa-Dorian PV, Perkins P. An experimental demonstration of echolocation behavior in the porpoise, *Tursiops truncatus*, Montagu. Biol Bull, 1961; 120: 163–176

Nummela S, Reuter T, Hemilä S, Holmberg P, Paukku P. The anatomy of the killer whale middle ear (*Orcinus orca*). Hearing Res 1999a; 133: 61–70

Nummela S, Wägar T, Hemilä S, Reuter T. Scaling of the cetacean middle ear. Hearing Res 1999b; 133: 71–81

Odend'hal S, Poulter TC. Pressure regulation in the middle ear cavity of sea lions: A possible mechanism. Science 1966; 153: 768–769

Oelschläger HA. Comparative morphology and evolution of the otic region in toothed whales, *Cetacea, mammalia*. Am J Anat 1986; 177: 353–368

Oliver JC, Slattery PN, O'Connor EF, Lowry LF. Walrus *Odobenus rosmarus* feeding in the Bering sea: A bentic perspective. Fishery Bull 1983; 81: 501–512

O'Loughlin BJ, Moore BCJ. Off-frequency listening: Effects of psychophysical tuning curves obtained in simultaneous and forward masking. J Acoust Soc Am 1981; 69: 1119–1125

Osen KK, Jansen J. The cochlear nuclei in the common porpoise *Phocaena phocaena*. J Comp Neurol 1965; 125: 223–258

Pack AA, Herman LM. Sensory integration in the bottlenosed dolphin: Immediate recognition of complex shapes across the senses of echolocation and vision. J Acoust Soc Am 1995; 98: 722–733

Palmer E, Weddel G. The relationship between structure, innervation and function of the skin of the bottlenosed dolphin (*Tursiops truncatus*). Proc Zool Soc Lond 1964; 143: 553–568

Pantev C, Lagidze S, Pantev M, Kevanishvili Z. Frequency-specific contributions to the auditory brain stem response derived by means of pure-tone masking. Audiology 1985; 24: 275–287

Pantev C, Pantev M. Derived brainstem responses by means of pure tone masking. Scand Audiol 1982; 11: 15–22

Pardue MT, Sivak JG, Kovacs KM. Corneal anatomy of marine mammals. Can J Zool 1993; 71: 2282–2290

Parker DJ, Thornton ARD. Cochlear traveling wave velocities calculated from the derived components of the cochlear nerve and brainstem evoked responses of the human auditory system. Scand Audiol 1978a; 7: 67–70

Parker DJ, Thornton ARD. Frequency-selective components of the cochlear nerve and brainstem evoked responses of the human auditory system. Scand Audiol 1978b; 7: 53–60

Patterson RD. Auditory filter shapes derived with noise stimuli. J Acoust Soc Am 1976; 59: 640–654

Patterson RD, Henning GB. Stimulus variability and auditory filter shape. J Acoust Soc Am 1977; 62: 649–664

Patterson RD, Moore BCJ. Auditory filters and excitation patterns as representations of frequency resolution. In: *Frequency Selectivity in Hearing*, BCJ Moore, ed. London: Academic, 1986

Patterson RD., Nimmo-Smith I. Off-frequency listening and auditory filter asymmetry. J Acoust Soc Am 1980; 67: 229–245

Patterson RD, Nimmo-Smith I, Weber DL, Milory R. The deterioration of hearing with age: Frequency selectivity, the critical ratio, the audiogram, and speech threshold. J Acoust Soc Am 1982; 72: 1788–1803

Payne R, Webb D. Orientation by means of long range acoustic signaling in baleen whales. Ann NY Acad Sci 1971; 188: 110–141

Peers B. The Retinal Histology of the Atlantic Bottlenose Dolphin *Tursiops truncatus* (Montagu, 1821). Thesis. Guelph Univ, 1971

Peichl L. Topography of ganglion cells in the dog and wolf retina. J Comp Neurol 1992; 324: 603–620

Peichl L, Berhmann G. S-cones are absent in the retina of the pilot whale. Invest Ophthalmol Vis Sci, 1999, 40: S238

Peichl L, Moutairou K. Absence of short-wavelength sensitive cones in the retinae of seals (Carnivora) and African giant rats (Rodentia). Eur J Neurosci, 1998; 10: 2586–2594

Penner MJ. Detection of temporal gaps in noise as a measure of the decay of auditory sensation. J Acoust Soc Am 1977; 61: 552–557

Pepper RL, Simmons JV. In air visual acuity of the bottlenosed dolphin. Exp Neurol 1973; 41: 271–276

Perez JM, Dawson WW, Landau D. Retinal anatomy of the bottlenosed dolphin (*Tursiops truncatus*). Cetology 1972, N 11: 1–11

Perry VH The ganglion cell layer of the mammalian retina. In: *Prigress in Retinal Research*, N Osborne, G Chader, eds. Vol 1. Oxford, Pergamon, 1982, pp. 53–80

Pick GF. Level dependence of psychophysical frequency resolution and auditory filter shape. J Acoust Soc Am 1980; 68: 1085–1095

Pick GF, Evans EF, Wilson JP. Frequency resolution in patients with hearing loss of cochlear origin, In: *Psychophysics and Physiology of Hearing*, EF Evans, JP Wilson, eds. New York: Academic, 1977, pp. 273–282

Picton T, Ouellette J, Hamel G, Smoth A. Brainstem evoked potentials to tone pips in notched noise. J Otolaryngol 1979; 8: 289–314

Picton TW, Skinner ChR, Champagne SC, Kellett AJC, Maiste AC. Potentials evoked by sinusoidal modulation of the amplitude or frequency of a tone. J Acoust Soc Am 1987; 82: 165–178

Piggins DW, Muntz RA, Best RC. Physical and morphological aspects of the eye of the manatee *Trichechus inunguis* Natterer 1883 (Sirenia: mammalia). Marine Behav Physiol 1983; 9: 111–130

Pilleri G. Über die Anatomie des Gehirns des Ganges Delphins, *Platanista gangetica*. Rev Suisse Zool 1964; 73: 113–118

Pilleri G, Gihr M. On the brain of the Amazon dolphin *Inia geoffrensis* de Blainville (Cetacea, Susuidae). Experientia, 1968; 24: 932–934

Pilleri G, Wandeler A. Ontogenese und functionelle Morphologie der Auges des Finnwals *Balaenoptera physalus L.* (*Cetacea, Mysticeti, Balaenopteridae*). Acta Anat 1964; 57: Suppl 50, 1–74

Plack CJ, Moore BCJ. Temporal window shape as a function of frequency and level. J Acoust Soc Am 1990; 87: 2178–2187

Plomp R. Rate of decay of auditory sensation. J Acoust Soc Am 1964; 36: 277–282

Plomp R. Bouman MA. Relation between hearing threshold and duration for tone pulses. J Acoust Soc Am 1959; 31: 749–758

Popov VV, Klishin VO. EEG study of hearing in the common dolphin, *Delphinus delphis*. Aquatic Mammals 1998; 24: 13–20

Popov VV, Ladygina TF, Supin AYa. Evoked potentials in the auditory cortex of the porpoise, *Phocoena phocoena*. J Comp Physiol A 1986; 158: 705–711

Popov VV, Supin AYa. Determination of the hearing characteristics of dolphins by measuring evoked potentials. Fiziol Zh SSSR (Physiol J USSR) 1976a; 62: 550–558

Popov VV, Supin AYa. Responses of the dolphin auditory cortex to complex acoustic stimuli. Fiziol Zh SSSR (Physiol J USSR) 1976b; 62: 1780–1785

Popov VV, Supin AYa. Quantitative measurement of auditory resolving power in man. Dokl. Biol Sci 1984; 278: 630–633

Popov VV, Supin AYa. Determining the hearing characteristics of dolphins according to brainstem evoked potentials. Dokl Biol Sci 1985; 283: 524–527

Popov VV, Supin AYa. Induced potentials of the auditory cortex in the dolphin brain, recorded on the body surface. Dokl Biol Sci 1986; 288: 353–356

Popov VV, Supin AYa. Characteristics of hearing in the beluga *Delphinapterus leucas*. Dokl Biol Sci 1987; 294: 370–372

Popov VV, Supin AYa. Diagram of auditory directionality in the dolphin *Tursiops truncatus* L. Dokl Biol Sci 1988; 300: 323–326

Popov VV, Supin AYa. Auditory brain stem responses in characterization of dolphin hearing. J Comp Physiol A 1990a; 166: 385–393

Popov VV, Supin AYa. Electrophysiological studies of hearing in some cetaceans and manatee. In: *Sensory Abilities of Cetaceans: Laboratory and Field Evidence*. JA Thomas, RA Kastelein, eds. New York: Plenum, 1990b, pp 405–415.

Popov VV, Supin AYa. Electrophysiological investigation of hearing of the fresh-water dolphin *Inia geoffrensis*. Dokl Biol Sci 1990c; 313: 488–491

Popov VV, Supin AYa. Localization of the acoustic window at the dolphin's head. In: *Sensory Abilities of Cetaceans: Laboratory and Field Evidence*. JA Thomas, RA Kastelein, eds. New York: Plenum, 1990d, pp. 417–426.

Popov VV, Supin AYa. Interaural intensity and latency difference in the dolphin's auditory system. Neurosci Lett 1991; 133: 295–297

Popov VV, Supin AYa. Electrophysiological study of the interaural intensity difference and interaural time-delay in dolphins. In: *Marine Mammal Sensory Systems*. JA Thomas, RA Kastelein, AYa Supin, eds. New York, London: Plenum, 1992, pp. 257–267

Popov VV, Supin AYa. Detection of temporal gaps in noise in dolphins: Evoked-potential study. J Acoust Soc Am 1997; 102: 1169–1176

Popov VV, Supin AYa. Auditory evoked responses to rhythmic sound pulses in dolphins. J Comp Physiol A 1998; 183: 519–524

Popov VV, Supin AYa. Contribution of various frequency bands to ABR in dolphins. J Acoust Soc Am 2000;

Popov VV, Supin AYa, Klishin VO. Electrophysiological study of sound conduction in dolphins. In: *Marine Mammal Sensory Systems*. JA Thomas, RA Kastelein, AYa Supin, eds. New York, London: Plenum, 1992, pp. 269–276

Popov VV, Supin AYa, Klishin VO. Frequency tuning curves of the dolphin's hearing: Envelope-following response study. J Comp Physiol A 1995; 178: 571–578

Popov VV, Supin AYa, Klishin VO. Frequency tuning of the dolphin's hearing as revealed by auditory brain-stem response with notch-noise masking. J Acoust Soc Am 1997a; 102: 3795–3801

Popov VV, Supin AYa, Klishin VO. Paradoxical lateral suppression in the dolphin's auditory system: weak sounds suppress response to strong sounds. Neurosci Lett 1997b; 234: 51–54

Popov VV, Supin AYa, Klishin VO. Lateral suppression of rhythmic evoked responses in the dolphin's auditory system. Hearing Res 1998; 126: 126–134

Popper AN. Behavioral measures of odontocete hearing. In: *Animal Sonar Systems*, RG Busnel, JF Fish, eds. New York, London: Plenum, 1980, pp. 469–481

Poulter TC. Sonar signals of the sea lion. Science 1963; 139: 753–755

Poulter TC. The use of active sonar by the California sea lion, *Zalophus californianus* (Lesson). J Aud Res 1966; 6: 165–173

Poulter TC. Systems of echolocation. In: *Les Systemes Sonars Animaux, Biologie et Bionique*, R-G Busnel, ed. Jouy-en-Josas, France: Laboratorie de Physiologie Acousticue, 1967, pp. 157–185

Prince JH. *Comparative Anatomy of the Eye*. Springfield, IL: Thomas, 1956

Prince JH, Diesem C, Eglitis I, Ruskill G. *The Anatomy and Histology of the Eye and Orbit in Domestic Animals*. Springfield, IL: C. Thomas, 1960

Provis JM The distribution and size of the ganglion cells in the retina of the pigmented rabbit: A quantitative analysis. J Comp Neurol 1979; 185: 121–137

Pubols BH, Pubols LM, DePette DJ, Sheely JC. Opossum somatic sensory cortex: A microelectrode mapping study. J Comp Neurol 1976; 165: 229–246

Purves PE, Pilleri G. Observations on the ear, nose, throat, and eye of *Platanista indi*. Inv Cetacea 1974; 5: 13–57

Pütter A. Die Augen der Wassersaugetierre. Zool Jahrb Abth Anat Ontog Thiere 1903; 17: 99–402

Ramprashad F, Corey S, Ronald K. Anatomy of the seal's ear (*Pagophilus groenlandicus*) (Exleben, 1777). In: *Functional Anatomy of Marine Mammals*, RJ Harrison, ed. Vol. 1. London: Academic, 1972, pp. 264–305

Rees A, Green G, Kay RH. Steady-state evoked responses to sinusoidally amplitude-modulated sounds recorded in man. Hearing Res 1986; 23: 123–133

Rees A, Møller AR Response of neurons in the inferior colliculus of the rat to AM and FM tones. Hearing Res 1983; 10: 301–330

Renaud DL, Popper AN. Sound localization by the bottlenose porpoise *Tursiops truncatus*. J Exp Biol 1975; 63: 569–585

Renouf D, Davis MB. Evidence that seals use echolocation. Nature (Lond) 1982; 300: 635–637

Repenning CA. Underwater hearing in seals: functional morphology. In: *Functional Anatomy of Marine Mammals*, RJ Harrison, ed. Vol. 1. London: Academic, 1972, pp. 307–331

Reysenbach de Haan FW. Hearing in whales. Acta Otolaryngol Suppl, 1956; 134: 1–114

Rhode WS, Greenberg S. Lateral suppression and inhibition in the cochlear nucleus of the cat. J Neurophysiol 1994; 71: 493–514

Rickards FW, Clark GM. Steady state evoked potentials to amplitude-modulated tones. In: *Evoked Potentials II. The Second International Evoked Potential Symposium*, RH Nodar, C Barber, eds. Boston: Butterworth, 1984, pp. 63–168

Ridgway SH. Electrophysiological experiments on hearing in odontocetes. In: *Animal Sonar Systems*, RG Busnel, JF Fish, eds. New York: Plenum, 1980, pp. 483–493

Ridgway SH, Bullock TH, Carder DA, Seely RL, Woods D, Galambos R. Auditory brainstem response in dolphins. Proc Nat Acad Sci USA 1981; 78: 1943–1947

Ridgway SH, Carder DA. Tactile sensitivity, somatosensory responses, skin vibrations, and the skin surface ridges of the bottlenose dolphin, *Tursiops truncatus*. In: *Sensory Abilities of Cetaceans. Laboratory and Field Evidence*, JA Thomas, RA Kastelein, eds. New York: Plenum, 1990, pp. 163–179

Riese W. Formprobleme des Gehirns. Zweite Mitteilung: Über die Hirnrinde der Whale. J Psychol Neurol 1925; 31: 275–278

Rivamonte A. Eye model to account for comparable aerial and underwater acuities of the bottlenosed dolphin. Neth J Sea Res 1976; 10: 491–498

Robineau D. Morphologie externe du complexe osseux temporal chez les sireniens. Mém Mus Nat d'Historie Natur, Nouv Sér, A, Zool 1969; 60(1): 1–32

Rochon-Duvigneaud A. L'oeil des cétacés. Archives Museum National Historie Naturelle. T. 58. Paris: 1939

Rochon-Duvigneaud A. *Les Yeux et la Vision des Vvertebres*. Paris: Masson, 1943

Rodenburg M, Vervei C, Van den Brink G. Analysis of evoked responses in man elicited by sinusoidally modulated noise. Audiology 1972; 11: 283–293

Roffler SK, Buttler RA. Factors that influence the localization of sound in the vertical plane. J Acoust Soc Am 1968a; 43: 1255–1259

Roffler SK, Buttler RA. Localization of tonal stimuli in the vertical plane. J Acoust Soc Am 1968b; 43: 1260–1266

Rolls RT, Cowey A. Topography of the retina and striate cortex and its relationship to visual acuity in rhesus monkeys and squirrel monkey. Exp Brain Res 1970; 10: 298–310

Rose JE, Woolsey CN. The relation of thalamic connections, cellular structure and evokable activity in the auditory region of the cat. J Comp Neurol 1949; 91: 441–466

Rose M. Der Grundplan der Cortextektonic beim Delphin. J Psychol Neurol 1926; 32: 161–169

Salt AN, Garcia P. Cochlear action potential tuning curves recorded with a derived response technique. J Acoust Soc Am 1990; 88: 1392–1402

Salvi RJ, Ahroon WA, Perry JW, Gunnarson AD, Henderson D. Comparison of psychophysical and evoked potential tuning curves in the chinchilla. Am J Otolaryngol 1982; 3: 408–416

Sauerland M, Denhardt G. Underwater audiogram of a tucuxi (*Sotalia fluviatilis guianensis*). J Acoust Soc Am 1998; 103: 1199–1204

Schevill WE, Lawrence B. Auditory response of a bottle-nose porpoise, *Tursiops truncatus*, to frequencies above 100 kc. L Exp Zool 1953; 124: 147–165

Schevill WE, Watkins WA. Underwater calls of Trichechus (manatee). Nature 1965; 205: 373–374

Schevill WE, Watkins WA, Ray C. Underwater sounds of pinnipeds. Science 1963; 141: 50–53

Schreiner CE, Urbas JV. Representation of amplitude modulation in the auditory cortex of the cat. I. The anterior auditory field (AAF). Hearing Res 1986; 21: 227–241

Schusterman RJ. Reception and determinants of underwater vocalization in the California sea lion. In: *Les Systemes Sonars Animaux, Biologie et Bionique*, RG Busnel, ed. Jouy-en-Josas, France: Laboratorie de Physiologie Acousticue, 1967, pp. 535–617

Schusterman RJ. Experimental laboratory studies of pinniped behavior. In: *The Behavior and Physiology of Pinnipeds*, RJ Harrison, RC Hubbard, RS Peterson, CE Rice, RJ Schusterman, eds. New York: Appleton-Century-Crofts, 1968, pp. 87–171

Schusterman RJ. Auditory sensitivity of the California sea lion to airborne sound. J Acoust Soc Am 1974; 56: 1248–1251

Schusterman RJ. Vocal communication in pinnipeds. In: *Behavior of Captive Wild Animals*, H Markowitz, VJ Stevens, eds. Chicago: Nelson-Hall, 1978, pp. 247–308

Schusterman RJ. Behavioral capabilities of seals and sea lions: A review of their hearing, visual, learning and diving skills. Psychol Rec 1981; 31: 125–143

Schusterman RJ, Balliet RF. Conditioned vocalization technique for determining visual acuity thresholds in the sea lion. Science 1970a; 169: 498–501

Schusterman RJ, Balliet RF. Visual acuity of the harbour seal and the Steller sea lion under water. Nature 1970b; 226: 563–564

Schusterman RJ, Balliet RF. Aerial and underwater visual acuity in the California sea lion (*Zalophus californianus*) as a function of luminance. Ann NY Acad Sci 1971; 188: 37–46

Schusterman RJ, Balliet RF, Nixon J. Underwater audiogram of the California sea lion by conditioned vocalization technique. J Exp Anal Behav 1972; 17: 339–350

Schusterman RJ, Barrett R, Moore PWB. Detection of underwater signals by a California sea lion and a bottlenose porpoise: Variation of the payoff matrix. J Acoust Soc Am 1975; 57: 1526–1532

Schusterman RJ, Katsak D, Reichmut CJ, Southal L. Why pinnipds don't echolocate. J Acoust Soc Am 2000, 107: 2256–2264

Schusterman RJ, Moore PWB. The upper limit of underwater auditory frequency discrimination in the California sea lion. J Acoust Soc Am 1978; 63: 1591–1595

Shailer MJ, Moore BCJ. Detection of temporal gaps in band-limited noise: Effects of variations in bandwidth and signal-to-noise ratio. J Acoust Soc Am 1985; 77: 635–639

Shailer MJ, Moore BCJ. Gap detection and the auditory filter: Phase effect using sinusoidal stimuli. J Acoust Soc Am 1987; 81: 1110–1117

Shaw NA Central auditory conduction time in the rat. Exp Brain Res 1990; 79: 217–220

Shibkova S. On structure of inner layers of the dolphin eye retina. Arch Anat Histol Embriol (Russ) 1969; 57(10): 68–74

Sivak JG. Accommodation in vertebrates: contemporary survey, 1980. In: *Current Topics in Eye research*, JA Zadunaisky, H Davson, eds. Vol. 3. New York: Academic, 1980, pp. 281–330

Sivak JG, Howland HC, West J, Weerheim J. The eye of the hooded seal, *Cystophora cristata*, in air and water. J Comp Physiol 1989; 165: 771–777

Skinner BF. *Cumulative Records*. New York: Appleton-Century-Crofts, 1961.

Sliper EJ. Whales. London: Hutchinson, 1962.

Smiarowski RA, Carhart R. Relations among temporary resolution, forward masking and simultaneous masking. J Acoust Soc Am 1975; 57: 1169–1174

Snell KB. The effect of sinusoidal amplitude modulation on gap detection in noise. J Acoust Soc Am 1995; 98: 1799–1802

Snell KB, Ison JR, Frisina DR. The effects of signal frequency and absolute bandwidth on gap detection in noise. J Acoust Soc Am 1994; 96: 1458–1464

Sokolov VE, Ladygina TF, Popov VV, Supin AYa. Somatotopic projection to cerebral cortex of the jerboa *Allactaga jaculus*. Dokl Biol Sci 1986; 288: 364–367

Sokolov VE, Ladygina TF, Supin AYa. Localization of sensory zones in the cerebral cortex of the dolphin. Dokl Akad Nauk SSSR (Proc Acad Sci USSR) 1972; 202: 490–493

Solntseva GN. Formation of an adaptive structure of the peripheral part of the auditory analyzer in aquatic, echo-locating mammals during ontogenesis. In: *Sensory Abilities of Cetaceans: Laboratory and Field Evidence*, JA Thomas, RA Kastelein, eds. New York: Plenum, 1990, pp. 363–384

Sonoda S, Takemura A. Underwater sounds of the manatees, *Trichechus manatus manatus* and *Trichechus inunguis* (Trichechidae). Rep Inst Breeding Res Tokio, Univ Agricult 1973; 4: 19–24

Spong P, White D. Visual acuity and discrimination learning in the dolphin (*Lagenorchinchus obliquidens*). Exp Neurol 1971; 31: 431–436

Stone J. The number and distribution of ganglion cells in the cat's retina. J Comp Neurol 1978; 180: 753–772

Stone J. *The Wholemount Handbook. A Guide to the Preparation and Analysis of Retinal Wholemounts*. Sydney: Maitland, 1981

Stone J. *Parallel Processing in the Visual System*. New York: Plenum, 1983a

Stone J. Topographical organization of the retina in a Monotreme: Australian spiny anteater *Tachyglossus aculeatus*. Brain Behav Evol 1983b; 22: 175–184

Stone J, Halasz P. Topography of the retina in the elephant *Loxodonta africana*. Brain Behav Evol 1989; 34: 84–95

Stone J, Keens J. Distribution of small and medium-sized ganglion cells in the cat's retina. J Comp Neurol 1980; 192: 235–245

Supin AYa, Ladygina TF, Popov VV. Properties of responses of the somatosensory cortex of the fur seal to vibrissae stimulation. In: *Electrofiziologiya Sensornykh System Morskikh Mlekopitayushikh (Electrophysiologi of Sensory Systems of marine Mammals)*, VE Sokolov, ed. Moscow: Nauka, 1986, pp. 158–170

Supin AYa, Mukhametov LM, Ladygina TF, Mass AM, Polyakova IG. Electrophysiological Study of the Dolphin's Brain. Moscow: Nauka, 1978.

Supin AYa, Pletenko MG, Tarakanov MB. Frequency resolution capability of dolphin hearing. Dokl Biol Sci 1992a; 323: 151–153

Supin AYa, Pletenko MG, Tarakanov MB. Frequency resolving power of the auditory system in a bottlenose dolphin (*Tursiops truncatus*). In: *Marine Mammal Sensory Systems*, JA Thomas, RA Kastelein, AYa Supin, eds. New York: Plenum, 1992b, pp. 287–293

Supin AYa, Popov VV. Recovery cycles of the dolphin brainstem evoked potentials for paired acoustic stimuli. Dokl Biol Sci 1985; 283: 535–537

Supin AYa, Popov VV. Curves of tonal masking of hearing in bottlenose dolphins. Dokl Biol Sci 1986; 289: 461–464

Supin AYa, Popov VV. Auditory frequency-resolving power of the dolphin. Dokl Biol Sci 1988; 300: 329–332

Supin AYa, Popov VV. Frequency-selectivity of the auditory system in the bottlenose dolphin, *Tursiops truncatus*. In: *Sensory Abilities of Cetaceans: Laboratory and Field Evidence*, JA Thomas, RA Kastelein, eds. New York: Plenum, 1990, pp. 385–393.

Supin AYa, Popov VV. Direction-dependent spectral sensitivity and interaural spectral difference in a dolphin: Evoked potential study. J Acoust Soc Am 1993; 96: 3490–3495

Supin AYa, Popov VV. Envelope-following response and modulation transfer function in the dolphin's auditory system. Hearing Res 1995a; 92: 38–46

Supin AYa, Popov VV. Frequency tuning and temporal resolution in dolphins. In *Sensory Systems of Aquatic Mammals*, RA Kastelein, JA Thomas, PE Nachtigall, eds. Woerden, The Netherlands: De Spil, 1995b, pp 95–110

Supin AYa, Popov VV. Temporal resolution in the dolphin's auditory system revealed by double-click evoked potential study. J Acoust Soc Am 1995c; 97: 2586–593

Supin AYa, Popov VV. Frequency-modulation sensitivity in bottlenose dolphins, *Tursiops truncatus*: Evoked-potential study. Aquatic Mammals 2000; 26(1): 83–94

Supin AYa, Popov VV, Klishin VO. Electrophysiological study of interaural sound intensity difference in the dolphin *Inia geoffrensis*. Experientia 1991; 47: 937–938

Supin AYa, Popov VV, Klishin VO. ABR frequency tuning curves in dolphins. J Comp Physiol A 1993; 173: 649–656

Supin AYa, Popov VV, Milekhina ON, Tarakanov MB. Frequency resolving power measured by rippled noise. Hearing Res 1994; 78: 31–40

Supin AYa, Popov VV, Milekhina ON, Tarakanov MB. Frequency-temporal resolution of hearing measured by rippled noise. Hearing Res 1997; 108: 17–27

Supin AYa, Popov VV, Milekhina ON, Tarakanov MB. Ripple density resolution for various rippled-noise patterns. J Acoust Soc. Am. 1998; 103: 2042–2050

Supin AYa, Sukhoruchenko MN. Characteristics of the acoustic analyzer of the dolphin *Phocoena phocoena*. In: *Morfologiya, Fiziologiya y Acustica Morskikh Mlekopitayushikh (Morphology, Physiology, and Acoustics of Marine Mammals)*, VE Sokolov, ed. Moscow: Nauka, 1974, pp. 127–135

Sur M, Nelson RJ, Kaas JH. The representation of the body surface in somatosensory area I of the grey squirrel. J Comp Neurol 1978; 179: 425–450

Sur M, Nelson RJ, Kaas JH. Representation of the body surface in somatic koniocortex in the prosimian *Galago*. J Comp Neurol 1980; 189: 381–402

Sur M, Nelson RJ, Kaas JH. Representation of the body surface in the cortical area 3b and 1 of squirrel monkeys: Comparison with other primates. J Comp Neurol 1982; 211: 177–192

Swets JA. *Signal Detection and Recognition by Human Observers*. New York: Wiley, 1964

Szymanski MD, Bain DE, Kiehl K, Pennington S, Wong S, Henry KR. Killer whale (*Orcinus orca*) hearing: Auditory brainstem response and behavioral audiograms. J Acoust Soc Am 1999; 106: 1134–1141

Szymanski MD, Supin AYa, Bain DE, Henry KR. Killer whale (*Orcinus orca*) auditory evoked potentials to rhythmic clicks. Marine Mammal Sci 1998; 14(4): 676–691

Tancred E. The distribution and sizes of ganglion cells in the retinas of five Australian marsupials. J Comp Neurol 1981; 196: 585–603

Tarakanov MB, Pletenko MG, Supin AYa. Frequency resolving power of the dolphin's hearing measured by rippled noise. Aquatic Mammals 1996; 22(3): 141–152

Tarpley RJ, Gelderd JB, Bauserman S, Ridgway SH. Dolphin peripheral visual pathway in chronic unilateral ocular atrophy complete decussation apparent. J Morphol 1994; 222: 91–102

Teas DC, Eldredge DH, Davis H. Cochlear responses to acoustic transients. J Acoust Soc Am 1962; 34: 1428–1489

Tedford RH. Relationship of pinnipeds to other carnivores. Syst Zool 1976; 25: 363–374

Terhune JM Directional hearing in a harbor seal in air and water. J Acoust Soc Am 1974; 56: 1862–1865

Terhune JM. Detection thresholds of a harbour seal to repeated underwater high-frequency, short-duration sinusoidal pulses. Canad J Zool 1988; 66:1578–1582

Terhune JM. Underwater click hearing thresholds of a harbour seal, *Phoca vitulina*. Aquatic Mammals 1989; 15 (1): 22–26

Terhune JM. Masked and unmasked pure tone detections of a harbour seal listening in air. Canad J Zool 1991; 69: 2059–2066

Terhune JM, Ronald K. The harp seal, *Pagophilus groenlandicus* (Erxleben, 1777). X. The air audiogram. Can J Zool 1971; 49: 385–390

Terhune JM, Ronald K. The harp seal, *Pagophilus groenlandicus* (Erxleben, 1777). III. The underwater audiogram. Can J Zool 1972; 50: 565–569

Terhune JM, Ronald K. Masked hearing thresholds of ringed seals. J Acoust Soc Am 1975a; 58: 515–516

Terhune JM, Ronald K. Underwater hearing sensitivity of two ringed seals (*Pusa hispida*). Can J Zool 1975b; 53: 227–231

Terhune JM, Ronald K. The upper frequency limit of ringed seal hearing. Can J Zool 1976; 54: 1226–1229

Terhune J, Turnbull S. Variation in the psychometric functions and hearing thresholds of a harbour seal. In: *Sensory Systems of Aquatic Mammals*, RA Kastelein, JA Thomas, PE Nachtigall, eds. Woerden, The Netherlands: De Spil, 1995, pp. 81–93

Thomas J, Chun N, Au W, Pugh K. Underwater audiogram of a false killer whale (*Pseudorca crassidens*). J Acoust Soc Am 1988; 84: 936–940

Thomas JA, Kasnelein RA, eds. *Sensory Abilities of Cetaceans. Laboratory and Field Evidence.* New York: Plenum, 1990

Thomas JA, Kastelein RA, Supin AYa, eds. *Marine Mammal Sensory Systems.* New York: Plenum, 1992

Thomas J, Moore P, Withrow R, Stoermer M. Underwater audiogram of a Hawaiian monk seal (*Monachus chauinslandi*). J Acoust Soc Am 1990; 87: 417–420

Thomas JA, Pawloski JL, Au WWL. Masked hearing abilities in a false killer whale (*Pseudorca crassidens*). In: *Sensory Abilities of Cetaceans. Laboratory and Field Evidence*, JA Thomas, RA Kastelein, eds. New York: Plenum, 1990, pp. 395–404

Thompson RK, Herman LM. Underwater frequency discrimination in the bottlenose dolphin (1–140 kHz) and the human (1–8 kHz). J Acoust Soc Am 1975; 57: 943–948

Tiao YC, Blakemore C. Regional specialization in the golden hamster's retina. J Comp Neurol 1976; 168: 439–458

Tremel DP, Thomas JA, Ramirez KT, Dye GS, Bachman WA, Orban AN, Grimm KK. Underwater hearing sensitivity of a Pacific white-sided dolphin, *Lagenorhinchus obliquidens*. Aquatic Mammals 1998; 24: 63–69

Turl CW, Penner RH, Au WWL. Comparison of target detection capabilities of the beluga and bottlenose dolphin. J Acoust Soc Am 1987; 82: 1487–1491

Turl CW, Skaar DJ, Au WWL. The echolocation ability of the beluga (*Delphinapterus leucas*) to detect target in clutter. J Acoust Soc Am 1991; 89: 896–901

Turnbull SD, Terhune JM. White noise and pure tone thresholds of a harbour seal listening in air and underwater. Can J Zool 1990; 68: 2090–2097

van Buren KM. *The Retinal Ganglion Cell Layer*. Springfield, IL: Charles Thomas, 1963

van der Pol KM, Worst JGF, van Andel P. Macro-anatomical aspects of the cetacean eye and its imaging system. In: *Sensory Systems of Aquatic Mammals*, RA Kastelein, JA Thomas, PE Nachtigall, eds. Woerden, The Netherlands: De Spil, 1995, pp. 409–418

Varanasi U, Malins DC. Unique lipids of the porpoise (*Tursiops gilli*): Difference in triacylglycerols and wax esters of acoustic (mandibular and melon) and blubber tissues. Biochim Biophys Acta 1971; 23: 415–418

Varanasi U, Malins DC. Triacylglycerols characteristics of porpoise acoustic tissues: Molecular structures of diisovaleroylglycerides. Science 1972; 176: 926–928

Vel'min VA, Dubrovskiy NA. Auditory analysis of sounds pulsed in dolphins. Dokl Biol Sci 1975; 225: 562–565

Verhaart WJC. The brain of the sea cow, *Trichecus*. Psych Neurol Neurochir (Amst) 1972; 75: 271–292

Viemeister NF. Temporal modulation transfer function based on modulation thresholds. J Acoust Soc Am 1979; 66: 1346–1380

Voigt HF, Young ED. Evidence of inhibitory interactions between neurons in dorsal cochlear nucleus. J Neurophysiol 1980; 44: 76–96

Voronov VA, Stosman IM. Frequency threshold characteristics of subcortical elements of the auditory analyzer of *Phocoena hocoena*. Zh Evol Biokhim Foziol (J Evol Biochem Physiol) 1977; 13: 619–622

Wainwright WN. Comparison of hearing thresholds in air and underwater. J Acoust Soc Am 1958; 30: 1025–1029

Wakakuwa K, Washida A, Fukuda I. Distribution and some size of ganglion cells in the retina of the eastern chipmunk (*Tamais sibiricus asiaticus*). Vision Res 1985; 25: 877–885

Waller G. Retinal ultrastructure of the Amazon river dolphin (*Inia geoffrensis*). Aquatic Mammals 1982; 9: 17–28

Waller G. The ocular anatomy of cetacea: An historical perspective. In: *Investigation of Cetacea*, GV Pilleri, ed. Vol. XVI. 1984, pp. 138–148

Walls GL. The vertebrate eye and its adaptive radiation. Can Inst Sci Bull 19. Michigan: Cranbrook Press, 1942

Wang D, Wang K, Xiao Y, Sheng G. Auditory sensitivity of a Chinese river dolphin, *Lipotes vexillifer*. In: *Marine Mammal Sensory Systems*, JA Thomas, RA Kastelein, AYa Supin, eds. New York: Plenum, 1992, pp. 213–222

Wartzok D, McCormick MG. Color discrimination in a Bering Sea spotted seal, *Phoca larga*. Vision Res 1978; 18: 781–784

Wartzok D, Schusterman RJ, Gailey-Phipps J. Seal echolocation? Nature 1984; 308: 753

Wässle H, Grunert U, Rochrenbeck J, Boycott BB. Retinal ganglion cells density and cortical magnification factor in the primate. Vision Res 1990; 30: 1897–1911

Wässle H, Hoon Chun Myung, Muller F. Amacrine cells in the ganglion cell layer of the cat retina. J Comp Neurol 1987; 265; 391–408

Watkins WA The activities and underwater sounds of fin whales. Sci Rep Whales Res Inst 1981; 33: 83–117

Watkins WA, Tyack P, Moore KE, Bird JE. The 20 Hz signals of finback whales (*Balaenoptera physalus*). J Acoust Soc Am 1987; 77: 1091–1101

Watkins WA., Wartzok D. Sensory biophysics of marine mammals. Marine Mammal Sci 1985; 1: 219–260

Webb SV, Kaas JH. The size and distribution of ganglion cells in the retina of the owl monkey *Aotes trivigratus*. Vision Res 1976; 16: 1247–1254

Weir C., Jesteadt W, Green D. Frequency discrimination as a function of frequency and sensation level. J Acoust Soc Am 1976; 61: 178–184

West LA. Sivak JG, Murphy CJ, Kovacs KM. A comparative study of the anatomy of the iris and ciliary body in aquatic mammals. Can J Zool 1991; 69: 2594–2607

Weston DE, Black RI. Some unusual low-frequency biological noises underwater. Deep Sea Res 1965; 12: 295–298

Wever EG, McCormick JG, Palin H, Ridgway SH. The cochlea of the dolphin *Tursiops truncatus*: general morphology. Proc Nat Acad Sci USA 1971a; 68: 2381–2385

Wever EG, McCormick JG, Palin H, Ridgway SH. The cochlea of the dolphin *Tursiops truncatus*: The basilar membrane. Proc. Nat Acad Sci USA 1971b; 68: 2708–2711

Wever EG, McCormick JG, Palin H, Ridgway SH. The cochlea of the dolphin *Tursiops truncatus*: hair cells and ganglion cells. Proc Nat Acad Sci USA 1971c; 68: 2908–2912

Wever EG, McCormick JG, Palin H, Ridgway SH. Cochlear structure in the dolphin *Lagenorhynhuchus obliquidens*. Proc Nat Acad Sci USA 1972; 69: 657–661

White D, Cameron N, Spong P, Bradford J. Visual acuity in the killer whale (*Orcinus orca*). Exp Neurol 1971; 32:230–236

White Jr MJ, Norris JC, Ljungblad DK, Barton K, di Sciara GN. Auditory thresholds of two beluga whales (*Delphinapterus leucas*). In: *Hubbs/Sea World Research Institute Technical Reports* San Diego, CA: Hubbs Marine Research Institute, 1978, pp. 78–109

Williams RW, Cavada C, Reinoso-Suarez0 F. Rapid evolution of the visual system: a Cellular assay of the retina and dorsal lateral geniculate nucleus of the Spanish wild cat and domestic cat. J Neorosci 1993; 13: 208–228

Wilson G. Some comment on the optical system of *Pinnipedia* as a result of observations on the Weddell seal (*Leptonychotes weddell*). Brit Antarct Surv Bull 1970; 23: 57–62

Wong ROL, Hughes A. The morphology, number and distribution of a large population of confirmed displaced amacrine cells in the adult cat retina. J Comp Neurol 1987; 255: 159–177

Wong ROL, Wye-Dvorak J, Henry GH. Morphology and distribution of neurons in the retina ganglion cell layer of the adult Tammar wallaby *Macropus eugenii*. J Comp Neurol 1986; 253: 1–12

Woolsey CN. Pattern of sensory representation in the cerebral cortex. Federat Proc 1947; 6: 437–441

Yablokov AV, Belkovich VM, Borisov VI. *Whales and Dolphins*. Moscow: Nauka, 1972.

Yost WA. The dominance region and ripple-noise pitch: A test of the peripheral weighting model. J Acoust Soc Am 1982; 72: 416–425

Yost WA, Hill R. Models of the pitch and pitch strength of ripple noise. J Acoust Soc Am 1979; 66: 400–410

Yost WA. Hill R, Perez-Falcon T. Pitch discrimination of ripple noise. J Acoust Soc Am 1977; 63: 1166–1173

Young EC. Response characteristics of the cochlear nuclei. In: *Hearing Science*, CI Berlin, ed. San Diego: College-Hill Press, 1985, pp. 423–460.

Young ED, Brownell WE. Responses to tones and noise of single cells in dorsal cochlear nucleus of unanaesthetized cats. J Neurophysiol 1976; 60: 1–29

Young NM, Hope GM, Dawson WW. The tapetum fibrosum in the eyes of two small whales. Marine Mammal Sci 1988; 4: 281–290.

Yunker MP, Herman LM. Discrimination of auditory temporal differences by the bottlenose dolphin and by the human. J Acoust Soc Am 1974; 56: 1870–1875

Zaytseva KA, Akopian AI, Morozov VP. Noise resistance of the dolphin auditory analyzer as a function of noise detection. Biofizika (Biophysics) 1975; 20: 519–521

Zook JM, DiCaprio RA. A potential system of delay-lines in the dolphin auditory brainstem. In: *Sensory Abilities of Cetaceans. Laboratory and Field Evidence*, JA Thomas, RA Kastelein, eds. New York: Plenum, 1990, pp. 181–193.

Zook JM, Jacobs MS, Glezer I, Morgane PJ. Some comparative aspects of auditory brainstem cytoarchitecture in echolocating mammals: Speculations on the morphological basis of time-domain signal processing. In: *Animal Sonar: Processes and performance*, PE Nachtigall, PWB Moore, eds. New York: Plenum, 1988, pp. 311–316

Zvorykin VP. Morphological basis of ultrasonic and echolocation features in dolphins (in Russ.). Arkhiv Anat Histol Embriol 1963; 45(7): 3–17

Zwicker E. Die Grenzen der Horbakeit der Amplituden-modulation und der Frequenzmodulation eines Tones. Acustica 1952; 2: 125–133

INDEX

ABR. *See* auditory brainstem response

accommodation 231, 265, 282

acoustic impedance 5, 6, 23, 55, 56, 159, 181, 200, 209, 218

acoustic window 181, 182, 184–189, 203

ACR. *See* auditory cortical response

adaptation (auditory) 51, 52, 93, 94, 97, 113, 201

adaptation (visual) 269, 271, 283

adaptive procedure 11, 15–18, 53, 141, 152, 210, 221

aerial myopia 233, 234, 256, 282

AM. *See* amplitude modulation

Amazon river dolphin 31, 54, 56, 92, 154–156, 159, 176, 183, 185, 233, 237, 239, 241, 254, 255, 257, 260, 283

Amazonian manatee 3, 222–225, 279

amplitude modulation 50, 85, 132, 135, 147

ANR. *See* auditory nerve response

anterior chamber (of the eye) 7, 230, 232, 264, 266, 279, 280

Arctocephalus australis. See southern fur seal

Arctocephalus pusillus. See south-

ern fur seal

area centralis 245–249, 272–277, 282, 283

audiogram 52–57, 136, 206, 207, 221, 223–227

auditory brainstem response 27–51, 56, 61–72, 75–86, 89–97, 103–105, 109, 111–113, 116–125, 145, 154, 155, 157, 164, 165, 168, 170, 171, 174, 181, 184, 187, 189, 196, 200–203, 223–227

auditory cortex. *See* auditory projection area

auditory cortical response 35, 120, 121, 145, 146, 147, 200

auditory filter(s) 73, 89, 97–101, 108, 109, 112, 119, 121, 127, 129, 130, 142, 195, 203, 211, 213

auditory nerve 25, 90, 157, 195, 285

auditory nerve response 38, 157–162, 176, 177

auditory projection area 26, 27, 35, 39, 121, 190–193, 203, 216, 217, 222

averaging technique 17, 18, 28, 30, 105, 116, 262

backward masking 60